T0156053

THE HISTORY OF HUMAN FACTORS AND ERGONOMICS

THE HISTORY OF HUMAN FACTORS AND ERGONOMICS

David Meister

CRC Press is an imprint of the
Taylor & Francis Group, an **informa** business

Copyright © 1999 by Lawrence Erlbaum Associates, Inc.
All rights reserved. No part of this book may be reproduced in any form, by photostat, microfilm, retrieval system, or any other means, without prior written permission of the publisher.

First published by Lawrence Erlbaum Associates, Inc., Publishers
10 Industrial Avenue
Mahwah, New Jersey 07430

Reprinted 2008 by CRC Press

CRC Press
6000 Broken Sound Parkway, NW
Suite 300, Boca Raton, FL 33487

270 Madison Avenue
New York, NY 10016

2 Park Square, Milton Park
Abingdon, Oxon OX14 4RN, UK

Cover design by Kathryn Houghtaling Lacey

Library of Congress Cataloging-in-Publication Data

Meister, David.
The history of human factors and ergonomics / Davis Meister.
p. cm.
Includes bibliographical references and indexes.
ISBN 0-8058-2768-4 (cloth : alk. paper) — ISBN 0-8058-2769-2 (pbk. : alk. paper).
1. Human engineering—History. I. Title.
TA166.M369 1999
620.8'2'09—dc21 98-49609
CIP

Printed in the United States of America
10 9 8 7 6 5 4 3 2 1

This book is dedicated to the memory of
Shirley Davis Meister
and to
The first generation of HFE professionals

Contents

Preface

In the past, many people, knowing my interest in history and the fact that I had studied for an advanced degree in this subject, have suggested that I write the history of HFE. I always resisted the idea because my concept of history was too narrow—history was, I conceived, mainly chronological events, and I did not see many of these in HFE (although there are).

History is events, but it is also much more. In a scientific discipline like HFE, events play a subordinate role to *concepts*. The most fascinating aspect of HFE is its intellectual character, which has been largely ignored in the one or two histories of the discipline I and a few others have written (Meister, 1995b, 1996c; Meister & O'Brien, 1996; Moroney, 1995). That is because an intellectual history is not immediately obvious; it must be teased out from the reminiscences of its professionals from their published writings—from the monthly Bulletins of the Society, which they established to support their work organizationally.

The present work is largely a history of ideas. Looking back over 45 years of HFE professional work, I find that this was an early interest of mine. There are two overarching themes in the intellectual history of HFE. The first is the need to transform technology, physical phenomena (equipment characteristics), into their behavioral (human performance) equivalents and to translate behavioral principles into technology (equipment characteristics). One might say that this is the modern equivalent of the mind–body problem that bedeviled the philosophers of the 17th and later centuries.

What was then only a philosophical conceit is now in the 20th century a major practical concern. The body is now the technology that arguably encases and drives us unmercifully. The mind is the corpus of behavioral principles that HFE seeks to apply to technology to ameliorate the rigors of an excessively mechanistic civilization. This may seem like old wine in new bottles, but the wine is still very potent.

The second theme, hovering bashfully behind the first, is the omnipresence of technology. What makes a history of HFE important in the general intellectual and cultural history of the 20th century (and presumably subsequent centuries) is the tremendous importance of technology. One does not need to be a historian of science to recognize the overwhelming growth of technology in our times. By any index, technology is growing exponentially. Hence, any discipline that seeks to influence the effect of that technology is almost as important as the technology itself.

Many HFE professionals may not agree with that last statement, although it massages their egos, because HFE, being both a basic and applied science, has directed much of its attention to the practical problems of experimentation and system development.

This is another reason that an intellectual history of HFE is so important. Starting almost from HFE's inception, HFE professionals have asked these questions: What is HFE? Where are we going? What should we be doing? No intellectual history can answer these questions, but it can describe to its professionals what others have thought and written about their discipline. If every discipline has a history, and I maintain that it has (to prove the point, I need only refer to the many histories of the discipline of history that have been published over the past 150 years; e.g., Novick, 1991), it is important if not essential that those who practice the discipline know what that history is.

I understand that in universities training HFE professionals, a course in its history is mandatory. Such a history is important because a discipline is never static; it is always in the process of development. If one does not know what that development consists of, the discipline is likely to wander in its course. As I have occasionally explained to skeptical colleagues, if one does not know where one has been, one cannot know where one is presently and one cannot then begin to steer a new course.

Historians have long argued about whether a knowledge of history enables a people to make more informed choices. I believe it was Santayana who made the oft-repeated statement that "Those who do not know history are doomed to repeat it."

In contrast to general history, such as that of the Napoleonic era (a knowledge of which may or may not have specific effects on those who are aware of that history), the history of a discipline will have practical and long-lasting effects on its practitioners because there is a much closer relationship between the content of the history and the quotidian concerns of the practitioners. At any rate, it is with this hope that the present volume has been written.

This book has several goals: It is intended as a contribution to cultural history because (a) ours is a technological civilization, and (b) one cannot understand technology outside of the various disciplines that make up that technology. A disciplinary history is of course highly specialized, but I maintain that HFE is distinctive in being the only discipline that relates humans to technology. Other behavioral disciplines like anthropology have little connection with technology. This is what makes HFE important in the present historical era.

HFE as a discipline has grown by accretions rather than having been developed systematically and deliberately. Hence, one goal of the book is to create a formal conceptual structure for HFE.

One might think that a discipline with only 50 years of history behind it would be relatively easy to analyze. Nothing could be further from the truth. Many times during the analysis that produced this book, I was astounded at the complexity of the subject matter and the variety of avenues with which one can approach the problems implicit in HFE. That is because I view HFE not merely as an adjunct to the design of equipment but as an attempt to understand fully the relationship of the human to technology and how to translate that relationship into physical terms.

I have also been judgmental in my evaluation of the strengths and weaknesses of the discipline. Some readers may feel that this is an unwarranted liberty. My vision of the discipline is that it has weaknesses as well as strengths, and that as long as those weaknesses are not addressed, HFE cannot achieve its full potential. Consequently, I have tried to indicate where I think HFE is and where it should go. For those who disagree with me, the book offers at least an intellectual challenge.

History for the sake of history is not the point in this book. The history of HFE (in the expanded definition of *history*, which includes the characteristics of the discipline) serves as a means of understanding more clearly what the discipline is and to emphasize by the changes it reports that the discipline today may not be what it will be 50 years from now.

List of Acronyms

APA—American Psychological Association
BCPE—Board of Certification in Professional Ergonomics
BIO—Biomechanics
CAD—Computer-assisted design
DOD—Department of Defense
DM—Decision making
HF—Human Factors
HEP—Human error probability
HFE—Human factors ergonomics
HFES—Human Factors and Ergonomics Society
HFS—Human Factors Society
HR—Human reliability
ICBM—Intercontinental Ballistic Missile
IE—Industrial ergonomics
IEA—International Ergonomics Association
M—Macroergonomics
MDS—Multidimensional scaling
NATO—North Atlantic Treaty Organization
NIOSH—National Institute of Occupational Safety and Health
NPP—Nuclear power plant
ODAM—Organizational development and management
OE—Operational environment
OST—Operational system testing
PSF—Performance shaping factors
PRA—Probabilistic risk assessment
RFP—Request for proposal
SME—Subject matter expert
S–O–R—Stimulus–organism–response
SOP—Standard Operating Procedure
S–R—Stimulus–response
WW—World War

1

Introduction to HFE

The structure of a discipline consists of its elements, its parameters, and the assumptions underlying it. These are global concepts. At a more detailed level, there are also a large number of variables whose interaction creates the phenomena with which HFE is concerned. There are also (or should be) hypotheses about these variable interactions that drive the research performed to investigate HFE phenomena.

Chapter 1 defines the HFE elements and parameters; chapter 2 deals with its assumptions; chapter 3 deals with the system as a fundamental HFE construct. Subsequent chapters describe the formal and informal history of the discipline, its research characteristics, what HFE practitioners do on the job, and some of the more important specialty areas.

Of course, the critical HFE elements are the human and the technology with which he or she interacts. These serve as the basis for the discipline, which is considered a science and possesses certain characteristics that are of interest to us. This originally homogeneous discipline progressively subdivides into specialties. The discipline/science has utility because its application to practical questions results in usable products. As a dynamic process, the discipline also has a history, which is the subject of this book. The building blocks of HFE are its systems, which are organizations of hardware (machines, tools, equipment, software) and humans. A critical HFE element is the system development process. The principles that guide or should guide the behavioral design of these systems are created by research. HFE professionals include both those who perform that research and those who assist in the system development process; these latter are called *practitioners*.

As part of research and system development, measurement of phenomena produces data and conclusions; when combined correctly, these produce the database on which the discipline relies for its knowledge. This knowledge is represented primarily in the form of publications.

HFE parameters include the discipline's basic elements, the human and technology, but also (a) the scope of the discipline (what it includes and excludes), (b) its purpose and subject matter, (c) its desired effects, (d) the functions the discipline performs, (e) how the discipline is conceptualized, (f) the factors affecting the discipline, and (g) the specialties into which it progressively subdivides.

Some of what is said in this chapter may seem obvious to the experienced professional. However, a little more analytic consideration may suggest that each of the topics treated has some previously unexplored dimensions.

Table 1.1 presents a taxonomy of the HFE discipline. The table is the beginning of an attempt by the author to develop a formal conceptual structure for the discipline. Later the reasons for developing such a structure are addressed, but it is necessary to first say a few words about the taxonomy. All of its categories are described in one way or another throughout the book, but not necessarily entirely within this or any other chapter. Nor is each topic addressed in the same amount of detail or in the order listed in the table. Nevertheless, it is hoped that the taxonomy serves as a road map for the reader.

TABLE 1.1
A Taxonomy of the HFE Discipline

The HFE Concept Structure	*HFE Functions*
Definitions	Research (analysis, measurement)
Elements	Application (system development, industrial ergonomics)
Constructs	
Assumptions	
Hypotheses	*HFE Disciplinary Concepts*
Variables	Subspecies of other disciplines
	Interdisciplinary approach to problem solution
HFE Elements	
Technology	Autonomous, distinctive science
The human in interaction with technology	
HFE professionals	*HFE Organization*
	The core discipline
HFE Purposes	HFE specialties
Ultimate purposes	
Improve system productivity	HFE Research Methods
Increase human comfort and safety	Objective
Add to the knowledge base of human–technology interactions	Subjective
Implementation purposes	*Factors Affecting HFE*
Analysis	Technology changes over time
Translation	Training of HFE professionals
Prediction	Professional organizations (HFES, IEA, etc.)
	Funding of research
HFE Scope	HFE job availability
Narrow	Relationships with other disciplines
Broad	

It may seem obvious to the reader that HFE does possess a formal conceptual structure that has been described many time in many textbooks. Why then introduce the topic anew? First, it is necessary because the standard texts do not treat these concepts as an integrated, formal structure. Moreover, the author believes that the standard treatment of HFE concepts overlooks many aspects. The reader will eventually decide on his or her own whether Meister's treatment is better or different than anyone else's. Although still looking for reasons to justify what he had intended to do anyway, the author suggests that his formal conceptual structure may serve as a source of stimulation to the reader in the selection of research topics.

Second, as is seen later, HFE professionals have their own conceptual structure (what and how they think about their work; chap. 5), and a formal conceptual structure can serve as a contrast with the one held by individuals. Third, it provides the author with a framework on which he can hang a number of admittedly speculative ideas.

In discussing constructs, assumptions, hypotheses, and so on, we attempt to describe the concept structure of the discipline because HFE history is bound up with the concepts it has developed over time. However, it is insufficient to talk about the HFE construct structure alone because there are in fact two: (a) that of the discipline that is formal and logical, and (b) the unformed, diverse, sometimes illogical, and incomplete, which is the concept structure of the HFE professionals themselves. These two structures are not the same, although they are linked, because the formal one would never have been developed except through the informal one. If HFE is a developing science, it is because the informal concept structure is also developing.

Unfortunately, the professionals' concept structure is never described in textbooks and, to the author's knowledge, has never been investigated and reported, although the Society has supported surveys of its members (Hendrick, 1996a; Kraft, 1970). However, the author contends that it is impossible to understand HFE without examining how its professionals think. The more informal concept structure, elements of which are described in chapter 5, is perhaps even more important in the early developmental stages of a discipline, and HFE has years to go before it can be considered completely mature. Because the personal concept structure of HFE is incomplete, the reader must not be repelled by its contradictions and incomplete ideas.

HFE DEFINITIONS

Infrastructure

The infrastructure of HFE is composed of the building blocks that form the conceptual and operational structure of the discipline. That infrastructure consists of elements, constructs, parameters, assumptions, variables, and hypotheses.

Elements are the entities that make up the HFE discipline. The two basic elements are the human and technology; the relationship between these two is what HFE research endeavors to determine. One can add HFE professionals to this because what they think and do becomes the subject matter of the discipline. For example, technology, considered as a whole, is an abstraction, but obviously there are concrete aircraft, automobiles, computers, and so on, all of which in combination form the construct that we call *technology*. The individual automobiles, computers, and so on are instances of the technology abstraction. Constructs also have dimensions that can be varied experimentally, which makes them variables.

Thinking of these dimensions as variables enables us to study them and derive conclusions that describe the construct. For example, the construct *system* has dimensions of size, complexity, organization, existence over time, and so on. Obviously one can study any individual system (e.g., a particular automobile, with its particular size, complexity, organization, etc.), but it would not be possible to generalize results to systems in general unless one thought in terms of the abstract system. HFE constructs include the discipline, technology, the human, the system, measurement, application, and system development.

Parameters are lower level elements that make up the more fundamental constructs. For example, the HFE discipline as a construct has parameters that include the purpose and subject matter of the discipline and the functions the discipline performs. Note that parameters are not dimensions of constructs; the purpose or scope of HFE as a parameter cannot be varied like dimensions of a construct.

Assumptions are givens about the relationships among elements and parameters. Assumptions are taken for granted, and therefore need not or cannot be tested. For example, the religious person assumes the existence of a divinity; this assumption, despite great efforts by many philosophers, cannot be tested. In Euclidean mathematics, an assumption is that the sum of the angles of a right triangle equals 180 degrees. In HFE, there are many assumptions (see chap. 2); one begins with the most fundamental assumption that there is a relationship between the human and his or her technology. Assumptions are necessary to the conceptual structure in which they are included. For example, if we did not make the assumption of a human–technology relationship, there could not be an HFE discipline.

Once an assumption is accepted by those who practice the discipline (e.g., that there is a relationship between the human and technology), it becomes a given. This means that it need not be tested, although in some cases it is possible to collect evidence that tends to validate the assumption. However, lack of supporting evidence does not disprove the assumption.

Although it seems logical that one cannot accept contradictory assumptions, the reality is that humans (who are not completely logical) can accept contradictions in their concept structure, particularly if they do not pay much attention to their assumptions. This is the case with many HFE professionals.

One cannot hold an assumption and its specific contradiction in the same concept space, but one can hold assumptions that, although not explicitly contradicting each other, do have a contradictory effect. It would be hard to hold the assumption that there is a relationship between human behavior and technology while maintaining that there is no such relationship. However, if one assumption is the general scientific notion that the goal of science is to predict phenomena quantitatively and another, held by many professionals, is that human behavior cannot be predicted even probabilistically, the two assumptions contradict each other, but both can be held because general scientific principles are relatively weak as compared with those that are specific to a discipline. Another reason is that, like most people, HFE professionals tend to segregate viewpoints into individual sections of their concept structure and do not permit any interaction among these sections (this is the notion of *logic-tight compartments*).

Hypotheses are imagined deductions from assumptions. For example, if we postulate a relationship between the human and technology, we can hypothesize that the relationship changes its form as the technology becomes more complex. Thus, there would be a different relationship between the human and his or her tool, like a knife, and the relationship between an operator and a command/control work station. Although there are only a few constructs and a slightly larger number of assumptions, there can be (depending on one's creativity and patience) an almost infinite number of hypotheses. Hypotheses are not givens; they must be empirically tested and may be accepted or rejected as the evidence suggests. The hypotheses dealt with in chapters 1 and 2 relate to the discipline as whole. The more common use of hypotheses by the professional is at a more specific human behavior level and involves much more molecular hypotheses.

The preceding may appear to the reader to be highly abstract, and it is. The dividing line among constructs, assumptions, and hypotheses may also sometimes appear unclear. Certainly the quotidian operations of the HFE discipline can be performed without the HFE professional having constructs, assumptions, and hypotheses in the forefront of his or her mind. If that is the case, the reader may ask, so what? These building blocks represent the context in which the discipline as a whole functions. Like the material in a computer memory, they may at any time be retrieved from memory and pushed to the forefront of consciousness. Moreover, a scientific HFE discipline would be unimaginable without them. At the same time, the fact that it is possible to describe HFE in the same terms as those used of that most basic of sciences, mathematics, helps to bolster the point that HFE too is a science.

Constructs and assumptions that form the discipline's infrastructure exist in the human's concept space, but at a lower level, if one can think of concepts occupying space in a topological, Lewinian (Lewin, 1936) sense. These are not retrieved from the core of the concept space except on occasions when the professional reads a text such as this one. He or she is much more concerned about variables that may or may not be especially significant except as they relate to the professional's specialty. Variables can be divided into three general classes, and which class the researcher selects depends on the nature of the specialty and the researcher's personal interests. The three variable classes are: (a) those relating to characteristics of the human (e.g., age, strength, work experience); (b) those relating to the environment in which humans function (e.g., earth, outer space, underwater, lighting, heating, etc.); and (c) those relating to the type of technology with which the human interacts (e.g., tools, equipment, systems, tasks). The operation of these variables produces human behavior, which is the subject matter of the research to be described in subsequent chapters. However, the next three chapters are concerned with exploring the nature of the HFE infrastructure because without some appreciation of this the functioning of the discipline will remain shrouded in obscurity.

It may seem to the practicing professional that the attempt to construct a rigorous conceptual structure for HFE is essentially irrelevant in view of the many practical problems HFE personnel encounter. It is of course possible to understand HFE without considering the intellectual basis of the discipline, but this leads to an imperfect understanding, which may have negative consequences on a practical level. For example, if one does not fully understand the role of the system in HFE, it is likely that research on system variables will be ignored.

THE HUMAN

Introduction

The two most critical elements of HFE are (no surprises here) the human and technology; without these, there is no HFE. They are discussed in this order. One might think that the human is an objective element simply because he or she exists. The human is represented in HFE by his or her performance in specific contexts (in relation to individual technological variables). What one learns from HFE research presumably defines the human. However, what the research describes is also determined by the researcher's concept of the human. It is possible to think of the human as a *black box*, in which his or her performance is determined solely by stimuli as inputs to the box and responses as outputs. This concept says in effect that what the human thinks about the stimulus situation and his or her responses to that situation are of no importance.

This is an extreme behavioristic concept and is no longer popular among researchers and practitioners. The contrasting concept says that the human's conceptualization of the individual technological situation and of technology in general is important because it determines, in part, his or her responses to technological stimuli. The major change in 20th-century HFE is not that we recognize that the individual thinks, but that what and how he or she thinks are important. The researcher who is concerned about maximum objectivity but who is aware that objectivity is inevitably modified by the human's internal processes is practicing what can be called *moderate behaviorism.* The author suspects that most HFE professionals fall into this category.

The great increase in concern for cognition, as shown by HFE research topics and textbook analyses, shows a definite progression, which in behavioral terms is as great or greater than the technological changes produced by the introduction of the computer. Although HFE research has maintained its long-term interest in sensory, perceptual, and motor functions, the years have seen an ever-expanding interest in cognition, especially decision making and particularly naturalistic decision making (Klein, 1993). Allied to that, but less intensively, is an interest in the emotional factors affecting performance (e.g., Dandavate, Sanders, & Stuart, 1996).

As Russian ergonomists keep emphasizing (see Bedny & Meister, 1997), the presumably objective stimulus can no longer be considered completely objective. One may induce an auditory stimulus of 1800 hertz, but it is no longer merely a sound of a given frequency and intensity because the subject asks him or herself what this sound in this immediate environmental context means. Complete objectivity, even in the form of psychophysical stimuli, no longer exists. This is shown by, among other indexes, the increase in the use of subjective techniques (e.g., interviews) to secure relevant data. As subjectivity becomes important, uncertainty (to be discussed in detail in chap. 3) also becomes important. This creates further problems: How much does subjectivity and uncertainty determine human performance? The HFE questions raised today are much more complex than those of 50 years ago if only because they are more realistic.

There are also new questions (e.g., How important is the human in human–technology interaction?). It might appear logical that the human does not affect technology because presumably hardware and software cannot be changed by the human responses to them. The human leaves his or her impact on technology in terms of requiring the equipment to accommodate human thresholds for sight, sound, and motor strength, and by requiring software, in particular, to be organized in accordance with human cognitive patterns.

This allows the human to use the technology. If Martians exist and if they are different from humans, we would expect their technology to be different from ours. Ultimately the nature of our technology is determined by the fact that it is designed to be useful to humans. Hence, we *imprint* that technology in the same way that Lorenz imprinted himself on ducklings that subsequently acted toward him as if he were their mother.

This is a human-centered concept of technology. Technology is effective (which is to say that it is useful) to the extent that the human responds to that technology as the designer of the technology had intended the user to respond. Is it too philosophical to ask, if I do not use the technology, does it exist for me? If I do not use a personal computer (PC) at home or in the office, the PC as an expression of technology hardly exists for me. Thinking in terms of Gestalt figure-ground terms (Koffka, 1935), it becomes part of the ground for me, although for an enthusiastic user a PC is definitely part of the figure. Another analogy is that of a primitive tribesman being shown a computer terminal; it has size, shape, and changing displays, but that is all. As a piece of technology, it hardly exists because the tribesman cannot use it as intended. This is not to say that technology as a background element has no effect on the human, but that the effect is muted and may be somewhat different than it is for the user of that technology.

The whole point of this disquisition is to show that the human is far more complex as a partner in the human–technological interaction than we might think. The human as designer determines technological characteristics; our concept of the human in the human–technological relationship determines what we research and how we interpret human performance in relation to that technology. If this appears to be like the shifting sands of uncertainty, it is; the more we learn about the human as an element in the technological environment, the more complex and less completely objective the human becomes.

Human Elements

The human elements with which HFE must be concerned (see Table 1.2) are three: physical, cognitive or intellectual, and motivational. Each of these has connotations for technology. The physical aspects of the human that technology must accommodate are strength, anthropometry, and threshold limitations in sensory and perceptual qualities. For example, in developing a system, one cannot assign the human the responsibility for moving quantities 200 pounds or larger without mechanical assistance simply because few men and certainly no women would be able to move such weights unaided. Similarly, one would not ask any human to reach for an object 20-feet high without a mechanical aid simply because anthropometry prevents this. One would not devise a type font of less than 8 points, for example, because the human visual capability would be overly stressed without spectacles.

Physical provisions such as these are relatively easy to accommodate because we understand them readily, but cognitive capabilities pose a much

TABLE 1.2
Taxonomy of HFE Elements: The Human

Human Elements	*Effects of the Human on Technology*
Physical/sensory	Improvement in technology effectiveness
Cognitive	Absence of effect
Motivational/emotional	Reduction in technological effectiveness
Human Conceptualization	*Human Operations in Technology*
Stimulus–response orientation (limited)	Equipment operation
Stimulus–conceptual-response orientation (major)	Equipment maintenance
	System management
Stimulus–conceptual–motivational–response orientation (total)	Type/degree of human involvement
	direct (operation)
	indirect (recipient)
Human Technological Relationships	extensive
Controller relationship	minimal
Partnership relationship	none
Client relationship	
Effects of Technology on the Human	
Performance effects	
goal accomplishment	
goal nonaccomplishment	
error/time discrepancies	
feeling effects	
technology acceptance	
technology indifference	
technology rejection	
demand effects	
resource mobilization	
stress/trauma	

greater problem for system developers. Although it is obvious that one would not expect a human to perform hundreds of numerical calculations in his or her mind without some sort of calculating assistance, many of the cognitive requirements of technology are much more complex, less overt, and more difficult to satisfy. Information processing is much more than threshold-limited capabilities, such as memory. For example, if decision making is required of the operator of a new system, one must ask: Does the human have certain preferred or stereotypical ways of processing information that should be incorporated in design? Does the new system impose special demands on the information-processing process? As automation takes us into an interactive role with the machine and as we progress toward artificial intelligence (AI), it will be necessary to determine the special ways in which humans think as opposed to relatively simple threshold-limited thinking processes.

Several questions are involved here: Are there distinctive ways with which humans think, and can one identify these? Does the system require the human to think in certain ways that are critical for system efficiency? Given that one can answer the preceding two questions positively, do we know how to translate the parameters of the human thinking process into physical equivalents or mechanisms that will permit the human as part of the system to think in these ways effectively and without stress? Will we allow the system to exercise those same human thinking qualities independently as a product of physical processes? In other words, can we transform our knowledge of intellectual activity from a verbal definition to a physical one? The third question is the most difficult one of the three to answer, but the preceding two also require HFE research in quantities and sophistication that we have never before attempted.

When we come to the motivational components of the human, one must begin by asking this question: Is it necessary to deal with a motivational component or can we explain and understand human performance without reference to the operator's motivation? An extreme behaviorist might say that if physical and cognitive factors are accommodated in research and system development, it is hardly necessary to be concerned about motivational problems especially because, if we include motivation as an area of interest, the transformation of knowledge about motivation into physical equivalents or structures that support human motivation is daunting in its difficulty.

However, a case can be made that to disregard the motivational aspect in HFE concepts and understanding is to avoid dealing with an important human process. In the normal routine of operating a system, the operator's motivation to perform may not be especially important, but if the system enters an emergency situation, motivation may become a much more serious problem. This is particularly the case if the emergency is a prolonged one. Similarly, if the situation is one of little stimulation over long periods of time, as in sonar/radar detection situations, some physical characteristic of the system may be needed to energize that motivation. Years ago, the U.S. Navy considered the possibility of providing sonarmen with false targets during simulated combat operations to stimulate their alertness. However, the notion was dropped because the false target might have become too real for the operator and the ship.

Ways of Conceptualizing the Human

It may not be exactly clear what is meant by conceptualizing the human. It is obvious that the system (at least as it is presently constituted) cannot function without a human within it. Even in highly automated systems like process control industries, there is still a role for the human, even if that role is greatly changed (Stubler & O'Hara, 1996). Nonetheless, questions arise. How far does one have to go in analyzing human performance to understand how the system functions? Is it enough to know whether the human has (or has not) made errors or performed to time constraints? Is it necessary to understand what the operator was thinking about when he or she performed the job or how he or she felt about job characteristics during this time? Is it necessary to determine modes of operator thinking as these may affect system operations?

A behaviorist might say that all such analyses are unnecessary, do not add anything to our understanding of system functioning, and do not suggest anything that will significantly aid human–system performance. Indeed, introducing human complexities into the analysis of system performance complicates the difficulty of arriving at that understanding. Thus, forget these human complexities.

The *catch-22* in this situation is that, to logically discard the complexities of cognitive or motivational functioning from our analysis of system functioning, we have to know just how important or unimportant these factors are to system performance. However, we cannot answer that question without first measuring those factors we should so like to discard. Not doing so might introduce an unwarranted assumption that they are unimportant. If one can measure these factors, presumably we can then introduce them into our analysis and development of systems.

Realistically, in any analysis of human performance, one can choose to ignore or include anything, but what happens if we study these factors and discover that they are actually important? Are we not then required to introduce them into the system development process?

If the theorist decides to ignore certain aspects of the human, there can be no objection to this because he or she operates solely on the level of theory. (Is this satisfactory? No, because theory without empirical measurement is simply, to use a vulgarism, *hot air.*) To bolster the position that one should ignore cognitive and motivational complexities in explaining system performance, it becomes necessary to measure these factors in the context of work performance. Unless these unknown factors have a zero or slight effect, we are then obligated to include them in system development. If we do not do so, after determining that their functioning is important to system operation, then we have deliberately ignored a factor that significantly affects system performance. Merely being able to conceptualize certain cognitive and motivational characteristics ultimately forces us to investigate them and include these factors in development.

This creates still other problems. If we have determined that certain cognitive and motivational functions are important to system performance, then as system designers we must include these factors in design, which means determining the hardware/software mechanisms needed for their functioning.

All of this may seem like vague philosophizing (the author seems to recall echoes of his graduate training in logical positivism), but it can be expressed more simply as a paradigm: If conceptual analysis suggests that a certain factor or quality exists, it is necessary to investigate that factor or quality empirically. If it appears that the factor or quality has a more than minimal effect on system functioning, it becomes necessary to incorporate that factor or quality into the system.

All of this is logical, but system development is only partially a logical process. Logic ceases to be effective when we lack the knowledge to implement the logic. If we recognize that lack of knowledge, it becomes incumbent on us to remedy the lack.

Human–Technological Relationships

It has been pointed out that until highly automated, computerized systems were developed, the operator controlled an equipment, energized it, directed its performance by attending to equipment status displays, and modified that performance by manipulating controls in a specified sequence at particular times until the equipment was turned off. The human–technological relationship in this case makes the operator a controller.

With the advent of computerized systems, the operator's role changed. He or she monitored system status, but it was the computer software that directed individual equipments to perform in specified ways. The operator's role was now to monitor, interpret a sequence of signals, and take remedial action if the system deviated from preset parameter values. In this situation, the human is in a partnership relationship with the machine.

The reason for going into so much discussion of the human–technological relationship is that so many HFE concerns are now changed. The operator's role now emphasizes perceptual recognition of stimuli and cognitive interpretation of the meaning of those stimuli. Because the system situation is much more fluid than it was when the operator controlled a proceduralized equipment, the operator must now manifest much greater flexibility in his or her performance and this emphasizes decision making. The implications of the new situation are that the nature of what we define as *error* changes; because of this, the measurement process also changes.

It was noted previously that, for some people, technology becomes merely background because they do not use technology or at least some part of technology (because it is difficult to avoid interaction with all machines). There are variations in technology usage. Most people do not operate all the technology that is available. When one is not the pilot of an aircraft but only a passenger in flight, one is a client in the human–technological relationship. The client is the beneficiary of services provided by the technology, and the major effect resulting from that technology is satisfaction or lack of it. If one never flies as a passenger, as a few people in industrialized countries do not, then technology as represented by an aircraft in the sky is only background stimuli for them. They are spectators in the human–technology relationship.

Two questions: Does technology as background or context affect people strongly? If so, what is that effect? Technology as background certainly affects people because it serves as a source of stimuli to which the human responds. The response is almost never in terms of overt performance because in this situation one is not an operator and one need do nothing. The response may be a change in the spectator's attitude or his or her concept structure.

None of this affects the system developer (who is probably saying, Thank God, or we would have even more tasks), but this is part of the human–technology relationship and the HFE researcher should be aware of it. He or she may think that the investigation of the effect of technological stimuli on human attitudes is more properly the province of the psychologist, anthropologist, or sociologist. However, if one of the purposes of HFE is to make people more comfortable with their technology, then all responses to technology, even of people who do not use a particular technology, should be considered.

TECHNOLOGY

Introduction

The major difference between the human and technology is that the latter is a construction of the former. This means that the human impress on technology is everywhere, and the more complex the technology the greater the impress. For example, the variations one can design into a penknife are limited, but those one can design into a computerized expert system are much greater.

Complex technology has existed for thousands of years (James & Thorpe, 1994), but only recently have people become concerned about its effect on the human. Early industrial engineers like Taylor and the Gilbreths (see chap. 4) were largely interested in improving the productive outputs of the worker. Only in the last few years has a humanistic influence developed that views the human as more than a *unit of production*. HFE—in its sociotechnical, macroergonomic, and organizational development interests—has become to a limited extent the ombudsman in the effort to restrain the potentially negative effects of unbridled technology.

What we have come to understand in more recent years is that there is a reciprocal relationship between technology and the human. The human, as represented by design engineers, managers, and capitalists, determines the shape and extent of technology. Once developed, the technology controls to some extent (sometimes greatly) how the human performs.

The Elements of Technology

Table 1.3 begins with a taxonomy of technology elements. There is obviously a progression of complexity, varying from individual components like microchips at the simplest end of the continuum (although obviously a computer chip is a complex device) to systems at the most complex end. Manifestly, molecular technology, such as resistors, transistors, and circuit boards, func-

TABLE 1.3
Taxonomy of HFE Elements: Technology

Technology Elements	Effects of Technology on the Human
Components	Changes in human role
Tools	Changes in human behavior
Equipments	
Systems	*Organization–Technology Relationships*
	Definition of organization
Degree of Automation	Organizational variables
Mechanization	
Computerization	
Artificial intelligence	
System Characteristics	
Dimensions	
Attributes	
Variables	

tions without human involvement unless the component fails and must be replaced or repaired. Simple tools require human involvement that is equally simple, although in the case of craftsmen performing tasks of great subtlety (such as the creation of jewelry) the precision of human functioning is greater. Equipments (e.g., printing presses) require continuing human intervention—notably continuous operation of the equipment. Systems, which are combinations and integrations of individual equipments, require the greatest and most complex amount of human involvement.

There are qualitative differences in the nature of human involvement with each technological element. As noted previously, there is little further human involvement once the component is created. Tools require a narrowly focused human involvement; the worker's attention is restricted to the tool and what he or she is doing with that tool. The range of human attention is greatest when the human operates, maintains, and monitors the system.

Such distinctions are somewhat arbitrary, and human interaction with the technological element is more of a continuum than that represented by discrete categories. Nevertheless, it is probable that as technology becomes more sophisticated, the human's involvement also becomes more sophisticated by calling on more human functions expressed in more differentiated ways. This is particularly so when one is interacting with systems. The topic of the system, its engagement with the human, and its relation to the discipline as a whole require much greater attention, which is provided in chap. 3.

Degree of Automation

The most significant dimension in technology is its degree of automation. We can distinguish three levels: (a) mechanization, (b) computerization, and (c) artificial intelligence. Mechanization merely replaces a human performance with that of a machine. For example, we use a steam engine as power to replace human muscles. In mechanization, the human is the controller of the equipment—starting it up and making it run through its paces.

Computerization also replaces human performance, but the mechanism is now a computer and, in particular, its software. The replacement of man by the machine is now much more extensive, more precise, faster, and so on. Mechanization reaches its highest level with replacement of the human brain by what has been termed *artificial intelligence* (AI; Fu, 1987). We have not presently achieved genuine AI but its development is being actively pursued and it seems likely that we shall some day achieve it. However, until that day occurs, HFE does not need to concern itself with AI except possibly on an experimental basis. The marriage of the system with progressively increasing automation gives the human a tremendous tool for control of technology.

System Characteristics

The system is on the technological continuum, but is so qualitatively different than equipment and tools that it is necessary to devote an entire chapter (see chap. 3) to its discussion. This is because it has dimensions, attributes, and variables that render it in some way *sui generis*. The author regards the system and system concept as fundamental to HFE, but little thought has been given in the HFE literature to this topic. This means that more effort must be given in this book to compensate for the lack of attention in other documents.

Obviously technological systems like the aircraft, automobile, or computer have been studied in the literature. However, as is seen in the chapter 6 discussion of HFE research characteristics, the system in HFE research serves merely as a source of stimuli to which the human responds. The author sees a great deal more in the system.

Effects of Technology on Human Performance

Much of what can be said about this topic has already been said in previous sections. In a technological civilization, technology is (next to sex, perhaps) the most potent stimulus for human responses. It not only changes the human role in immediate interaction with physical objects, but serves as a backdrop for almost everything the human does.

This is something the HFE professional is not likely to think about perhaps because it is so obvious. The notion that everything one does is somehow affected by a ubiquitous technology is a concept more likely to occur to the historian, anthropologist, and sociologist than to the HFE specialist. That is because the former are observers of the human condition, whereas the HFE professional is an activist who has been trained to intervene in the human–technology relationship by measuring performance, creating a stimulus, and observing its effect. Such an act of intervention is more likely to produce relevant data. At the same time, the emphasis on intervention as opposed to observation is likely to constrain the HFE specialist's mental purview to the specific act of intervention. A physical analogy might be that he or she observes in a 90-degree quadrant with high resolution, whereas the more passive anthropologist or sociologist observes in terms of 180 or 270 degrees, but with less resolution. The HFE specialist sees only the results of the immediate experiment and its relationship to experiments on the same subject matter.

The HFE obsession with experimental intervention has led to a general neglect of passive measurement techniques, such as observation, and a reluctance to use subjective techniques in part because there is less intervention in the use of subjective techniques than in experimentation. It is possible for someone who is not an experimentalist to take a wider view of phenomena. This is not a rejection of experimentation as a research technique, but a recognition that excessive concern with such a technique has a few undesirable side effects.

The preceding is only a side comment, but the thoughtful HFE professional might well consider how his or her role narrows the view he or she takes of the phenomena being observed. The reader may well believe that all of this is useless philosophizing (the author can imagine him or her saying, get on with the theme; what can one do with these concepts?). However, to the extent that a discipline is the product of a conceptual viewpoint, the concentration in HFE on measurement by means of experimentation (which is an act of intervention) inevitably colors what one sees and concludes from the experiment. In any event, technology has tremendous effects on human behavior, which is why it is a subject for scientific consideration.

The increase of automation means that a qualitatively different and widely expanded range of human functions are now required of system operators. In more primitive mechanized equipment, the human was required to follow a preestablished set of procedures and thus to make use largely of perceptual functions, to read dials and gauges, and simple motor behaviors by throwing switches and pushing buttons to force the equipment into preset procedures.

In automated systems, although perceptual functions are still important to receive individual signals, much greater use must be made of cognitive functions to integrate a flow of symptomology and interpret its meaning. Whereas before symptoms of deviant equipment performance were largely individual, although they could occur sequentially over time, now symptoms can manifest themselves concurrently from multiple subsystem sources and must be treated in relation to each other, in much the same way that a physician interprets a number of physiological indexes of sickness to form a diagnosis of what is wrong with the patient. In the presence of a malfunctioning device, the operator now acts as the physician would act (see Sanderson, Reising, & Augustiniak, 1995).

What this means practically for HFE is a shift of attention from routine, proceduralized equipment activities to those of trouble shooting and, concurrently, a greater need to examine cognitive and, in particular, the operator's decision-making (DM) behaviors. Although in the past (see Edwards, 1987) there has been much attention given to DM, this has been in the context of an experimental laboratory environment. Attention to what has been termed *naturalistic DM* (Klein, 1993) has been relatively recent. In HFE research, attention to maintenance and trouble shooting performance in particular has been much less, as is seen in chapter 6. Again these are significant impacts on HFE research.

There are also changes in human behavior, as differentiated from performance, but these are much more tenuous. For example, there are intimations that use of computers by humans creates a feeling of increased power and control in the human and a progressively increasing reliance on larger and larger databases as a prerequisite for DM. In addition, technology can be enthusiastically accepted by younger people and just as enthusiastically rejected by a number of older people. We should not stereotype the older cohort because there is a continuum of acceptance and rejection of technology. The latter is manifested by a dislike of automated devices, increased difficulty in utilizing them, and so on (see Davies, Taylor, & Dorn, 1992). Although such variations in human responses to technology are important, we are only in our infancy in understanding them; they appear to be less important than the changed operator role.

Organizational–Technology Relationships

The development of complex automated systems inevitably brings with it what can be called an *organization*. The simplest definition of an organization is that it is a set of rules for humans to interact with and within systems. Such rules are necessary because the interaction process often determines the efficiency of system performance.

In simpler equipments, particularly those that are proceduralized, the organization consists of the operator and procedure for running the equipment. When operators begin to function in teams, where there is interaction among them in terms of communication, and, in particular, where contingency is permitted in their response to stimulus events, and hence system flexibility, one may speak of a true organization.

An increase in the complexity of technological entities invariably creates a need for an organization to manage these entities. An individual equipment does not require an organization, but a system consisting of *N* equipments and the humans to run them does require some sort of organization to coordinate equipment and personnel operations.

Regardless of whether one likes it (and it is the author's impression that many HFE professionals probably do not because it is difficult to intervene in organizations experimentally and because social factors enter much more into organizations than in the strictly human–machine entities that HFE professionals study), the organization is an integral part of technology. It is not even a concern of modern technology because one supposes that the Egyptian bureaucracy responsible for raising the pyramids was probably much concerned about the organization of their peasant work groups. The organization is human-centered because it makes relatively little use of equipment (e.g., primarily simple communication devices such as telephones, FAX machines, etc.). The organization is not fixated on machines as such except as aids to human processes. It is a system, or rather a subsystem of the total system, because a system is in essence only an organization; but it is a human or procedural system rather than a human–machine system. It does not produce outputs (other than paper), but aids the processes that do produce physical outputs.

Therefore, the organization has much more in common with industrial and social psychological concerns than with more hardware-oriented equipment operations. However, regardless of whether one likes it, if the HFE specialist is going to work with large, complex systems, he or she is going to encounter organizations as an inevitable correlate of systems. Organizational/system relationships are discussed in chapters 3 and 7.

THE DISCIPLINE

If the reader refers to Table 1.1, he or she will see that, in terms of the taxonomy described, we have dealt so far only with the HFE concept structure and its elements. The following section follows the organization of the remainder of Table 1.1, but does so briefly. Additional related material is described in subsequent chapters.

HFE Professionals

The characteristics of any discipline depend greatly on what its professionals do and how they conceptualize the nature of their discipline. Professionals also have a concept structure that is coordinate with but less formal than the more formal disciplinary one that is reported in textbooks.

The formal concept structure of HFE is expressed mostly in terms of theories such as that of the human–machine system, the signal-detection theory of Green and Swets (1966), and the attentional resources theory of

Wickens and his colleagues (e.g., Wickens & Goettle, 1984). Because the concept structure of the professional (which is almost never discussed in texts) is far less formal than that of the discipline, it includes many more things, such as the nature of the discipline, its relationship to other disciplines, what the discipline should encompass (its scope), the role the discipline should play in research and application, and what the products of the discipline should be. The concept structure of HFE professionals is discussed in more detail in chapter 5 as part of what the author calls the *informal history* of HFE.

Purposes of HFE

The first thing to note is that the heading of this section is multiple. Some purposes have to do with productivity, others with comfort, and still others with adding to the knowledge base. Moreover, one can approach any single purpose of HFE at two levels: one, the ultimate effects one hopes to achieve with the discipline; the other, at a somewhat less remote distance from its operations, what one hopes that HFE will be able to do to implement its ultimate purposes.

Among the ultimate purposes of HFE one would hope that, because HFE is a science, it would add to the knowledge base describing human–technology relationships. Then because one assumes that HFE should have utility to people, it would be desirable for the discipline to (a) increase the productivity of humans interacting with machines (and thus to increase the productivity of the overall system—an assumption in itself that increasing human productivity will also increase the productivity of the systems of which humans are a part); and (b) increase the comfort/safety that personnel interacting with machines feel while operating these machines (or, at the least, preventing machines from imposing excessive stress on their personnel).

This last purpose divides into two. Increasing comfort for operators requires much more than merely preventing machines from imposing unacceptable stresses on operators. Comfort means adding something to technology it did not originally possess; preventing stress means avoiding certain technological characteristics that impose on people. How does one build comfort and satisfaction—its corollary (both motivational and idiosyncratic in nature)—into machines?

As was said previously, these are ultimate goals for the discipline. To achieve these, it is necessary for the discipline to have certain implementation goals. These are: analysis of the human–technology relationship, translation of human (behavioral) variables into technological (physical) variables and vice versa, and prediction of the direction such relationships will take.

It is necessary to distinguish between purpose and function. Purposes (when accomplished) create certain desired effects; functions are performed to achieve these purposes. Sometimes purposes and functions are difficult to distinguish. For example, research and application are major (top level) functions of the discipline. Analysis, translation, and prediction, while helping to implement the overall purposes of the discipline, also assist the functions of research and application.

In all three of the implementation purposes, the knowledge aspect is most important. We want to know what the human–technology relationship is and so we analyze that relationship largely through the research function. Because technology and the human inhabit quite different domains (human = behavioral, technology = physical), we want to know how the domain effects are manifested. A more concrete way of putting this is that the physical characteristics of technology (characteristics of equipments and systems) produce certain behavioral (performance) effects in the human and we want to know what these are. Similarly, the characteristics of the human (e.g., anthropometry, intelligence, skills) are most effectively utilized by certain physical (equipment/system) characteristics. This describes the translation purpose, which can be more inelegantly expressed as which technology characteristics are equivalent to which behavioral characteristics. Knowledge of this type is of tremendous utility in the application of behavioral principles (e.g., design guidelines) to the more effective development of systems.

A subpurpose of translation is prediction, presumably in a quantitative manner. If one knows what a technology–human relationship is (e.g., knowledge gained by experimentation in a laboratory), one should be able to make certain predictions as to how this relationship will manifest itself in a different environment and under different conditions.

The interesting thing about these purposes is that their validity depends on whether professionals think that these purposes are desirable and possible. If they do not maintain these beliefs as part of their concept structure, all the positive statements in texts are meaningless. One can only estimate what the extent of the beliefs among professionals is because surveys do not ask these questions and individuals are likely to conceal their true feelings for the sake of scientific correctness. The reason for going into such length about purposes is that belief in these purposes has practical implications.

In reverse order, if HFE professionals believe that prediction is an important purpose for the discipline, they will work at developing predictive databases. If they believe translation is important, then major deliberate efforts will be made to develop design guidelines specifically applicable to system development. If they believe analysis is a critical element of the disciplinary purpose, they will endeavor to perform more research.

If one were to conduct a questionnaire survey of the strength of belief in each of the three purposes, one would probably find that almost all professionals believe in analysis because they see analysis and research as essentially equivalent. Most, but certainly not all, HFE professionals would consider that translation was a critical purpose. Relatively few professionals would consider prediction (at least in quantitative terms) to be a critical purpose.

Considered as abstractions, they might nod their agreement to all of the prior notions. Considered in terms of what they do or would not do in relation to these, there would be far less agreement. For example, almost no researchers seek to develop design guidelines from their research (a function of the translation purpose), and even fewer are concerned about the development of predictive databases (prediction purpose).

The purposes are important if we keep them in mind, but few professionals do. These purposes that are nested in the remote recesses of the mind are rarely brought forth to daylight and dusted off. Professionals of all disciplines rarely like to be brought back to consider first principles; they feel they have gone beyond these, but sadly how much they lose by not thinking of the discipline as opposed to their daily activities.

Scope of the Discipline

The scope of the discipline is the conceptualization of its professionals of what the discipline represents or encompasses. That scope can be narrow or broad, and the scope one adopts can have discernable effects on aspects of the discipline (such as what is researched) and on the professionals in the discipline. If the narrow scope is adopted, this limits the discipline to research about those variables that affect the human when the human interacts with technology and to application of research to system development. This is a fairly broad palette and will include almost all the activities in which HFE engages. However, it is less encompassing than the broad scope.

The broad scope or definition of the discipline says, in effect, that everything relating the human to technology is within the purview of the discipline, although for practical and interdisciplinary reasons HFE professionals may not wish to address the full scope of the discipline. For example, the attitudes of technology users to the technology they use is within the broader scope of the discipline. Except for an occasional article about this aspect, not much is done with this subject because many HFE professionals view it as a subject for urban anthropology, for example, or some other *softer* discipline. The sociological effects of technology, as it relates to the effects of urban overcrowding, crime, morality, and so on, can be considered part of HFE in the broad scope, but almost all HFE professionals would allow sociologists to deal with these phenomena (except perhaps those like Moray, 1993).

The author adopts the broadest possible definition of the discipline, but he does so theoretically, in the sense that he believes nothing in the technology–human interaction is foreign to HFE. However, certain topics are more important and more practically achievable in HFE than are others and he would prefer that we work on these, but theoretically he can admit of no restrictions on disciplinary scope.

In particular, although the discipline concerns itself with factors that affect the application of HFE to equipment/system development, it is entirely proper for those interested in the topic to examine how people use their technology once it is developed and made available for operation. Thus, for example, it is perfectly appropriate to ask how frequently automobile users clean their cars, how frequently office workers use elevators, and so on. The utility of such research would vary, of course, but that is true of all research topics. Anything is grist to the HFE mill, which should encourage innovation.

Functions Performed by the Discipline

Closely related to the purposes of the discipline are the functions the discipline performs. The two primary functions are research and application, each of which subsumes a number of more molecular implementing functions. For example, research implies the selection of a research topic, the venue or environment in which the research is performed, the selection of subjects, and so on. This is followed by performance of the research, analysis of the data, writing of a research report, submission of the report to a publication medium, presentation of the results to an interested audience, and so on.

The application function has even more variegated subfunctions. These include application for a job; performance of the job, which, if in system development, includes analysis of a development problem; development of alternative problem solutions; selection of one solution; testing of the solution; modification of the solution product; development of personnel training; and so on. Each of the preceding subfunctions has subfunctions of its own, but a description of these would require a book of its own.

The discipline, considered as an organizing mechanism, plays a primary role in the research function by providing concepts that will eventually become research topics and publication/presentation outputs. The discipline has little or no influence on application because application is energized and organized by nondisciplinary and nonbehavioral (e.g., engineering or commercial) entities.

Even in the research function, the discipline has limited effectiveness because it has almost no control over funding sources, although these sources may call on the discipline (i.e., its professionals) for suggestions as to what to fund.

The discipline has a strong influence on the training provided to potential HFE professionals. This occurs not only through more general functions of communication and publication, but more specifically through its accreditation of university training programs. It indirectly controls training content by specifying what it will or will not accept as accreditation criteria.

Regardless of whether one considers it as a specific function, the discipline organizes its subfunctions like communication and publication. It does so through the efforts of its societies, which represent the effort to take abstract ideas such as the concepts the discipline deals with and to transform these into concrete activities. The discipline is a construct, an abstraction, but its functions transform these abstractions into concrete activities. Hence, although it might be difficult to pinpoint the abstractions, it is possible to specify their transformed, concrete representations.

At the same time, the discipline has limited power to influence these functions. The discipline operates not only directly (through its publication mechanisms), but more so as a context within which certain functions are implemented. All of these functions make it possible to deal with the discipline as a semiphysical entity.

Analysis of human–technological relationships enters more immediately into the design or application function, over which the discipline has relatively little control. However, the discipline does have the capability (which it almost never exercises) to analyze and evaluate its own activities and thus to provide a certain amount of feedback to its members as to how well the discipline pursues its goals. Occasional evaluative letters to the editors of publication media such as the Society's monthly bulletin or papers published in journals or proceedings documents permit individual professionals to reflect on this evaluation activity. The discipline's analysis/evaluative capability can serve as a sort of homeostatic mechanism to keep the discipline, using a somewhat fanciful term, *on course*. To perform such analysis/evaluation, the discipline through its professionals must have some concept of a goal and progress to that goal. Unfortunately, that goal is often not well articulated and the evaluative function is often not well performed, when it is performed at all.

Research and application are the two primary activities with which the discipline (as an overarching construct) gets involved. Inevitably, the question arises of the actual and desired relationship between these two. Does research stand apart from application? Can application (i.e., system development construed in its most general sense) function without relating to research?

It is often implicitly assumed that the outputs of research should be utilized in application. Looking at the history of science over the past 400 years or so, it seems that, reasoning from its consequences, one of the major purposes of research is to provide usable inputs for more utilitarian purposes. For example, ship navigation made use of astronomical knowledge. The knowledge produced by Harvey about the circulation of the blood had a tremendous effect on surgery. In the case of HFE, the utilization of research in system development would seem all the more necessary because the latter requires a transformation from one (behavioral) domain to another (physical) domain.

Such a transformation requires a deliberate research effort of its own, and there are difficulties involved—most notably, developing the energy to undertake the transformation process. Despite adjurations, a substantial percentage of HFE researchers may well have no desire to get involved in the possibly tedious and difficult effort to make use of their research. As is seen later, the need to make research results useful may have implications for the selection of research topics (thereby interfering with the self-interest of professionals). If one sets as an overall research goal the development of useful information, large areas of research interest may be put out of bounds.

Of course, the application arena—primarily the development of systems—may have research questions of its own that are worthy of research investigation on their own. More is said about this later.

Conceptualization of the Discipline

A central theme of this book is that how professionals conceptualize their discipline has a significant effect of the way in which the discipline operates. One aspect of their structure is how they view their discipline in terms of three possible options:

1. HFE is a subspecies of another discipline. Other disciplines have made significant contributions to HFE, the most important ones being psychology, engineering, and physiology. Many professionals feel that what they do is a part of psychology (engineering or industrial psychology) and others that it is a special aspect of either engineering or physiology.
2. HFE is an interdisciplinary discipline—an awkward phrasing that means that what is distinctive about it is its orientation or approach to solution of certain problems.
3. HFE is a distinctive, hence autonomous, discipline with features peculiar to itself that differentiate it from other disciplines.

The concept that professionals have of their discipline (whatever that concept is) has significant effects on the manner in which they practice their profession. If HFE is a subordinate part of another discipline, it is likely that those who have this concept will utilize concepts and methods derived from the superordinate discipline. These concepts and methods may be more or less congruent with the purposes and requirements of the subordinate discipline (HFE).

If HFE is considered interdisciplinary, it is likely that those who possess this notion will lack the will to develop special concepts and methods that distinguish their work from that of any other discipline. There may be borrowing from other disciplines, but it is just as likely that HFE methodology will be fortuitous and heuristic, or that those who maintain this option will not seriously endeavor to create distinctive disciplinary features. If HFE is considered a distinctive discipline, it is more likely that professionals will look for new methods with which to attack problems. Indeed, it would be assumed with this attitude that the problems encountered are distinctive and require novel solutions.

Consequently, the different attitudes will have different effects on HFE analysis, measurement, and research in general.

HFE as a Disciplinary Subspecies. HFE developed primarily as an outgrowth of psychology. The initial application of experimental psychology to the solution of weapons-related problems in World War II was not HFE as we know it today. Indeed, that is why HFE has a history that we endeavor to describe. Other disciplines have had (and probably still have) an influence on HFE, but that influence was and is minor compared with that of the predecessor discipline—psychology.

At some point in its history, HFE divorced itself from the parent discipline. The divorce is incomplete, however. Moreover, because much of the training given to HFE professionals was (and still is) of a psychological nature, the concepts and methods of the predecessor have carried over to the new discipline.

One might also conceptualize HFE as a branch of engineering because certainly HFE has in certain respects modeled itself on engineering (its action orientation, primarily). One might also think of HFE as a species of physiology or safety because it is certainly concerned with some of the problems raised by these disciplines.

HFE as an Interdisciplinary Discipline. This point of view is perhaps only a variation of the preceding one. Early in the history of HFE (HFS Bulletin, April 1963), the Netherlands Ergonomics Society declared that it had decided that ergonomics is only an attitude or approach, which of course it is—although if that is all it is, the effects can be disastrous. Such an orientation dismisses the entire concept of HFE as a discipline, therefore attempts to develop a formal, distinctive conceptual structure and methodology are pointless. Obviously HFE was and is influenced by a number of other disciplines; this fact and the apparent inability to discover that there are any distinctive features in HFE lead a number of professionals to this option.

HFE as an Autonomous Science. The point of view maintained in this book is that HFE is a distinctive discipline that utilizes the methodology of science (e.g., empirical investigation, experimentation, measurement, derivation of conclusions) and therefore is an autonomous science.

The reasons for maintaining this point of view is that no other discipline is concerned with the relationship between the human (behavior) and technology (engineering). Psychology is not concerned with technology; industrial psychology, which is most similar to HFE, is concerned primarily with working conditions and schedules, worker selection, health and safety, and worker attitudes toward the job. A central component of HFE—system development—is not even mentioned in psychology texts. However, if one confines oneself to HFE and psychological research, the identification is much closer because, as is demonstrated in chapter 6, HFE research has retained its psychological orientation (which is one reason for its comparative ineffectiveness vis-à-vis application).

If one looks at engineering, one finds nothing behavioral in it. Of course, HFE has been influenced by engineering, but that influence has been exerted on professional attitudes (e.g., pragmatism, lack of concern for laboratory research, preference for heuristics), not in any research (few HFE research studies have been initiated by technological problems; see chap. 6).

Physiology is self-centered; it is concerned only with the individual's physiological responses to stimuli, among which are environmental conditions like temperature, noise, lighting, and acceleration, which are also the concern of some HFE specialists. No one would seriously think of HFE as a species of physiology simply because HFE is concerned with many things that do not involve physiology.

If, as has been demonstrated, the three major disciplines that have influenced HFE do not account for certain HFE aspects, then logically HFE must be a distinctive discipline. Whatever points of overlap exist between HFE and other disciplines the overlap is small enough so that HFE remains distinctive. If HFE is distinctive, then any attempts to develop distinctive concepts and methods for it, as suggested in this book, are worthwhile.

HFE ORGANIZATION

The functioning of a discipline over time results in the progressive fractionation of the discipline into specialties based on the personal interest of the professional in a part of the total discipline and the practical impossibility of being equally conversant with all aspects of the discipline. One finds fractionation in all disciplines, and it is impossible to prevent it (nor would one wish to do so). One becomes then a specialist in one or two fields and a generalist in all others, just as one may become a specialist in cardiology while still being able to do simple surgery, general diagnosis, and so on. The same applies to HFE. The author thinks of himself as a specialist in system development as well as test and evaluation, but a relative novice in cognitive engineering, aviation psychology, and so on.

This question may be asked: Is there a fundamental set of core knowledges in HFE? There is, and this core, which applies primarily to research, has been largely taken over from experimental psychology. It includes a basic knowledge of human anatomy and physiology; certain key concepts, primarily the S–O–R paradigm, experimental design (including how to perform an experiment or conduct research generally); and a basic knowledge of statistics. To this might be added in the present era a knowledge of computers and software.

What is probably a HFE core knowledge is the ability to analyze problem situations and extract the variables that impact the problem and how they do so. This ability is to be found in both research and application, but is particularly marked in the latter, where it is the initial phase of system development.

Whatever their idiosyncratic interests are, specialties have many points of similarity, these being part of the core HFE knowledge (see chap. 7). What is important about the specialties is the effect they may have on the overall discipline by diverting their professionals' attention away from problems of the discipline as a whole to those of the specialty area.

This assumes that the discipline as a whole has problems that the specialist may not recognize as problems because, in the context of a specialty, they are not recognized as such. For example, specialties that emphasize the idiosyncratic functioning of the individual (e.g., aging, cognitive engineering) may not recognize the importance of systems to HFE. The prediction of performance will seem to the specialist in visual performance to be of little importance. To the extent that each specialty seems to have problems of its own that need solution, the importance of problems that affect other specialties (and thus the discipline as a whole) will appear to be remote and hence unimportant. Specialists in a particular field define themselves at least in part by the problems that are peculiar to their specialty. Therefore, they are less likely to recognize the problems of other specialties, which in toto are the problems of HFE as a whole. General problems are obscured by the details of the specialties. The rejoinder to this by the specialist is that the solution of the particular problems of the specialty contributes to the solution of HFE problems as a whole. However, the solution of specialty problems may be essentially irrelevant to the discipline as a whole. For example, if one of the major problems of industrial ergonomics is the development of a model for strength (e.g., Kim, 1990), how will the development of such a model assist in the general HFE problem, of, say, developing design guidelines?

Considering that specialization is inevitable, what can one do not to retard the advance of specialization, but to encourage professionals to think in disciplinary as well as specialty terms? An attempt to find the common elements in the various specialties would help because these would presumably reflect the discipline as a whole. Chapter 7 describes one such attempt. However, if the specialist defines him or herself solely by the characteristics of the specialty, it will be difficult for the professional to recognize the claims of the discipline.

HFE Methodology

A discipline is distinguished primarily by its subject matter, which, in the case of HFE, is the relationship between the human and technology. It is also distinguished, but in much lesser degree, by its methodology.

The *hard* sciences (e.g., physics, biology, chemistry) need not be concerned about the consciousness of its subject matter because that subject matter, physical forces, chemical, biological elements, has no consciousness. This is not true of HFE, which has this in common with all the *soft* sciences—that the subject matter of its measurement efforts is affected by that measurement. The Heisenberg effect does apply to the physical sciences, but much less so. HFE, along with psychology, has the problem of accommodating both objective measures of performance and subjective ones. The problem is that the measurement in two ways of the same phenomenon may produce discrepant results. At the same time, it seems to be illogical if one were not to give one element of the measurement, the human, an opportunity to provide data. If atoms could talk, would not physics utilize their communications?

The objection to subjectively derived data is not that they are subjective in origin, but that they are skewed in unknown ways by the variability inherent in the human. The HFE ethic derived from general science tends to denigrate subjective methodology because of this variability. Humans are assumed to be biased by self-interest and flawed by errors in perception.

At the same time, the discipline cannot refuse to face the fact that much of its phenomena are mediated (communicated to an observer) by subjective means. In psychology, this led in certain circumstances to a rejection of all subjective data and the selection of subjects who cannot communicate except by their performance: witness the running of rats. However, this too is illogical because using the rat or primate as a subject (e.g., in learning studies) assumes that there is some relationship between the animal and human performance. Thus, the inability of the subject to communicate becomes in a twisted way an advantage for the science. Rats cannot be used in HFE because they can interact with technology, only in the most primitive way, by bar pressing, for example; but if they could, some HFE researchers would probably espouse their use as subjects.

The philosophy of objectivity in data has led in HFE to a reluctance to deal honestly and fully with the subjective aspects of performance. That there are such subjective aspects is obvious, but there has been a continuing tendency to devalue those aspects in HFE research. The existence of the problem must be recognized and methods developed to solve the problem. To do this the discipline must recognize that a conscious object of measurement presents a problem as well as an opportunity to gain knowledge that would otherwise not become available. An entirely objective methodology is limited because there can be no communication between the researcher and the measurement object. Only an extreme behaviorism attached to the *black box* concept of the human would adopt such a methodology.

Another methodological problem the discipline faces is its preoccupation with experimentation to the exclusion of other measurement techniques. It has already been pointed out that the experiment is an artificial method and that, although it has advantages for examining the variables that affect a phenomenon, it also creates problems in terms particularly of validation of experimental findings. Because of this artificiality, there is a problem with testing their validity in real-world environments. In an introductory chapter, it would be unwise to discuss the problem of experimental methodology in great detail (this will be done later). However, it must be recognized that the experiment does present a problem to a hybrid discipline (i.e., both physical and behavioral), which has multiple test venues.

Factors Affecting the Discipline

Technological Changes Over Time. Because the HFE subject matter is the relationship of the human to technology, major shifts in that technology will have significant effects on the discipline. Indeed, HFE developed as a discipline in response to the technological changes occurring in World War II. More recently, the major shift to computers and automation has *forced* (although the word is somewhat inappropriate) many professionals to divert their attention to the new technology. The increase of automation in certain industries has changed the role of the human in those industries and has required changes in HFE in the definition of error and measurement methods. These technological changes are not unwelcome, but they obviously require changes to HFE. At the same time, as the technology changes to focus more on cognitive behaviors than was previous practice, the importance of the human and HFE specialists who study the human becomes greater. If designers can be pressured to recognize the new situation, HFE may be given greater opportunities by engineering to assist in system development. This may or may not result in increased respect for the discipline and a healthier relationship with other disciplines.

Personnel Training. The training that HFE professionals receive obviously impacts their performance, particularly in research. (In application, the nature of the design situation is much more important.) Until the 1950s, at least, that training was psychological. With time and the development of accredited university courses, the content of that training has expanded to include some engineering and a heavy emphasis on computers and statistics. A discussion of the nature of what the novice professional should know is left to a later chapter. The author's underlying premise is that training should be driven by what the novice will later be asked to do. Whether the university actually presents sufficiently realistic material is unclear.

The importance of training stems, in part, from the intellectual baggage that professionals carry away from that training. Those who in the past and at present received their training as a specialization within the psychology department of the university are likely to have received a number of courses in general and experimental psychology, and have been influenced by psychological concepts and methods. Such training was characteristic of the first generation of professionals. Later it would be possible to receive advanced degrees in human factors within the industrial engineering or the operations research department of the university, and one would expect much more emphasis on nonpsychological concepts and methods. Whatever other training they received, present professionals have been strongly indoctrinated in experimental psychology methods, advanced mathematics and statistics, and computer science.

Those who have been strongly influenced by their psychological training are probably more apt to think of HFE as a subspecies of psychology and to use psychological concepts and methods in their work. Those emerging from a nonpsychological training environment are perhaps more likely to think of HFE in nonpsychological ways. These attitudes almost certainly influence in some measure the research they perform.

The nature of psychological training for the future professional emphasizes highly sophisticated analysis of psychological concepts particularly when these are examined in an experimental context. Ph.D. examinations in psychology often emphasize close analysis and critique of experiments. This tends to produce a highly sophisticated researcher, one who "plays exquisitely on" molecular behavioral variables, but it is unlikely to produce people who can develop new concepts—particularly those that deviate from classical psychological ones. This inability to move beyond psychology has a tendency to retard HFE progress because exclusively psychological concepts are only partially relevant to HFE problems.

The inability to experience the application environment in a university setting, which only partially simulates that environment, may lead to a lack of understanding on the part of some HFE researchers of the genuine conceptual problems posed by application and particularly by system development. This is evidenced by the infrequency with which problems arising from the development process lead to published research studies (see chap. 6).

Professional Organizations. The HFE discipline, which is a construct in the minds of its professionals, is made manifest by two things: the work they perform and the societies developed to represent the discipline. The societies do many things, but the function that is most important to us is to help organize the concepts of the discipline and communicate information about these, primarily through research reports and symposia. How well the societies perform their functions significantly affects the well-being of the discipline.

The major American society (the Human Factors and Ergonomics Society [HFES]) is described in some detail in chapter 5. There is also Division 21 of the American Psychological Association (APA), together with the International Ergonomics Association (IEA). However, the last has little influence on American professionals because it is headquartered in Europe. One society task that is examined critically here is whether the societies examine and analyze critically how well the discipline is performing its functions.

Research Funding. Whatever HFE research is performed must be funded; there is no nonfunded research except possibly as masters' theses and doctoral dissertations. Even these, when they require exceptional physical resources and subjects, must be paid for by some agency. "Whoever pays the piper, selects the tune," therefore, funding agencies determine research topics and often the way in which the topics are addressed. Therefore, funding, which is largely provided by governmental agencies, drives the discipline in its research mode.

Of course, the bureaucrats in the funding agencies are often (possibly typically) HFE professionals. To that extent, the discipline, which has created these professionals, drives its own destiny, but not completely or indeed in large part. Extradisciplinary influences in the funding agencies are strong and may well be as strong as any disciplinary determinants. The amount of money available to support research (never sufficient to address all possible problems) requires the selection of certain research topics and the rejection of others. The emergence of a new technology like computers will obviously drive much of this funding because the novelty of the new technology will raise questions that must be answered by research. Attitudes of nonprofessionals in charge of funding will also have an effect; to the extent that nonprofessionals have a positive or negative attitude toward HFE, or a particular aspect of it, will determine the amount of funding.

Job Availability. However much one might prefer the discipline to be concerned solely with conceptual processes, nonconceptual factors exercise a strong influence. For example, if there are no jobs for HFE professionals to support themselves, those unable to find jobs will drift away from the profession. Therefore, the discipline is affected by economic expansion and retraction and more so perhaps than are other disciplines, which are considered by *money men* to be more important than HFE. For example, the ending of the cold war produced a contraction within industries developing sophisticated weapons systems. To the extent that HFE professionals were working as part of the development teams for these systems, they felt the resulting contraction and had to scramble to find other jobs. Budgetary restrictions also cast a pall on research funded by government. When professionals are driven by economic circumstances, they divert whatever attention they have paid to the discipline to more mundane considerations. This inevitably shrinks the discipline, but by how much one cannot say.

The Relationship of HFE With Other Disciplines. Such relationships are of two types: formal and informal. An example of the former is when the HFES combines with other societies such as Safety Engineering to develop a standard that involves both safety and HFE considerations, or when they co-sponsor a meeting. Such relationships, although indicating official acceptance of HFE by others, does not seriously affect the way in which HFE performs its functions.

The informal relationship is much more important because it directly impacts on the effectiveness of the discipline in its application mode. Research is essentially an inhouse activity; application, particularly in system development, requires HFE professionals to work in an engineering context and as part of an engineering design team. Because engineering is concerned with physical variables, it has been (and still is) difficult to get engineers to give adequate consideration to behavioral concerns in design (indeed, if they did, there might be much less need for HFE, at least in its application mode). What makes the situation critical is that major engineering decisions are almost always made by engineers, so that if behavioral considerations are brushed aside by engineers, this makes HFE look weak and insufficiently cost-effective by system development managers. This problem is examined in greater depth later.

A CONCLUDING NOTE

This introductory chapter has attempted to develop in the reader a view of HFE broader than that acquired from university training. We live in a world that has been transformed by technology (at least in its industrial societies). HFE represents a new discipline that has developed out of that technology. It is a discipline that attempts to change that technology by introducing the influence of the human and humanistic concerns without losing its scientific credentials. Consequently, one cannot correctly view HFE without considering its wider sociological, philosophical, and historical contexts. This means that for an HFE professional to view what he or she does as merely a job (although it is that also) is to demean not only him or herself, but the discipline as a whole. It is necessary to think of what we, as representatives of the discipline, do in broad as well as particular terms.

2

The Conceptual Structure of HFE

The history of a discipline should logically begin with the conceptual structure (CS) of the discipline, which is the assumptions on which the discipline depends. A discipline cannot be fully understood without examining its foundations (i.e., the assumptions and their implications). Assumptions have a number of attributes, each of which must be examined in turn. Some assumptions are entirely logical, which means that they are required by the logic of the situation. For example, the assumption that HFE is an autonomous discipline is required by logic, or else why are we describing HFE? Other assumptions do not depend on logic, but are personal beliefs of individual professionals. These assumptions may be positive or negative and may have emotion associated with some of them. When a large enough number of professionals (what has been termed a *critical mass*) develop a set of common personal assumptions, one can speak of them as a professional CS.

The personal beliefs of professionals logically precede those assumptions that characterize the discipline as a whole. A small group of HFE professionals, whom one can term an *elite* (but only because they are a subset of all professionals), write textbooks and papers; this elite takes the raw material of the professional CS and refines it to form what one can call a *disciplinary* CS. In the process of this refinement, many (possibly most) professional assumptions are eliminated as being too specific or too negative. The process of refinement is not necessarily deliberate; it may occur subconsciously.

Before progressing further, it is necessary to define what an assumption is. An *assumption* is a statement of a belief that describes some aspect of the discipline. An assumption can be operationally further defined as a statement that cannot be broken down into more detailed substatements. However, this does not mean that an assumption cannot have a host of implications or other associations related to it.

As indicated previously, some statements are required by the nature of the discipline; others represent personal viewpoints that result largely from experiences of HFE professionals and the interpretations they have made of those experiences.

It is important to note that not all assumptions are held by all professionals. There is also a great deal of variability about these assumptions. For example, some assumptions are more strongly held than are others. Because it is impossible for the author to peer into the minds of HFE professionals (except possibly his own), much about these assumptions must remain obscure. This cloak of invisibility is intensified by the fact that many of these assumptions (probably those that are not written down in textbooks) function below the level of consciousness. This does not mean that we must conjure up something like the psychoanalytic concepts of id, ego, and superego. It is merely that most professionals do not often think about the assumptions on which they act. However, assumptions can be retrieved and consciously examined, which is what we are doing in this chapter.

Assumptions are predispositions to action or inaction. Sometimes a professional's assumptions will lead him or her into a positive action of some sort; sometimes (unfortunately, more often than not) they predispose him or her to a nonaction (i.e., failure to take an action that is required by the logic of the discipline). This enables us to make a further differentiation between positive assumptions (those leading to a desirable HFE consequence) and negative assumptions (those leading to actions or inactions that produce negative consequences for HFE).

When negative assumptions are revealed, they are displeasing to the professional because they are a threat to his or her ego; it is at this point that one encounters the psychiatric phenomenon of *denial*. That is, when a professional is presented with HFE negative assumptions, he or she will strenuously deny that he or she harbors any such attitudes or beliefs. It is akin to the response most people evince if one were to accuse them of being racist, anti-Semitic, or sexist.

There is of course no objective method of proving that professionals harbor negative assumptions because, as indicated previously, such assumptions are largely unconscious and can only be inferred from performance (more correctly, from nonperformance). Moreover, some assumptions the author might call negative would be considered by some professionals to be positive. An example of this is the assumption that basic research is superior to applied research.

Hardly anything of what has so far been described (particularly as it relates to negative assumptions) will be found in textbooks and papers. The CS of HFE is *terra incognita*. Perhaps those who write texts describing HFE simply assume that all professionals are aware of the concepts that serve as the foundation for their actions. It is also possible that, because the writing elite are also individual professionals, they are not consciously aware of all their own assumptions.

Until now, nothing has been written specifically about the CS of HFE. Thus, it was necessary for the author to develop it by inference from the logic of the discipline, from the characteristics of HFE research (see chap. 6), the activities of professionals, the manner in which they behave, and from his own self-scrutiny. The assumptions described in this chapter represent an attempt to formalize much less structured ideas; the author would never suggest that the CS as described in this and other chapters is complete and impervious to revision.

We spoke of the assumptions as being predispositions to action or inaction. These predispositions arise because every assumption has a number of implications. The implications are important because they are what one may call the *drivers* that animate the discipline. An HFE assumption without implications for the practice of the discipline is meaningless.

In summary, there are assumptions in an informal state that are believed by individual professionals. When a large enough number of professionals believe the same assumptions, we can call these the professional CS. When some professionals write about HFE as a discipline, they extract some (usually only a few) of the professional assumptions; from this they create what can be termed a disciplinary CS. Some assumptions have positive consequences for HFE (these are called *positive assumptions*), whereas other assumptions have negative consequences for HFE (these are termed *negative assumptions*).

The CS of a discipline is thus composed of the refined beliefs/assumptions of its professionals. The belief structure of the professional is like that of any individual: a grab bag of hardly conscious beliefs, some with and some without emotional associations, some more conscious than others, some with positive effects for the discipline, and others with negative effects. This belief structure is influenced by experience and therefore can be manipulated to positive uses by training and discussion. The reader may ask, is it possible that certain assumptions are neutral (i.e., that they have no positive or negative consequences for the discipline)? An assumption that had no consequences would have no implications, and assumptions cannot exist without implications.

If all of this (and the remainder of the chapter) seems sort of inchoate and less structured than what one finds in the typical HFE textbook, it is because we are investigating minds, and minds are rarely highly ordered structures. One can look at a discipline as an attempt to supply the missing order. However, one must also recognize that the discipline is the product of this disorder, which makes the effort to impose order more laborious. Tables 2.1 and 2.2 list respectively the positive and negative assumptions.

TABLE 2.1
Positive Assumptions

(1) HFE is an autonomous science, distinct from its predecessors. Its subject matter is the relationship between the human and technology.
(2) HFE is in continuous development; because of this, it has a direction in which it moves.
(3) Purpose drives all HFE elements. Three major functions of HFE, derived from its (multiple) purposes, are research, application, and prediction of human performance.
(4) The natural tendency of any developing discipline is to organize its elements.
(5) Tasks are developed from purposes. The task is the organizing principle within the human–technology relationship and thus brings the human and technology together in a meaningful relationship.
(6) The purposes of HFE as they pertain to research are to: (a) gather knowledge and explain phenomena describing the human–technology relationship; from a scientific standpoint HFE exists to contribute to the scientific knowledge base, and (b) assist in system development and other practical applications.
(7) The purposes of HFE as they apply to practical ends are to: (a) assist in system development, (b) improve system productivity, (c) ensure operator safety and health, and (d) increase operator comfort in performing tasks.
(8) The human–technology relationship presents difficult research problems.
(9) The S–O–R paradigm is adequate to understand individual human–machine interactions but is insufficient for higher order entities.
(10) As HFE matures, it develops a conceptual structure (CS) that influences and guides all other HFE elements.
(11) HFE as a discipline is distinct from its professionals while still including them as an element in the discipline.
(12) HFE exists in relationship with a construct called the *operational environment* (OE).
(13) The human and technology co-exist in independent domains (behavioral and physical) whose differences cause a dynamic tension in which the human influences and is influenced by technology.
(14) The interaction of human and technology creates a new entity, the system, whose properties are distinct from either entity alone.

The remainder of the chapter describes each assumption in greater detail and its implications for the discipline.

IMPLICATIONS OF THE POSITIVE ASSUMPTIONS

This section examines the implications of the assumptions listed individually in Table 2.1.

HFE as an Autonomous Discipline

A great deal is made in this book of the professional's view of what his or her discipline actually is. There are three general viewpoints: (a) HFE is a branch of psychology and/or engineering, (b) HFE is a multidisciplinary approach to the solution of a certain class of problems, and (c) HFE is autonomous—an entirely distinctive science.

TABLE 2.2
Negative Assumptions

(1) In most of its concepts, HFE is really part of psychology.
(2) (a) Research and application are two distinctly different aspects of HFE.
 (b) HFE research is infinitely more desirable as an occupation than work in HFE applications.
 (c) Basic research is superior to applied research.
 (d) HFE research is intended solely to add to scientific knowledge; it is not intended to assist application nor can it do so.
 (e) Problems inherent in applying HFE to system development or other industrial applications are not appropriate subject matter for HFE research.
 (f) The immediate purpose of performing research is publication.
(3) (a) The human is so variable in his or her responses that quantitative prediction of human performance is impossible.
 (b) Because personnel performance is so strongly influenced by contextual factors, it is useless to look for general HFE principles.
(4) (a) The only research that is truly scientific is the experiment when conducted in a laboratory.
 (b) Research performed under nonlaboratory conditions is almost always contaminated by its environment.
 (c) Because of this, validation of research results can be achieved only in the laboratory.
(5) The HFE database consists of the research papers it publishes.
(6) Any difficulties HFE has had in interacting with other disciplines or non-HFE elements is the product of the ignorance and hostility of those who are not behavioral scientists.

One cannot say unequivocally that one point of view is correct and the others are wrong. The author knows and demonstrates which conceptual focus he prefers. The important point, however, is that each concept has different consequences and some consequences are better (for HFE as a whole) than others.

Of course, it is possible to have no attitude at all toward HFE as a discipline. However, this alternative is not acceptable because then the consequences are essentially random. Moreover, it is sheer ignorance not to have a view about one's work. The matter is discussed at greater length so that the reader can decide which choice to make.

The assumptions that were developed for Table 2.1 are the result of logical deduction. That, at least, was the attempt. Logically one must begin with the assertion of HFE as an autonomous discipline because to do otherwise is to consider it as a branch of psychology or engineering, or as merely a composite of techniques derived from psychology, physiology, medicine, and so on. The result of maintaining these other points of view would be to delegitimize the discipline. This statement is made despite that fully half the members of the HFES consider HFE as a part of psychology (Hendrick, 1996a). Remember the statement made previously that logically derived assumptions need not be believed by professionals to be valid; only their logic confers validity.

That HFE is a behavioral science is unquestionably true, but the author maintains that it is a unique type of behavioral science. If one examines the two major candidates as predecessor sciences, psychology and engineering, it is immediately apparent that the human is the immediate focus of psychology; technology is the immediate focus of engineering. The same objection can be made to any other possible predecessor discipline (e.g., physiology or safety). However, HFE is exclusively concerned with the relationship between the human and technology—a concern not to be found in any other discipline. One can sympathize with those professionals who see the relationship between HFE and psychology as overwhelmingly close, but that is because they have failed to see the importance of the human–technology interaction. The same is true for those who see HFE as a species of engineering.

Some readers may feel that the importance of the assumption is exaggerated, and it is true that one can continue to perform HFE regardless of how one approaches it conceptually. However, there is a loss of efficiency in adopting an inadequate viewpoint. If the nature of the relationship between the human and technology is the issue for HFE, then any research performed must seek to explicate the nature of the relationship. Although it is unlikely that HFE research would overemphasize the technology or engineering part of the relationship, it is entirely possible to ignore this aspect of the research and overemphasize the purely human part of the relationship. That in effect is what the author claims in chapter 6. For example, it is possible to concentrate one's research on the human and make the technology merely a context for human performance. It is entirely possible for the questions that initiate research to focus unduly on the human. This is what happens when the S–O–R paradigm is applied as a governing principle for one's research. Technology becomes solely a source of stimuli to which the human responds and the human–technology relationship, which may be largely if not exclusively in the cognitive faculty (O), is ignored.

The existence of a human–technology relationship suggests that there must be some sort of translation of one into the other, and the research consequences of this suggests certain types of research that would otherwise be ignored. For example, what is the behavioral equivalent of some aspect of the machine? Not merely a performance response to that aspect, but an equivalence. How would the equivalence be expressed (e.g., in an error probability)?

The importance of the human–technology relationship is twofold: First, it validates the autonomous nature of the discipline because no other discipline is so focused on the relationship. Second, its importance is that every human performance measured in a study must be interpreted in terms of its effects on technology. For example, what does a particular performance mean in terms of the machine design required or implied by that performance? For example, does it mean that interactive displays are appropriate? The relationship dictates a prediction (qualitative if not quantitative) of the impact of that performance on the design of the equipment from which the human performance was derived. In other words, HFE research has meaning only if the results of the research can be read in terms of a translation between X (the human) and Y (the machine). If, as happens all too often, research conclusions cannot be interpreted (i.e., translated) in terms of some effect on technology (the machine), the research has psychological meaning but not HFE meaning.

The author is aware that some readers may suggest that he is *straining at a gnat*, but the professional's attitude and orientation are critically important in HFE, if not in other sciences. How one phrases one's research questions will largely determine the kind of conclusions one will derive. That is why the focus on the human–technology relationship is so all-important; this focus is difficult for the researcher to achieve if he or she thinks that HFE is *only* psychology or *only* something else. The questions implied by the assumption are subtle ones, but they have profound effects.

The importance of the relationship and what that relationship means in terms of research operations are critically important for HFE and must never be simply assumed to be only part of the professional's conceptual baggage.

HFE Is in Continuous Development and Because of This It Has a Direction in Which It Moves

Why is this assumption a logical one? Because any discipline that does not develop is moribund and HFE is certainly not that; it constantly receives new inputs from its professionals. All disciplines develop; the rate at which they mature and the direction in which they move are (along with their subject matter) two aspects that vary among disciplines.

More important is the corollary to the assumption: Because HFE is in development, it is moving in a direction either known or unknown to its professionals. Whether that direction is on course is a matter for analysis by its professionals. Whether that direction is purposively directed and influenced by deliberate examination of HFE activities by professionals is a fundamental question to be answered. Such an examination and analysis represent another implied assumption: that it is possible for professionals to affect the activities of the discipline as a whole simply by attempting to influence these.

Some readers may suggest that something as complex and unwieldy as a discipline cannot be moved in particular directions. However, it would be illogical to assume that a discipline is anarchic (i.e., is out of any control by its professionals). Recommendations for specific methods of maintaining oversight of the discipline are made later, although one can anticipate by referring to chapter 6, which describes an analysis of research characteristics.

The logical development of a discipline sometimes does not correspond to its actual development. Theoretically, a discipline begins when people (the number cannot be specified) note that problems exist with certain entities in which these people are interested. In the case of HFE, the entities were the human and technology and, in particular, the relationship between them. Interest in these occurred prior to World War II (WWII), but the critical mass prior to the war was insufficient to cause HFE to develop as a discipline. (For illustrations of what can be called *engineering psychology research* before WWII, see chap. 4.) Critical mass, as suggested by Wilson (1996), is essentially the volume of observations of and concern for the phenomena of interest to the new discipline. In WWII, the number of new technological developments increased so rapidly and the need, in terms of the numbers of men and women required to interact with these technological developments, was so great that a new discipline emerged, although its emergence was recognized only after the war ended.

One performs research to answer the questions posed by the new problems. In a feedback relationship, the WWII research produced a purpose for HFE; this purpose was to perform HFE research. One does the research in WWII and then realizes the purpose that stimulated the research and formalizes that purpose. The deed produces the goal to be accomplished. This may seem illogical, but it often happens; the purpose is inherent in the activity, but that purpose is not recognized as such until later.

The original intent of WWII HFE research was application—application to the solution of practical problems. It was only later that some first-generation professionals tried to divorce application from research perhaps by making the somewhat artificial distinction between basic and applied research.

One might say that having performed the applied research, the researchers then began asking what that research meant in terms of more fundamental principles—principles oriented to their psychological training. People trained in basic science would always approach any problem, however oriented to practical ends the research was, with a concurrent bias to discover its meaning in terms of more fundamental questions.

Other reasons for developing a more fundamental research program might be related to the venue in which the research was performed. Some HFE research laboratories were established at universities, and academia looks down on application from the high standard of basic research.

Logically, research suggests application unless one cares to redefine the term *application* to mean supplying material to the science knowledge base. If one does so, such vulgarisms as the solution of practical problems can be quietly shelved.

A more important but less clearly understood reason for divorcing the research–application contract, which is inherent in the human–technology relationship, is that application is (contrary to almost universal opinion) inherently more difficult than research. One can find many topics and ways to perform research (see chap. 6), but to find a research conclusion that tells one something about how technology should be modified in the light of that conclusion is a great deal more difficult; researchers, even the most accomplished of them, abhor difficulty.

It is not suggested that application in the form of system development fell into disuse after the war ended; activity in the form of behavioral system development was more intense than ever. However, the connection between research in laboratories and universities and its application in industry began to fall apart. One aspect of this divorce was that problems arising out of the system development and use process gradually initiated fewer and fewer research studies. HFE professionals working in industry lost the capability to do research because industry had its own imperatives, which insisted there was no time for research during system development and certainly no opportunity to perform that research unless the government paid for it. On the whole, the government preferred to put its research money into its own facilities.

The human–technological relationship, which is the unique characteristic of HFE, logically requires research to apply its results because technology is made manifest by application. However, the tendency to view HFE as merely another behavioral discipline tends to downgrade the application aspect of research. This in turn tends to break down the technology half of the human–technology equation.

The previous material refers to early development of the discipline. Later development involves a phenomenon known as *fractionation*, in which professionals adopt specialities because of individual interests and opportunity to work in particular areas, such as aviation or computers. Moreover, the scope of HFE is so large that it becomes a practical necessity for the professional to specialize in one or two areas.

Specialization has had both positive and negative effects. It permits the professional to make a more efficient contribution to the speciality by concentrating his or her mind and efforts in a focused manner; this is positive. However, by narrow concentration, he or she may lose sight of concerns that relate to the discipline as a whole. There are problems that are HFE-wide, which may not be recognized because the professional is unaware of them or less aware than he or she should be. Problems specific to the specialty may compete for the professional's attention relative to disciplinary problems. One might suppose that the disciplinary problem, like the development of a database or validation, would apply just as much to the speciality. Specialists are aware of these problems, but they make less impression on the specialist because he or she is so concerned about the specifics of the specialty. Such discipline-wide problems remain general; although their solution is desirable to the specialist, they rarely become concrete and specific enough to engage his or her full attention. It is impossible to resist the notion of specialization because it is inherent in the nature of every discipline. However, there is much more to say about this (see chap. 7).

HFE Purpose and Functions

Purpose drives all aspects of HFE functioning; if there is no reason for doing something, why do it? Therefore, any analysis of HFE functions must begin with their purpose.

The three functions are research, application, and prediction. The first two obviously derive from the necessity to solve the problems that initiated the discipline in the first place. Therefore, research and application are immediately accepted by the professional as necessary. The third function, prediction, has met with continuing resistance probably because it has been misunderstood by the generality of HFE professionals.

The prediction function derives logically from the human–technology relationship. If one is going to translate the characteristics of the human into those of the machine by adopting a particular design configuration, one is making a prediction that the design will aid operator performance. In other words, any application of a research conclusion to technology involves a prediction. Once an equipment design is decided on, there is an inherent prediction that that design will produce the operator performance desired.

A factor contributing to the professional's unwillingness to consider prediction in the same light as research and application is that equipment specifications for new designs almost never include requirements for a certain level of operator performance. There is an implicit prediction in the specification that the system, once designed, will produce an acceptable level of operator performance, but this is almost never articulated verbally and certainly never quantitatively.

If one asks the reason for this, the professional would answer that a quantitative predictive technique and database for HFE does not yet exist and one cannot then set down explicit performance requirements. One of the negative assumptions in Table 2.2 [Item 3(a)] is the belief that human performance is too variable to make any predictions about it, much less quantitative ones. Many years ago, a high-ranking government functionary responded to the author's proposal to perform research on prediction by producing the same negative argument. The feeling against quantitative prediction of human performance is so great that continuing efforts to initiate this kind of research have been largely frustrated (see chap. 7).

However, prediction is inherent in any research conclusions made. The specification of the conclusion says, in effect, that researchers believe with reasonable confidence that the following conclusion represents usual human behavior and will produce the effects described in the conclusion when applied to technology or when observed in the exercise of an equipment by an operator. Even statistical interpretation of research data implies a prediction: We bet that these results validly represent human performance in all operational situations to which these data apply.

Prediction, then, is an inherent part of general scientific principles. Moreover, prediction implies quantification. A science that is merely descriptive hardly merits the name of a science. From that standpoint, there should be a much greater impetus to quantitative prediction in HFE. Several reasons may explain the resistance to quantitative prediction: the great amount of effort that is required to implement it and the fact that primary attention has been paid to experimental statistics that prove the validity of experimental results.

The Natural Tendency of Any Developing Discipline Is to Organize Its Elements

A discipline in development must organize itself into certain elements to accomplish its functions. One would like to imagine that this organization is systematic in its development, but often it is not, with professionals developing individual pieces and then interrelating them.

In HFE, these elements are as follows: (a) an HFE conceptual structure (CS), which includes all other HFE elements; (b) HFE principles generally accepted by professionals; (c) unresolved problems that are recognized as problems; (d) knowledge of facts and methods; (e) HFE professionals, including their skills; (f) the operational environment (OE) in which HFE functions; (g) nontechnical factors that impact HFE; and (h) processes like publication that implement HFE elements. Each of these elements is discussed hereafter.

The Conceptual Structure (CS). The organization of the discipline begins with the development of a conceptual structure that incorporates all other HFE elements, but that also stands alone. The CS is a construct that has, like all others, concrete manifestations. The CS begins with questions such as these: What is the purpose of the new discipline? What is its scope? What are the assumptions that underlie the new discipline? What are their implications? What theories, paradigms, or facts generalize from the predecessor to the new discipline? Chapter 4 describes how these questions were used by the first generation of professionals to formalize the new discipline by organizing the Human Factors Society (HFS) in 1957.

General Principles. The result of the research process is the search for general principles that can explain not a single research result, but an entire class of such results. A general principle can be explanatory; an example is the skills, rules, and knowledge theory of Rasmussen (1983), which attempts to explain a variety of learned behaviors. A general principle can also be instrumental, such as a methodology like task analysis (R. Miller, 1953), which can be applied to the solution of many HFE problems.

The search for general principles (a search inherent in the research process) results in the progressive formalization of the discipline, which is seen most clearly in its publication process. Chapter 6 describes how HFE research studies have become progressively more formal in terms of the type of research HFE publications are willing to publish. Formalization is strongly impelled by academic influences, which insist on formalization as a signifier of their status.

Unresolved Problems. The research process that leads to the solution of problems that initiated the discipline produces a number of more or less general principles. At the same time, the research is unsuccessful in solving all of its problems or, just as likely, uncovers during the research additional problems. It is probable that merely thinking about a discipline, what is known and unknown, will lead to the discovery of unresolved problems. Initial research is likely to be directed at specific problems. However, after some critical mass of research has been performed, the relationship of any individual specific research question to all the questions being investigated leads to the recognition that general principles are involved. What are they?

The problems must be recognized as such or there will be no great effort to solve them. HFE has a number of unresolved problems, one of which—quantitative prediction—is largely unrecognized by the mass of professionals. Hence, it does not receive the attention it deserves. Another problem type, validation, or rather the lack of it, appears to be so difficult that professionals refuse to attempt to solve it even while recognizing its importance (see chap. 7). Unresolved problems continue to remain unresolved.

Knowledge of Facts and Methods. Obviously research produces facts and conclusions and new methods to conduct research. In general, there are few questions about facts, but there may be many about conclusions. To answer these questions, these conclusions must be organized into explanations of phenomena. This is aided by the development of theories. Progressively the storehouse of facts increases with research. This makes the explanatory effort more difficult because the number of theories and their quality do not keep pace with the accumulation of facts and conclusions.

Elsewhere, the author (Meister, 1992) proposed a distinction between explanatory and instrumental knowledge: the first explaining post hoc what and how certain phenomena occur, the second making use of that knowledge to apply to constructive activities like the development of new systems and to prediction of their operations. This distinction parallels the human–technology relationship in the sense that technology requires application of research results and application requires what has been termed *instrumental knowledge.*

Professionals. The formal structure of a discipline tends to make one forget that the discipline is ultimately the professionals who work at it. Professionals are human, but they are not part of the human–technology equation because they are observers and those who measure. The humans in the equation are the people whose performance one wishes to explain and predict—the equipment operators, system personnel, experimental subjects, and ordinary people.

Nevertheless, one cannot divorce HFE from its professionals. Because of this, there is a necessity to explore what professionals think about HFE and how they feel about its various aspects. The notion that one must, as part of the discipline, examine the professionals who are part of that discipline is a fairly new idea; it will not necessarily find favor among most HFE professionals because it smacks of anthropology or sociology. Nevertheless, one cannot understand or control the discipline without examining its personnel.

In any event, in accordance with the Heisenberg principle that the act of measurement affects the measurement results, we assume that the professional is part of the measurement process. In that light, it is just as meaningful to examine the professionals' thinking and acting processes (of course, only while doing his or her job) as it is to calibrate a physical measuring instrument.

This may appear to violate the hard science ethic that pervades HFE, but even the hard sciences like physics or biology have written about their professionals (e.g., Heisenberg, 1971). The history of a CS requires examination of the people who create the CS. For example, what the professional considers to be important determines what he or she will research and even the kinds of interpretations of research results. One way to control HFE is to understand the professional more fully and then try to change his or her point of view.

The Operational Environment. Humans and machines function in an operational environment (OE), which serves as a context for HFE activities. The OE, like HFE, is a tremendously important construct. As the term suggests, it is a special environment—the one in which the equipment and system were designed to function. It is not a knowledge, fact, or process. It is something that envelops HFE, is context for HFE, but is at the same time part of the HFE discipline and must therefore receive serious attention, which, until now, it has not received.

The OE is important because it affects the human–technology relationship and because it serves as a context for HFE operations. The OE has an obverse: the non-OE. Although the human–technology relationship can be studied in a non-OE, all such observations in the non-OE (and this includes the experimental laboratory) are automatically artificial, nonrepresentative of the OE, and false (to a certain extent) because they are nonrepresentative.

Therefore, the OE serves as the referent for all HFE elements. To use an abstract and imprecise term, the OE serves as the ultimate *reality* to which all HFE activities must be referred (compared). Later there is a much more detailed discussion of the OE as one of the HFE assumptions.

Nontechnical Factors. One aspect of OE is nontechnical factors; these impact HFE severely—without them the HFE as we know it could not exist. These nontechnical factors (because they do not deal with efforts to explain the human–technology relationship) include, most prominently, funding for research and jobs for professionals. Unless these two are in favor of HFE, the discipline will cease to exist.

At the heart of funding and jobs is the attitude of those *in charge* toward HFE. In the past, this attitude has been both positive and negative (positive, generally, from government; negative, generally, from industry). Because of the importance of these nontechnical factors, a great deal of HFE attention has been diverted from technical to nontechnical problems.

Processes. To make HFE function effectively, certain processes must be developed. Two of the most important processes are the development of societies, which help manage the discipline in certain respects, and publication, which is closely tied to societal activities, but more directly implements the HFE research process. More about these two later.

Each of the HFE elements described previously affects every other element and each requires continuing oversight to ensure that it performs effectively. There is an unfortunate tendency to concentrate HFE attention on research and societal/publication processes because these are the most evident to professionals. However, all HFE elements are worthy of attention and all receive further consideration in later chapters.

The Task As an Organizing Element

From a theoretical standpoint, this question arises: How is the human–technology relationship energized? Initially the two entities are motionless abstractions at one moment, and the next they are in full interaction.

As indicated previously, purpose is the ultimate energizer of all HFE operations. However, purpose alone either in its general or specific sense, does not energize the human–technology relationship. Purpose provides a goal, but does not contain instructions for how the interaction between the two entities is implemented.

Operating instructions are provided by the task, which is derived from the purpose and machine characteristics. The task, which may be in written or verbal form, in greater or lesser detail, describes the steps by which the human *shall* interact with the machine (the word *shall* implies nonrandomness). Of course, the task is deliberately man-made; before the human operates on the machine, he or she requires instructions. The task description is like the scenario of a two-character play. Like the play, the task description must contain a setting, a description of the environment, and conditions in which the action takes place.

The importance of the task for HFE is that, in measurement, the subject must be given a task to observe his or her interactions with the machine. The representativeness of the task determines the extent to which one can have confidence in the measurement of the behaviors occurring (in this connection, see Vicente, 1997). Synthetic tasks—those created deliberately to require a particular function or variable to be manifested—are not completely representative of actual tasks performed with an actual machine. The conclusions reached using such synthetic tasks must therefore be validated. The nature of the task in a research study provides a bridge to the reality with which one validates the measurement. The task encapsulates the reality inherent in the operational environment.

The task is at a higher hierarchical level than the variable or function that the task calls forth so that both of these latter must be inferred from the task. A variable or function is a molecular abstraction; the task is a concrete entity because, when integrated with the human and machine, it is now manifested in actions. The variable or function cannot be directly observed; their observation occurs only through the task and must be inferred from the task.

The task also contains all the behavioral problems that must be solved in the design of new equipment. If one can understand what is implied in the task in terms of the demands imposed on the human in his or her interaction with the machine, one can ameliorate these demands. The machine, and particularly the human–technology interaction, imposes certain demands on the operator that must not be excessive. Theoretically, task analysis will reveal the extent of those demands. The task description is in essence a shorthand description (hence somewhat inscrutable to the uninitiated) of reality. Because of this, all analyses of the human–technology situation begin with analysis of the task. If the task is not specified in detail (which may be the case initially in the design of a new system), it must be constructed.

The concept of the task and the techniques of task analysis are the most potent tools for analysis of the human–technology relationship. The importance of the task is so great that literally dozens of volumes (e.g., Goodstein, Andersen, & Olsen, 1988) have been written about it. Therefore, this chapter is not intended to be an exhaustive treatise of the subject. However, the task is a fundamental part of the theory that underlies HFE. If S–O–R is the fundamental paradigm of HFE, then its concrete representation is in the form of the task. The task is a necessary linch-pin in the theory of how the human–technology relationship is transformed into actual operations.

The Research Purpose of HFE

The important thing about this assumption is that it mandates both basic and applied research. The terms *basic* and *applied* are distasteful to the author because they imply the superiority of basic over applied work. This implication is demonstrably false and, as such, is one of the (implied) negative assumptions of Table 2.2. The effect of believing that basic research is superior to applied research is that less intense attention is paid to the latter.

The implication of the superiority of basic research is false because its definition suggests that it produces general, more comprehensive, or more abstract principles as opposed to the more specific, less comprehensive, and more concrete principles resulting from applied research. The implication is also false because it suggests that where one gets one's general principles from depends on the research venue. If the study is done in a university laboratory, it produces general principles; if it is done in industry or some other nonuniversity or use-environment, it cannot produce general principles.

The research assumption in Table 2.1 has no value judgment associated with it in the author's estimation. It merely says that HFE research has two functions: (a) to contribute to scientific knowledge, and (b) to aid system development and system use. Neither aspect says anything about the general or specific nature of the conclusions reached or their relative utility. The reader may say, well, the author has brought the matter of basic versus applied up on his own. However, this interpretation is made by many professionals—not only those in HFE, but also those in other disciplines.

It is a tenable argument that if the unique factor in HFE is the human–technology relationship, and if the most obvious expression of that relationship is the development of new equipment, then questions raised by equipment development are just as likely to produce general principles as questions posed by, for example, the effort to develop a model of color perception. The applied study has implicit in it all the fundamental questions related to its subject matter. In the CS of some HFE professionals, general principles are those that apply only to the human part of the human–technology equation; of course this is the essence of the psychological orientation of HFE. Their reasoning proceeds as follows: Technology serves merely as stimuli for human responses (S–O–R again) and general principles cannot be derived merely from studying stimulus characteristics. Therefore, let us concentrate on the human part of the equation.

Because system development is a unique aspect of HFE, it seems reasonable to suggest that questions arising out of system development should, with regard to HFE, lead to general principles and are therefore basic. HFE developed out of the WWII context of research specifically designed to solve system development problems. Prior to this time period, any behavioral research conducted with relation to industry was essentially industrial psychology research because it concentrated on the workers' environment, health, and safety. It took WWII to change the emphasis from these aspects (which are handled today by industrial ergonomics) to that of aiding system development. HFE professionals who propagate the notion that research stemming from system development is applied research are merely attempting to retreat to an earlier time period—a psychological womb. This viewpoint is intensified by academic influences, which abhor research that can actually help something or someone. (Is it too cynical to suggest that general principles are not designed to help anyone or anything specific unless it is another researcher?)

If, as the assumption says, HFE research is supposed to aid system development and other practical functions, the only way it can do so is by performing research based on questions arising from system development and system use (chap. 6 demonstrates that HFE research does not).

This brings up another facet of the basic versus applied controversy. Applied work has the connotation that it directly aids people and things (and that is its purpose), but basic research does not. The notion of aiding in the minds of some professionals somehow denigrates what they do. Regardless of whether one worries about the basic versus applied issue, questions about major aspects of HFE, when correctly asked and when research on them is properly performed, will produce general principles. That HFE questions have, until now, not produced general principles in any quantity is because the correct/meaningful questions have not been asked.

HFE Purpose As It Relates to Practice

This assumption provides a theoretical rationale for activities that were already being performed; HFE specialists were being accepted in industry once WWII ended. If HFE had not been put to practical (industrial) ends after that war, it would have been still born.

The human–technology relationship is given its greatest challenge in the effort of HFE to aid system development. The development of each new system always presents new problems, and system development as a process is a problem-solving situation. The transformation of the behavioral into the physical, by making hardware and software features fit a behavioral requirement, is far more difficult to accomplish than performing research. Experimentation is a process with explicit rules that do not exist in system development. Research is ultimately dependent on the interpretation of data in the form of conclusions that are generalities. The application of these conclusions to a specific instance of physical design is actually the ultimate validation of those conclusions. Without this application, one could say anything in these research conclusions. What could demonstrate the correctness or falsity of the conclusions other than application particularly because professionals avoid any systematic efforts at research validation? In this sense, application of research to system development is absolutely necessary to HFE not only on a practical level of providing HFE jobs, but by providing the opportunity to test the utility of research conclusions. One might say that utility is the ultimate validation test.

Unfortunately, this test has not turned out well. Applicability of research conclusions can only occur where the questions that initiate the research stem from problems produced by practical activities, primarily system development. As chapter 6 demonstrates, little research has been initiated over the years by system development and use questions.

Most HFE research has had explanatory value only. Its conclusions have been useful primarily by stimulating speculation, which we call *theory*, and further research. If research conclusions are only interpretations of data, there may be multiple interpretations particularly where the data are relatively inscrutable. The propensity to look for the effects of variables in research means that one must infer the actions of those variables from the data; they cannot be seen directly. This makes any research conclusion highly speculative unless the conclusion merely restates the data.

The struggle to find HFE principles (other than the most simple ones) that can guide design is a real struggle when the specialist faces the behavioral tasks inherent in system development. One would like to consult texts that provide definitive guidance, but there are almost none. There are principles to be found in MIL-STD 1472E (1994) and in Williges and Williges (1984), but most of these are heuristics based on the experience of HFE experts.

The author's impression, based on his own experience in system development over many years, is that system development activities are messy. This is not to say that they are disorganized because there is an internal unity to these activities, but the specialist is working in a problem-solving environment that has characteristics similar to those of a detective story, which features many unknowns and frustrating efforts to find a satisfactory answer. One hopes that HFE research would provide clues, but it provides few clues.

One way to justify HFE research as it relates to system development is to have the equipment design accommodate human limitations. Even if HFE research does not produce principles to guide design, it at least provides information about human characteristics and limitations—information that at least permits designers to avoid making the more obvious mistakes in their design as these relate to the operator. However, this justification of HFE research is only partially satisfactory because it ignores the need for a much more comprehensive HFE contribution to system design.

System development consists of design, production, and testing stages. The most difficult (intellectually) part of this is design, but production and testing also present problems, although not of the same difficulty. Behavioral problems in production are those that have a long history in industrial psychology as well as industrial engineering. Testing of the equipment that represents the design solution to development problems presents fewer difficulties to the HFE specialist because much of HFE is measurement, which generalizes without too much difficulty to system testing.

Another HFE purpose is to improve system productivity. This assumes that if one improves operator performance, technological performance will also improve. If one reduces error potential, for example, the likelihood that operator failure will degrade the system is also reduced, and this alone should enhance productivity. This is a negative way of improving productivity, and one hopes that somehow the human contribution could be positive as well as negative.

HFE has the additional goals of improving the operator's safety, health, and comfort in the work setting. The main practical value of these purposes is to stimulate research leading to ways of reducing workload, stress, cumulative trauma disorders, and so on. How successful research has been to satisfy these purposes is unclear. Probably much of it has been tied to efforts to understand related phenomena, but some of this may have rubbed off on actual practice.

The Human–Technology Relationship Presents Serious Research Problems

The human–technology relationship is a concept abstracted from a great number of actual human–machine interactions, each of which varies in certain respects from all the others. The purpose of HFE research in its scientific mode [assumption Table 2.1, 6(a)] is to make a number of general statements that summarize commonalities among these individual interactions. That is difficult to do because only a relatively few studies are involved to represent these commonalities. Because this research contains many lacunae and discontinuities, the general statements must be supported by theory, which in essence is supposition based on inductive and deductive logic and intuition.

The actual human–machine interactions can be thought of as the HFE reality. The general research statements are an attempt to reflect that reality in conceptual and verbal terms. Logically, it is obvious that success in representing reality is inevitably only partial because no general statement, no matter how detailed and comprehensive, can hope to represent reality as it occurs and is perceived *in situ*. The overall problem of HFE research is how one gets from these specific, concrete human–machine interactions in the OE to the general statements. HFE professionals are aware that their success in describing reality is incomplete, but this does not prevent them from maintaining the research effort. (Indeed, they must continue or the whole research effort grinds to a halt.)

Every professional accepts the experiment as the single most powerful research instrument available to HFE. However, the experiment is quite artificial in the sense that it is usually an attempt to discover underlying variables that are minor constructs that are invisible to the researcher's eye and must be inferred from human performance.

The existence of variables can be inferred only by arranging the experiment in specific ways that permit a variable to be inferred from subsequent human performance. For example, if one wishes to study the effects of skill level on a certain performance, one arranges two groups of performers: One is trained and the other is untrained. In a correctly conducted experiment, all other potentially confounding conditions are controlled by being made identical between the two groups (e.g., age, sex, etc.). This is the essence of experimental design.

In real-world human–machine interaction, no such arrangements occur; this is what makes the experiment artificial. Of course, it is assumed that the rationale for the experiment is to reveal something about the real world; no one cares what happens in the experiment except as the experiment is a surrogate for the real world. The local context in which the real interaction takes place may or may not be reproduced in the experiment; this is the fidelity with which the OE can be reproduced. The presumption is that the local OE performance context in some way influences the actual interaction. From this the problem of representativeness (retaining the operational context of the interaction in a nonoperational test environment) arises.

Representativeness can be ignored, of course. If one is dealing with molecular variables, such as human strength or body movements, the assumption can be made that at this level of scrutiny the local context in the OE is not significant; the behavior involved will not be greatly changed. As the nature of the human–machine interaction becomes more complex (e.g., more complex tasks), representativeness and OE fidelity become more important. In actuality, most professionals ignore the problem of the representativeness of their test situation; in effect, for them the problem disappears.

A further problem is presented by Heisenberg's paradigm: that the act of measurement influences the behavior being measured and thus distorts (to an unknown degree) the resultant human performance. For example, there is the well-known problem of bias. If a company desires to know how workers feel about its management practices and thus conducts a survey requiring workers to identify themselves on the response instrument, it is likely that only skewed responses will be received because workers fear reprisal if they are honest. If one conducted the survey on only a sample of other managers, the sample would obviously be biased.

Even under the best experimental conditions, the research conclusion that attempts to describe real-life human–machine interactions cannot adequately represent reality. Logically, then, the researcher wishes to know how closely the research conclusion represents that reality. This is the problem of validation. It becomes necessary to compare the reality of actual human–machine interactions with the research conclusion based on nonoperational experiments performed in a nonoperational environment. Usually such a comparison requires the gathering of comparison data in the real work world of the OE. The paradox of this *Catch 22* is that it is difficult to exert experimental control in the real world, which is why researchers initially retreated to the less real measurement environment of the test site, the simulator, and the laboratory.

Of course, it is possible to conduct research in the OE if one has the capability to control OE operations (occasionally possible) or if the research technique one uses does not interfere with OE operations. This last option involves observation, which is almost entirely passive.

If it is not possible (as it usually is not) to repeat in the OE the experimental procedure on which conclusions are based, how does the researcher validate his or her findings and the resultant general statement? (There must be a general statement because otherwise one has only a series of poorly connected experimental results.) There are several options:

1. The researcher can ignore the problem and implicitly assume that validation is irrelevant in the experimental process. This is what most HFE researchers do perhaps because they see the problem of validation as intractable.

2. Some researchers repeat individual studies in the same laboratory environment; if they receive the same results, they call this *validation*. However, it is not; it is merely the measurement of reliability. Further, they may perform the same study in a new test environment that is *not* OE and seek to find the same results. This too is not validation, but rather generalization. Both reliability and generalization are highly desirable; it is simply that they do not validate.

3. A third option, which is almost never attempted, is to make a prediction based on the research conclusion. The prediction is inherent in the conclusions described in the research report because the report implies that, given certain conditions, such and such performance effects will result. It is possible for the researcher to extract this prediction from the research conclusion explicitly and to look for the predicted performance effects in the real world. Why not test the prediction in the laboratory, simulator, or test site? Because the research conclusion refers to human–machine interactions in the real world and not in any substitute environment. This is why the prediction function of HFE is so important. Unless the prediction is made and verified, HFE general principles are merely speculation with some probability of correctness, but that probability is tied to a non-real-world environment.

Validation is not a matter of the determination of experimental measurement error or bias in running an experiment, as Kanis (1994) would have it. It is a prediction of performance in the operational environment (the real world)—a prediction based on a conclusion derived from a prior study. If the validation process must be accomplished in the real world (and it seems it must), the researcher encounters again the problem of measurement in that environment.

If one cannot (or can only rarely) perform experiments in the real world, then other techniques must be developed to observe real-world interactions. These nonexperimental techniques are more difficult to develop than the experiment; the experiment is essentially a set of man-made rules for arranging a performance—a bit like a director arranging a stage setting. Inevitably, because a nonexperimental technique involves some loss of control over variables, some other compensating feature is needed. It is in part because of the comparative ease with which one can design an experiment that the experiment has become a habit as much as a technique.

Prediction can be performed in two ways: (a) determine by analysis of the research conclusion the conditions under which a particular phenomenon occurs in the real world; (b) then go to the real world and look for situations in which the phenomenon of interest is manifested and determine whether the predicted performance actually occurs. If it does, the conclusion is validated; if not, it is not validated.

This observation of the real world is a messy process. In the first place, the prediction implicit in the research conclusion may be vague and difficult to extract. Researchers do not like to "put their money where their mouths are." Then actual observation of reality may produce only partial results: The validation may be somewhat tenuous. There may be dispute about the meaning of the observed performance particularly because the prediction in the conclusion is usually qualitative not quantitative. No matter; one does not expect perfection in the human condition. If the results of the observation merely suggest that the prediction is partially validated, this is probably as much as one can reasonably expect.

One technique that may be utilized in the real world is the use of subject matter experts (SMEs). These are people who, because of their expertise involving many years of experience with certain human–machine interactions, can be assumed to represent reality, although that reality is transformed or viewed through the prism of their intellect and motivations. However, it is possible to take advantage of their knowledge (even as it relates to the effects of invisible variables) by using them as experimental subjects. Because their only response is to questions, it should be possible to develop (arrange) the research questions to elicit statements from them that bear on the prediction issue. The questions need not be only verbal; they may be graphic and situational (e.g., what do you think is happening when such and such occurs).

Inevitably it will be objected that subjectivity contaminates their responses and, of course, it does. However, because everyone is human, any reality apprehended by humans must be tainted by subjectivity. Even experimental conclusions are tainted by the subjectivity of the researcher who writes them.

Therefore, it is necessary to accept the subjectivity of one's humans while working to reduce or control that subjectivity.

Use of the SME is inevitable unless the researcher is prepared to spend years becoming as much an expert as the SME. Armies of researchers all becoming experts in particular fields would be necessary to make a representative survey of all the different types of human–machine interactions in the OE.

The reality that expresses itself in cynicism makes it seem unlikely that HFE research will adopt the SME approach, although it has great potential for validation of predictions as well as gathering original data. Perhaps one could collect one's data first from SMEs, develop hypotheses based on those data, and then validate the data in the experimental laboratory. In the past, although SMEs have been used—usually in military applications to enable the researcher to become familiar with equipment details needed to conduct a study—they usually have been employed in an ancillary, tutorial position, not as potential instruments to be used to collect data.

Of course, one could attempt to develop quantitative predictions based on the raw data of the individual experiments performed in the laboratory and then try these predictions out by running new experiments in the laboratory, thus avoiding the necessity of going to the OE for validation. This process requires the researcher to (a) quantify his or her predictions in terms of the probability of error, (b) associate certain probability values with certain performance conditions, and (c) arrange experiments to determine if the quantitative predictions hold up. Because there is much effort involved in this procedure and the possibility of failure to validate the prediction always exists, it is unlikely that many researchers will follow this process.

Leaving aside the previous measurement problems and assuming that HFE research conclusions are valid, the question arises of what one does with one's research. Assumption 6(a) in Table 2.1 specified that one of the purposes of HFE research is to add to scientific knowledge. Every researcher would agree because, in their minds, nothing need be done—no effort need be exerted to make this contribution, merely publication of the research. As soon as this occurs, an increment of knowledge is presumably added to the knowledge stockpile. This is the ideal answer to the question because it involves no effort.

Is it as simple as that? The individual research report that describes the results of the individual experiment has little value until it is combined and integrated with other researches into a more comprehensive general statement that can be the basis of—what? After performance of the research, after publication of the research paper, what then? Almost all HFE professionals would say that there is no afterward, no what then; one simply returns to the further performance of research. This is nonsense of course, although it is generally accepted nonsense—that the scientific contribution to knowledge can be accomplished merely by publications suggesting that something is missing. One does research presumably to accomplish something.

Assumption 6(a) in Table 2.1 specified that HFE research was to be useful in applied activities. What happens after research is supposed to be application. Moreover, application of research results, if successful, is the ultimate validation of the meaning of those results.

If the author can summarize what he has been trying to say in this section, it is this: Research is not or should not be simply a habit. There are profound problems in the research process. One can perform research at various conceptual levels, some of which are simple and obvious, and this is probably what most researchers do. It is also possible to look at the research process with a greater degree of conceptual sophistication, and this is what the author has been attempting to do. Of course, in performing these more sophisticated analyses, one runs into further problems, some of which are difficult to overcome; this is why most researchers function on a somewhat superficial level. However, the complexities inherent in the research process are fascinating for those who can stretch their mental legs an additional mile.

The S–O–R Paradigm Is Adequate to Understand Individual Human–Machine Interactions, But Is Insufficient for Higher Order Entities

It is assumed that a reduction in the quality of human performance, such as error, may lead to a degradation in machine performance. This provides the ultimate justification for HFE as a discipline: the need to design systems in such a way that the frequency of human error and other behavioral inadequacies will be reduced or eliminated.

It also seems obvious that if the machine malfunctions in some manner, the operator's performance will be substantially impacted and may ultimately result in accident. The change in the performance of either entity must be great enough that it disrupts normal functioning of the relationship. Many operator errors have no influence on machine performance, and machine performance other than malfunction will have no effect on the operator until that performance unduly stresses him or her.

Because not every variation (e.g., error) in human performance affects the machine and every variation in machine operations does not affect the operator, it is a reasonable question for HFE research to determine the conditions under which the performance of one entity in the relationship exerts a significant effect on the other.

A change in machine operations outside of preset limits can lead to a malfunction that will require the operator to engage in compensatory behavior. In a period of increasing automation, when the human's role in the human–machine interaction is changing—from controller to monitor and diagnostician of machine performance—the machine can exert a significant demand on the operator's perceptual and cognitive capabilities as he or she struggles to interpret symptomology and respond appropriately.

That demand is a potential subject for HFE research because it is a source of tension that may degrade the entire human–technology relationship. (Research on workload is research on machine and situational demand.) At present, the human exerts no comparable demand on the machine because the machine has not been given the capability to observe and consciously respond to the human (although there are adaptive training and performance systems; Kelly, 1969). In the future, that capability may be designed into the machine, at which time the factors affecting the machine's consciousness will become important for research.

The variables discussed earlier are generally understood by professionals in terms of the traditional S–O–R paradigm carried over to HFE from psychology. The changed operator's role in automated systems and the notion of demand exercised by the machine on the operator can be conceptualized in S–O–R terms, although the paradigm must be expanded. This paradigm tends to fixate the professional's attention on the human in the relationship (because all the S–O–R elements are human) and to overlook the interactions between the human and machine.

The human–machine interaction is commonly thought of and phrased in terms of a single operator and a single machine. Engineers also design a single machine. However, the interaction between human and machine is only a part of a more general human–technology relationship, which is vastly more complex than the one-on-one interaction. A hierarchy is involved, which is depicted in Fig. 2.1.

As can be seen in Fig. 2.1, the human–technology relationship involves several levels: the interaction of systems, which are major components of a technology as a whole; of humans in relation to systems; and of humans in

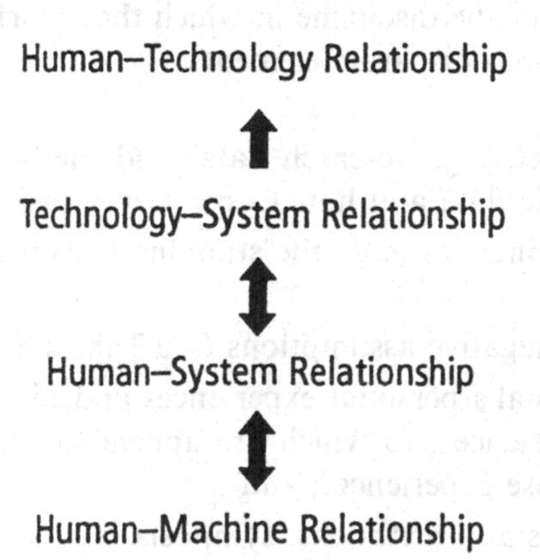

FIG. 2.1. An expanded view of human–technology relationships.

relation to individual machines. The traditional S–O–R paradigm applies to the single human–single machine interaction, but a new formulation is required for the human–system and technology–system relationships. This does not mean that the professional needs to discard S–O–R, but that this paradigm must be understood in more complex ways. The paradigm is sufficiently powerful to organize HFE concepts as long as one considers only the single operator–single machine configuration.

As HFE begins to confront the system, the S–O–R formulation becomes appropriate only for the single human–machine unit. When multiple humans and multiple equipments are organized into a system, that system develops dimensions and variables not to be found in the single operator–machine interaction (see Meister, 1991). These higher order dimensions and variables influence the S–O–R formulation even when it is correctly applied at the more molecular level.

The point of the previous discussion (and the reason for emphasizing it) is that the S–O–R paradigm is adequate to understand the nature of the human–machine interaction as long as it is restricted to the single human–machine unit. That unit is part of a larger entity—the system. When the professional raises his or her eyes from the unit level to the level of the system and its major components, the simple S–O–R formulation is no longer useful to understand system phenomena. The simple formulation must be expanded and modified in the minds of professionals to encompass many more dimensions and variables described in chapter 3.

As HFE Matures, It Develops a Distinctive Conceptual Structure (CS)

The conceptual structure (CS) of a discipline is, quite simply, what its professionals think about the discipline in which they work. The CS includes a multitude of sometimes disparate elements:

1. Matters of fact (e.g., research data) and methods (e.g., how to use anthropometric data and how to perform a task analysis);
2. Theoretical principles (e.g., the stimulus–organism–response [S–O–R] paradigm);
3. Positive and negative assumptions (see Tables 2.1 and 2.2);
4. The professional's personal experiences and his or her interpretation of those experiences, to which are appended attitudes and emotions relative to those experiences; and
5. Formal papers and textbooks, symposia, and so on.

How much any single individual's CS contains varies, of course, with the individual.

Because it includes the totality of the thoughts and feelings of individual professionals and professionals as a group, and the formal analysis of the discipline as published in documents, the CS is a remarkable melange of the formal and informal, the academic and the personal, the factual and attitudinal, the logical and irrational. The CS is important because it directs all other HFE elements and has particular impact on HFE research. Its influence on application is weaker because application, in particular system development, has its own principles and methods that may or may not coincide with those of the overall CS, but in any case supersede the latter.

From a developmental standpoint, the HFE CS began by the transfer from the predecessor disciplines, but primarily psychology, of specific matters of fact (e.g., principles of learning), specific principles, the S–O–R paradigm, general scientific principles (e.g., the need for objectivity), methodology (e.g., the method of performing an experiment), and experimental statistics. Manifestly, in the early days of HFE (1942–1945), the CS was largely that of the predecessor discipline. With the publication of the first documents about wartime research (chap. 4 describes these), HFE began to develop its own formal CS.

Obviously one must divide the CS into its formal and informal aspects and its personal (individual) from its disciplinary aspects. The CS is always an individual one (because it is always viewed through the individual); at the same time, its disciplinary aspects are being published. That is why publication is so vital to the discipline; without publication, disciplinary concepts will not be communicated. As soon as the individual begins to publish, a formal and disciplinary CS begins to be developed. When read by other professionals, this material is incorporated into the personal CS. Therefore, the professional's CS contains both formal and informal material and a great many personal concepts that are never published and may indeed hardly ever be expressed except possibly in casual conversation. If one could look into the mind of the HFE professional, one would find it a wondrous storehouse (some might call it a *junk pile*) of discordant elements, emotions shoved up against research findings, gross prejudices chock-a-block with theory, and so on. Naturally, when any professional decides to write about his or her discipline or some aspect of it, he or she selects with care among these elements because there is some rubbish that one will not wish to have exposed to the common gaze.

The extraordinary aspect of the CS is that much of it is essentially subconscious—not far below consciousness, however, and relatively easily retrievable. The formal factual and theoretical material is easy to fish out of the waters of the CS, but the fascinating thing about CS functioning is that so little of it is actually used in daily HFE activities. This is particularly true of application. For example, in analyzing an engineering drawing of a work station interface, only a few simple principles are applied (see Lund, 1997). The mass of what the specialist has learned is, like all background learning, hardly remembered except when needed. If the situation actually requires it, one knows where to go in one's library to find material to be used in the specific situation.

This is true of research as well, but the researcher makes much greater use of factual material. For example, no one is permitted to write a research report without first reviewing and analyzing research data found prior to one's own research effort. To the extent that what one studies in research is a matter of choice, the CS also contains inclinations to work in one or the other specialized fields like aviation psychology, to select certain topics like the glass cockpit to study, and so on. The professional's CS also includes material about funding possibilities, job availability, and so on—all the nontechnical factors that influence the discipline.

It is possible to view the CS in various ways. It can be classified, as the author did in his original thinking about the topic, into three main categories: concepts dealing with the human, technology, and the discipline as a whole. Most CS concepts seem to fall into the discipline category. It is also possible to think of the CS as an engine that organizes all the positive and negative assumptions and translates these into action when, for example, the professional selects a research topic or methodology. In large part, the CS is responsible for the formalization of the discipline and for emphasizing certain general principles of science, like objectivity. The CS is a unique, separate entity incorporating all other HFE elements while being one of the HFE elements. This is because, in the conceptual space of the Lewinian type (Lewin, 1936), different concepts can occupy the same space.

The individual's awareness of his or her own CS almost certainly varies. It is usually not articulated or expressed verbally as a CS even in formal writings, which is why one must tease it out by inference and deduction. The individual CS may contain discordant elements; that of the discipline cannot. It is assumed that the more aware the individual and discipline are of their CS, the more effective the CS is in performing its functions. The CS is a decision maker; if one is aware of one's CS, then decisions can be examined and made more efficiently.

The CS may be largely quiescent, in a state of stasis; it begins to function when a decision must be made or some action taken. It may seem as if only the individual can make decisions and take actions, but the discipline does so also, although only through the medium of the individual. The decision-making process in the disciplinary CS is evidenced by (a) selection of research to be performed, (b) selection of methods to be employed in that research,

(c) selection of research to be published, and (d) what is to be taught to novice professionals.

The influence of the CS in the application (system development) area is comparatively slight because system development has its own forces and rules that dominate the individual. That may be why there is a continuing state of tension between research and application: the CS is more operative in one and much less so in the other.

The individual CS is molded by experiences and has emotional characteristics. The disciplinary CS (that which is written about in texts) is affected by the individual CS in composite, but is refined by general principles of science—like the need for objectivity, experimental design, and so on. These act as a sort of censor. For example, standards of selection for publication demand an experimental design, statistics, objectivity, and so on so that the individual's CS must accept these also. The general principles of science that affect the CS are taught in university. The individual CS must accept the dictates of the disciplinary CS, but the latter can be expressed only in action through the individual because the individual is the one to decide what to research and how. This must be qualified, of course. If one performs research for money (e.g., in a consulting firm or on a university grant), the individual researcher does not have complete freedom in the research process. Rather, he or she must abide by general principles of science, but even more so by the preferences and interests of the funding agency. That agency is also influenced by the disciplinary concepts of those who work in the agency.

However, the individual CS may violate prescriptions of the disciplinary CS—not directly by contradicting them, but by ignoring them. For example, one of the concepts of the disciplinary CS is that, because there is an OE, all conclusions derived from venues other than the OE must be validated. This means validation with reference to the OE because the OE is the ultimate referent for HFE. However, little validation is performed in HFE because it requires great effort. There is presumably an assumption that research must lead to application; again, little effort is exerted in this area. In rejecting a logical requirement such as validation, the professional rationalizes as an excuse (e.g., one cannot experiment in the OE, there is no funding for it, etc.).

The individual CS often has emotion expressed in terms of preference: "I would much rather do research than work in system development" or "I want to work in a laboratory rather than in industry." The disciplinary CS also has preferences, such as for research over system development, but these are expressed in a textbook only indirectly, as in Proctor and Van Zandt (1994), by devoting only 5 pages out of 500 to system development. It is tricky to write about the CS because, as we see in the Proctor and Van Zandt example, so much of the CS is subconscious (concealed from view).

The individual CS contains certain disciplinary concepts and therefore overlaps the disciplinary CS. However, the individual CS also has elements dealing with self-interest, desires, prejudices, and so on that the disciplinary CS is not supposed to have but is influenced by anyway. This is because the disciplinary CS is verbalized through the writings of individuals who inherently function through personal concepts and motives of self-interest.

The process by which individual and disciplinary concepts interact is shown graphically by Fig. 2.2. Disciplinary and personal concepts are absorbed by the individual who may proceed to develop new disciplinary concepts, which, when published, are absorbed by other professionals who integrate these with their own personal concepts and the process repeats itself indefinitely.

Concepts are resources that are drawn on by the professional when they are required by the job. The term *job* includes research because research is also a job, however much the researcher may wish to idealize the activity. The nature of the job determines which concepts are demanded. When the concept is utilized, it interacts with and overlaps the working experiences of the professional and may be modified by those experiences. Theories and models may be modified by empirical test. Therefore, HFE concepts may change over time and their number may increase as the individual professional expands the range of his or her experiences.

For the most part, disciplinary concepts are independent of work experiences. However, as was pointed out previously, they overlap when concepts are called on by the professional at work. The use of an abstract concept in work may make the concept less abstract. Yet when the professional's work is described in a textbook, the reality of the work is formalized and becomes more abstract, thus changing its reality. For example, in some of his books, the author described various concrete system development processes in step-by-step form, to make them clearer to the reader. He became aware that, in the descriptive process, the actual concrete activities were being intellectualized and transformed into concepts because of the words that were being used to describe them. When written about, actual processes tend to lose their dynamic character and become static. For the professional who reads those words, the reality of the actual work with which he or she is familiar is transformed into a variation of that reality. When we look at real events with a conceptual focus, they become a different reality. This is true not only for words but also for graphics; when an actual task is decomposed

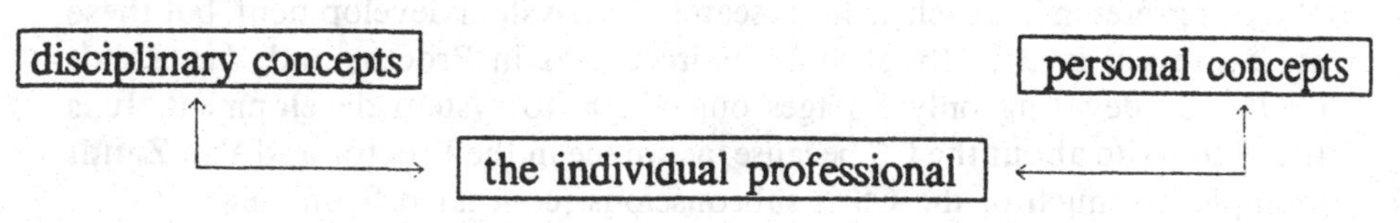

FIG. 2.2. The interaction of personal and disciplinary concepts.

into its elements by an operational sequence diagram or when a real work decision is described by a decision–action diagram, the actual task and decision have been subtly changed by the graphic descriptive process, which is at bottom an intellectualizing one.

It is somewhat surprising (or is it?) that so little about the HFE CS is to be found in HFE textbooks, despite their emphasis on research. There are great gaps in the formal descriptions of the CS (when these are provided), and of course there are no descriptions of the personal concepts in the CS. There may be good reasons for this: much of the CS, like much of the unconscious of the average individual, would be embarrassing if revealed.

Professionals do not talk much about the HFE CS because the topic is at once rather arcane and somewhat lacking in respectability: A discipline should be superordinate to its professionals. Also, because of the pragmatism inherent in HFE, professionals may feel awkward about what they call *philosophy*. Because the CS is inherently subjective, it does not seem to fit the concept of a highly objective discipline, and yet nothing is done in HFE without human mediation. Thus, to ignore the CS is to ignore what is arguably the most important part of the discipline.

Whatever is written about the formal concepts of a discipline must be presented in a positive manner. It is undesirable (because it does not fit the HFE image) to emphasize how far we are from the accomplishment of various HFE goals, although such recognition is necessary if progress is to made. Indeed, even HFE goals must be inferred from other material. CS material, which is part and parcel of any disciplinary oversight, is discussed only occasionally in less formal symposia and papers published in the *Proceedings* of the HFES.

The preceding should not be considered a criticism of HFE specifically; all disciplines, one suspects, are fundamentally the same in this respect. Formal descriptions of a discipline are almost always optimistic because they usually prefer not to dwell on disciplinary problems. If one assumes, as the author does, that the CS (at least its formal concepts) directs the progress of the discipline, the absence of much discussion of that CS hinders oversight of the discipline. If all that is discussed in textbooks and papers are research details, models, and theory, there is not much room for consideration of the direction in which the discipline is going.

In earlier pages, the desirability of some conscious oversight of the discipline was suggested. Unfortunately, where oversight material is to be found, as in the volumes published by the Human Factors Committee of the National Research Council (see Nickerson, 1995), the material is oriented only to specialty areas (e.g., the opportunities that will arise in virtual reality).

It is the author's opinion that the absence of a full-scale treatment of the conceptual foundations of HFE (at least those parts of the CS that may be respectably revealed) has retarded HFE progress because it has allowed negative assumptions and tendencies (to be described in detail later) to affect the course HFE is taking. Analysis and discussion of HFE problems can only aid the discipline.

The HFE Professional Is One of the Elements of the Discipline

As has been suggested by the preceding assumptions, nothing happens in HFE except through the agency of the professional. Although the professional stands apart from the discipline, in the sense of being able to view it somewhat dispassionately and apart from his or her own role in that discipline, he or she is at the same time one of its elements. It is this dual role and dual perspective that makes the professional so interesting a subject for study.

In describing HFE in textbooks, attention is given to almost every HFE element except the professional. One wonders why this is so. Is the discipline supposed to be an abstraction separate from its professionals? Such a concept would certainly aid the formalization of the discipline. Is it that the professional as man or woman represents a potentially disruptive, subjective HFE element? From a historical standpoint, it might be possible to trace the HFE effort to renounce subjectivity (and what could be more subjective than the individual professional?) to the influence of psychological behaviorism, which is reflected in the work of Watson and Skinner.

From that standpoint and that of the principles of science, a discipline should be a coldly calculating logic machine. Actually those who describe HFE in textbooks and papers tend to give this impression, as if the research professionals (those who apply HFE are usually not considered) were disembodied specters. This is of course nonsense, but it is nonsense hallowed by many years of tradition. A researcher is known only by the papers he or she has published. What precedes the research and writing of the paper is ignored, although arguably it is almost as important as the research product.

The inability to recognize the human dimension in HFE means that the discipline reflects only one or at most two dimensions. Of course, it is possible that the discipline, through its elite, prefers it this way.

Any description of the discipline ignores that there is such a thing as a personal CS and that, as a result, the discipline contains discordant elements, some of which are negative to HFE progress. It is likely that in this characteristic as in others, HFE is much like any other science, although the potential for disruption by these negative elements is perhaps greater because there is somewhat greater subjectivity in HFE. One argument for maintaining conscious oversight of HFE is the need to control these discordant elements, which is discussed in the next section.

Functioning in the context of these discordant elements is a principle of general science—that ideally the professional should be free to select the research he or she wishes to perform and write about anything he or she wishes. It is further assumed that this principle of scientific liberty will produce more positive effects than if there are constraints on research freedom. Behind this is the Enlightenment concept of the *free marketplace of ideas*, which is a foundation of popular democracy.

In actual fact, of course, there is no unlimited research freedom; those who supply funding for research are those who control the ideas in that research, although the degree of control ranges from the slight and indirect (where money is given for more general research purposes) to the direct and great, where contracts are let (often by government) to investigate specific research questions. Again, this is as true of medicine or nuclear physics, for example, as it is of HFE. Of course, the HFE professional is almost never an unwilling subject of this influence. For example, in a particular kind of research because of special interest, he or she becomes a willing participant in this funding–research relationship.

It is the author's contention that, although the individual professional should be allowed as much freedom as possible to be the best possible scientist he or she can be, the discipline as a whole, whose function is to organize and integrate the contributions of the individual professionals, requires some control simply to perform this organization and integration function efficiently.

One example of the control demonstrated by the discipline is the manner in which publication is performed—where control is exercised by the editors of refereed journals. Although isolated voices have been raised against particular instances of editorial control, by and large the discipline out of necessity accepts this. Moreover, this control—brutal perhaps because it rejects much more research that it accepts—is accepted because it is supposedly that of peers, although these peers are *primus inter pares*.

The freedom the individual professional may feel (in research, not in application) is in part an illusion. The elite, as represented by those who write texts and those who edit journals, exercise control, as do funding agencies. However, it is important for the professional to feel free. There are then two disparate forces: the freedom of the individual professional and the controlling forces exercised by funding agencies and editors. (There is also academia and the university, although this influence tends to wane after one graduates.) How is the discipline supposed to integrate these divergent tendencies? The answer is that every effort should be made to accord the individual professional as much freedom as possible, but that freedom should be exercised in a context or matrix of information about disciplinary tendencies—information gained by a formal oversight process that could include surveys of professional attitudes.

One could argue that editorial and funding control already perform an oversight function, but this oversight occurs informally and at the individual level. What the author suggests is that the oversight be directed at the overall discipline by examining trends in that discipline over time. For example, what are the most popular research themes in the published literature? What underlying concepts do these themes reflect? To what extent is there an attempt by researchers to find applications for their work? Questions of this type are examined empirically in chapter 6.

The Operational Environment (OE) Serves as the Ultimate Reality for HFE

The *operational environment* (OE) is defined as the situation in which equipments and systems normally and routinely perform their assigned operations; the OE is the environment for which they were designed. For behavioral design principles derived from HFE research, the OE, the ultimate referent and validation for these principles, is the system design project. Moreover, there are multiple concurrent OEs, which complicates the problem.

Presumably the designer of a system has in mind an environment in which that system will routinely be exercised. (The environment is specified as part of the system mission.) Simplistically put, it is in the sea for ships, the air for aircraft, and the highway or city streets for automobiles. It is a physical environment, but it is also a condition that may have any number of variations. For example, the sea in which ships sail may be calm or in a force 5 gale, or highways may be almost empty of cars or clogged with automobiles.

Therefore, the phenomena of system operations are normal to that environment and not normal to any other environment. Any environment in which the phenomena are recreated, other than the one for which it was intended, is artificial and unnatural.

The OE has an antithesis—the non-OE. It may seem strange that there is an environment other than the real one. This is the test environment, of which there are three forms: laboratory, special test site, and simulator. The laboratory is the non-OE par excellence because in it there may be no attempt at all to reproduce the characteristics of the OE. The simulator and test site attempt to reproduce these characteristics with varying success. The importance of the OE is that it serves as a referent for conclusions reached in HFE research. All research is meaningless unless it explains some facet of the OE.

We already discussed the nature of the experiment, which requires the decomposition of the task situation into experimental and controlled variables. If Variable X is studied to determine how it affects a certain kind of human performance, the intent is to relate the finding in some fashion to the actual functioning of Variable X in the OE. For example, if we study the effect in the laboratory of ambient illumination on the ability to read various sizes of type font, we do so to understand the effect of illumination on reading situations or tasks in which reading is ordinarily performed. In other words, it is assumed that one studies a phenomenon to understand how that phenomenon occurs under the conditions that are normal for that phenomenon. There may be those who think the laboratory is the real world because they are not interested in anything outside the laboratory. This may be acceptable for physics or chemistry, but not for HFE.

It is simplistic to assume that a non-OE test venue is representative of the OE simply because subjects utilize the same human functions (e.g., perception, motor responses, decision making) that are also utilized in the OE. The functions may be the same, but the different contexts in which they occur change the character of the human functions. Decision making in a computer game like Space Fortress (Mane & Donchin, 1989) is not the same as decision making in actual air warfare, although they are related.

The laboratory environment in which most HFE research is conducted (see chap. 6) is therefore an artificial environment unless it reproduces the features of the OE. Any findings in any test venue have no meaning unless they can be referred back to and are found to occur in the OE.

If one can think of a reality for a discipline, the OE is that for HFE. The concept of reality has been a fundamental problem for philosophy since Plato. It may seem to be inappropriate for a science, but reality as it is implied in the notion of representativeness is fundamental to HFE. For example, physically it is central to simulation. Of course, the question that immediately arises is: what are the characteristics of the OE that differentiate it from a specialized research site like the laboratory? This is a difficult question to answer because, until now, there has been no systematic study of the OE. (Anthropologists and sociologists study the OE as it applies to people, their roles, their customs, etc., but this study involves technology only peripherally and does not deal directly with systems. There is then a need for what can be termed *anthropological ergonomics*.)

If one thinks of the laboratory as the paradigm of research measurement (and it actually is because most HFE research is performed in a laboratory), its one outstanding characteristic is that it eliminates all factors that may interfere with the study of variables that have been decomposed from the normal task situation. In the OE, chance or other factors may enter into the task situation and affect that situation. This would be intolerable in a laboratory or any measurement venue that has been arranged so that only certain variables of interest are allowed to exert an effect on task performance.

If, in the process of ensuring that these variables are controlled to emphasize their effect on performance, it is necessary to simplify the measurement situation by changing equipment, task, or situational characteristics, the researcher will do so. How much simplification he or she will permit depends on whether he or she is searching for general principles (so-called *basic* research) and the actual equipment or task is simply an instrument to permit one to study these principles; or whether the measurement is designed to answer some question relating to the specific equipment, task, or situation employed in the measurement (*applied* research). All of this suggests that if one wished to reproduce OE characteristics in a laboratory test situation, it would be desirable to introduce one or more chance factors. However, this might interfere with the action of the experimental variables, and therefore would be undesirable.

The simulator comes closest to the OE, and it would be possible to introduce chance (i.e., unexpected) factors into it without disturbing the essential reality of the situation. For example, the introduction of malfunctions into a simulator test operation is common. The question is whether one can study basic principles in a simulator closely tied to a particular operational system or situation.

The OE also provides technological, sociocultural, and financial contexts for HFE (as it does for everyone living in a technological civilization). HFE is particularly susceptible to the financial and cultural aspects of the OE. Because the human–technology relationship is an inherent part of the OE, it is possible that that relationship cannot be completely understood except in relation to these other OE factors.

In a technological civilization, technology is a significant part of the OE. It must be understood not only in relation to a worker operating a machine, but also in relation to all those not working machines. Technology provides the physical, behavioral, and social environments of the OE in which all humans function.

It has been pointed out that, because of the differences between the laboratory and the OE, the necessity of validating research findings arises. If one observes that Variable X exerts a certain effect on performance in the test condition, then it is important to determine that Variable X exerts the same effect in the OE.

This last requirement produces difficulties because, in the real world of the OE, task performance is not normally decomposed. With some effort, task performance in the OE can be decomposed to assist measurement, but the paradox (the Catch 22) is that to do so is to render that task performance no longer normal, routine performance. Heisenberg's principle is particularly virulent for HFE because the human is tremendously sensitive to any manipulations of the task or other aspects in which he or she performs.

Validation provides a truth test. It is as close as one can come in HFE to ultimate reality. Therefore, it is not sufficient for professionals to ignore the problem, as they generally do. This is a failure of will—an unwillingness to exert the effort needed to validate. It is recognized that there are degrees of deviation from performance in the OE as regards measurement and that minor deviations are relatively unimportant. Much depends on the nature of the task being measured. If one is performing a psychophysical task that is likely to be invariant in almost all situations, then validity of the results can be assumed. Of course, it is not the business of the HFE researcher to perform psychophysical studies that have little relevance to technology.

We interpret the meaning of the OE broadly, which means that the number of conditions for task performance in the OE may be so many that it is impossible to include them all in the measurement situation. For example, all illumination conditions from absolute darkness to blinding glare may be too many to include in a perceptual viewing experiment, and some of these conditions, such as absolute darkness, may be irrelevant to a real-world viewing task. The researcher may measure points along a continuum or at the extremes of that continuum.

The criterion of truth also depends on the nature of the questions asked in the measurement. If one wishes to estimate the importance of a particular variable in affecting performance (the normal case in experimental design), the determination of the validity of findings relative to that variable is difficult because, as pointed out, it is almost impossible to decompose performances in the OE—or, if one does, the act of decomposition changes the performance from normal to artificial. The action of Variable X relative to overall performance is covert and cannot be immediately determined by simple observation of that performance in the OE; it is inferred from the arrangement of the experimental design. However, if one merely wishes to describe the observed performance of a task (e.g., the number of errors that operators make in reading a display), it is much easier to validate findings by observation of the same performance in the OE. That is because it is not necessary for observation to decompose performance into constituent variables.

There are (hypothetically) various ways to validate a research conclusion other than repeating the same task in the OE. The following are some examples:

1. *Prediction.* On the basis of the research findings or on a model of a certain phenomenon, one might predict that performance of a particular task in the OE would occur in certain ways. This is not the same thing as repeating an experiment in the OE. If the researcher can predict as a result of the original finding that certain observable effects will occur in the OE, one can attempt to find those effects by observation and nonexperimental measurement. If the predicted effect is found, the original research conclusion is validated.

2. *Simulation.* If the research is performed in a simulator that reproduces the major characteristics of the OE, the more faithful the simulator is to the OE, the more confidence one can have in the research conclusion. If no simulator is available, one can attempt to introduce certain characteristics of the OE into the original experimental situation. It is difficult to say what

these OE characteristics would be because there have been no systematic efforts to examine the OE.

3. *Robustness.* The term *robustness* has a statistical meaning in relation to a set of measures, but that is not what is meant here. Robustness here means that when one repeats the original measurement situation but introduces certain negative conditions that stress the subject, and if one has achieved the same results under these changed conditions, the conclusions based on the results appear strong enough to stand. Negative conditions might be less well-trained subjects, a high level of stress, or reduced context conditions like poor lighting or noise.

4. *Calibration.* The developers of the Data Store (Payne & Altman, 1962) adopted a procedure to refine the data they retrieved from the experimental literature (see chap. 7). They reasoned that the experimental situation tended to encourage the commission of errors (to demonstrate statistical significance of differences). On the basis of independent data, they applied a correction factor derived from the literature to the raw experimental data. Similarly, Swain (1967) calibrated the error data he used by applying a correction factor based on data from other field tests. This calibration feature does not apply to research conclusions but to raw data; what effect the refined data would have on their statistical qualities is unclear. The correction factor applied to the Data Store was a single value; whether it would be meaningful to apply a single correction to different task performances is a matter of judgment.

Prediction is a long-honored tradition in the physical sciences. Some of the phenomena that Einstein predicted in his special theory of relativity were later measured and the results confirmed the prediction. Simulation fidelity is an excellent answer to the possibility of invalidity, if the degree of fidelity is high. The procedure termed *robustness* does not validate directly, but suggests that the experimental conclusion is strong enough to be probably valid. The calibration method is probably the weakest of the possible methods of ensuring validity. In any event, although the previously described techniques are possible ways of ensuring validity, they are, in the present HFE climate, unlikely to be tried because there is no will to try them.

The OE stands as the antithesis of the laboratory experiment. In between, on a continuum of representativeness relative to the OE, we have the simulator and the test site. The special characteristics of the laboratory and the experiment as measurement situations have been extensively explored in textbooks, but nothing corresponding to this has occurred with regard to the OE. One would imagine that, because the OE is the situation that is the referent for all HFE tests, there would be attempts to explore the differences between the OE and other measurement settings. For example, even if one does not perform experiments in the OE, it is possible to measure human performance as it occurs there and to compare that performance on the same or a similar task in the non-OE environment. However, one gets the impression that the prestige of the laboratory in HFE is so great that collecting data in or about the OE must be very down-scale indeed.

The reluctance to validate HFE conclusions and methods leaves the discipline in a sort of limbo. The conclusions reached over 50 years of research may seem somewhat tenuous. One has to assume that they are valid, although they have not, except in a few cases, been validated. The alternative is to disbelieve almost everything in HFE research, which is an enormity no one wishes. However, if there is any one area one would point to as deserving significant additional research, it is that of the relationship among the OE, the laboratory, and validation.

Technology Makes Demands on the Human, Which Creates Tension Between Them

All technology, even if one is not in direct relationship with a machine and the technology only serves as a part of the operational environment, creates tension. That is because technology, as a set of stimuli to which the human must respond, makes demands on the human for behavioral resources to satisfy those demands. These resources, in the form, for example, of capability to attend to competing stimuli and to process masses of information, may or may not be sufficient at any one moment to satisfy those demands. This uncertainty about resource availability is at least part of what creates the uncertainty found in technological life.

Technological demand varies in intensity. Of course, it is greatest when one is interacting directly with a machine, but it also exists when technology is simply context in the OE. Technological demand is greatest when the human is initially introduced to the machine but tapers off after the human learns appropriate responses.

When technological requirements demand information-processing resources that are easily satisfied because the human need produce only well-learned responses, the technology-induced tension is minimal. When learned responses are not available or when resource allocation is critical (e.g., attention must be divided among several competing stimuli), tension rises correspondingly. The difference is between the requirement to turn a light on (which, even for young children, is easy to learn and perform) and the requirement to operate a complex personal computer (which many people never learn to do).

When technological demands are not excessive, they create a drive or motivation to perform. When these demands are excessive, they create an undesirable state of stress and reduce motivation (Schroder, Driver, & Streufert, 1967). In highly complex situations, technological demand creates uncertainty in the human because the meaning of stimuli and the responses they require may be unclear.

Technology creates a generalized tension in the OE because of the possibility that it may impose demands on people that cannot be satisfied. Railroads, aircraft, buses, subways, and even office buildings can make one uncomfortable until one becomes familiar with them. Think of an aborigine suddenly brought to Los Angeles. This may seem an extreme example, but think of anyone who faces daily trips in the New York City or London subway system; these people experience technological tensions in some measure. In most cases, this tension is low key; it acts as a sort of continuing background noise. When the human is more directly involved in technology, as when he or she must operate or maintain an equipment, the tension increases considerably and the human becomes very aware of it. Consider the common situation in which a driver cannot make his or her automobile start: The resultant tension is very evident.

Technology and the Human Form a New Construct—The System

The combination of technology and the human(s) in an organization designed to achieve a common goal creates a system that is an entity different than either the human or technology alone. A system is an organization that, because of its uniqueness, has certain properties that distinguish it from its constituent elements. This means that when humans interact with, react to, and control or monitor technological operations, the unit that describes this combination of human and technology has characteristics that require it to be treated and studied as a unique entity. The concept of the system is a higher order construct for HFE data. Because the combination of the human and technology is the fundamental object of concern for HFE, the system becomes an integral part of HFE and its study becomes a primary rationale for HFE research.

The assumption of the system concept as a higher order organizing construct is unfortunately not accepted by all HFE professionals; for others it is accepted only in a *pro forma* manner. For example, a review of the HFE research literature (see chap. 6) reveals almost no research directed at systems in general, although the term is used to describe types of systems such as automobiles and aircraft. The fact that so fundamental an assumption is not generally accepted in HFE suggests that the discipline is presently in a transitional stage of development.

The system formulation suggests that the human–technology relationship creates a new set of phenomena for which psychological concepts are no longer completely appropriate. A whole new set of theory and variables are required of HFE that cannot utilize solely psychological processes.

The question is what that new theory and model are. In the S–O–R formulation, technology is a stimulus to the human, but it is different from a nontechnological stimulus; the technological stimulus has many elaborate characteristics of its own. The human responds to the technological stimulus in diverse ways, but ways that are controlled and limited by the technology. Technology exerts control in the form of procedures that must be followed. Indeed, failure to follow these procedures means that the human is in error. Technology in the form of the system creates a concept of error that does not really exist in a purely psychological context.

The assumption that HFE is founded on a system formulation also requires a further assumption—that HFE phenomena cannot be completely explained by traditional psychological concepts. Such concepts may be adequate to explain human behavior as long as technology does not play a large role in human life, but are insufficient as soon as the individual proceeds to operate a machine or respond in any other way to technological stimuli.

Psychological explanations for how a human senses and attends to stimuli in general, analyzes these stimuli, and responds with a motor action may be appropriate as long as one does not consider the technological context in which the human functions. However, as soon as one includes the machine in the analysis, a psychological explanation for behavior may be correct at one level of analysis, but is insufficient at the system level of analysis. At this point, one must include the system as a necessary construct.

Analysis of HFE phenomena assumes that the human is breathing (a physiological level of analysis) and perceiving (a psychological level of analysis) while interacting with the machine (an HFE level of analysis). Different levels of analysis require different constructs, and selection of one set of constructs for a particular level of analysis does not imply that other analytic levels with their particular constructs are in some way wrong. One adopts physiological concepts to explain why the human is breathing, psychological concepts to explain why the human is perceiving, and HFE concepts (i.e., those of the system) to explain what the human–technology combination is doing.

The implication of the foregoing is that HFE in its research mode must seek for explanations of novel phenomena that cannot be found in psychological textbooks. New questions arise:

1. What characteristics of technology have the effect of exercising control over the human?
2. How is this control exercised (what are the control mechanisms)?
3. What are the criteria to which the human must perform?
4. If technology is manmade, how does one create a system that will utilize human performance efficiently?

These questions arise solely because of technology. The human performance resulting from the human–technology interaction is quite different from that which one finds in psychology texts. These issues and other questions that arise from the system formulation are discussed in chapter 3.

NEGATIVE ASSUMPTIONS

Introduction

Negative assumptions are simply strongly held views (with certain emotional overtones) on the part of some HFE professionals. It is not intended to suggest that all professionals hold these beliefs, but that a sufficiently large number of them do. The result is that they exercise a significant effect on the discipline. These assumptions are largely unconscious because they exist primarily as tendencies to action or, more often, *inaction*—or the unwillingness to do certain things. They are neither consciously viewed by those who possess these attitudes as being negative nor are they necessarily expressed as negatives. One will not find them described in HFE textbooks because they are entirely informal. Consequently, to discover these assumptions, they must be inferred from positive actions or, more frequently, actions that are not taken. These negative assumptions are listed in Table 2.2.

Some readers may feel that even to mention undesirable characteristics of a discipline is to denigrate it; even if such characteristics exist, it is best to ignore them. The author holds the viewpoint that a discipline must be known, *warts and all*, and that the only way to eliminate the inadequacies is to confront them and see what can be done to remedy them.

In Most of Its Concepts, HFE Is Really Part of Psychology. The author considers this as a negative assumption because, in a sense, it denies the legitimacy of the discipline. The falsity of the assumption was suggested earlier. Psychology is possibly *the* most important predecessor discipline for HFE, but psychology does not deal with technology as a critical factor and has no use for the system concept that underlies HFE. This would seem to disqualify psychology in claiming ownership of HFE, but many HFE professionals cling to psychology because they overemphasize the role of the human in the human–technology relationship. This is most apparent in HFE research (see chap. 6), which deals with technology almost as an afterthought.

That the assumption delegitimizes HFE is only a minor objection. The real problem is that if HFE is only a psychological speciality, there is no compelling need to find solutions for HFE-peculiar problems. The psychological model to which so many professionals cling helps sustain a lack of awareness of crucial HFE propositions and problems. It adds to the lack of concern about the inability to apply HFE research conclusions to system development, the failure to develop a quantitative predictive database, and the indifference to the lack of validation of HFE research conclusions, which renders these conclusions tenuous, if not dubious.

Research and Application Are Two Distinctly Different Aspects of HFE. Research and application are different, but not so different that one cannot feed into the other in a loop arrangement. The concept that these entities are inherently different, when allied with the psychological preoccupation with the human in the human–technology relationship, is largely responsible for the fact that HFE research has little to offer application and that significant problems in application are not themes for HFE research.

One of the purposes of HFE research as specified previously is to assist in application. This purpose has been largely ignored by researchers in part because of the differences they see between what they do and what practitioners do. Consequently, there are two HFE cultures: that of the researcher and that of the practitioner (Meister, 1985). Attempts to bridge these two have barely been attempted.

Research Is Infinitely More Preferable as an Occupation Than Work in HFE Applications. This assumption is closely tied to the one prior to it. Undoubtedly, if one surveyed the HFE professional population, a small majority would support the concept, and it may be true. However, the effect of this concept is to denigrate HFE application, which in turn results in few efforts to apply HFE research and hardly any use of application problems to initiate research.

Basic Research Is Superior to Applied Research. This is a widely held conception, in other disciplines as well as HFE. It has a long history going back to the rise of Western science in the Renaissance. The ethos of science requires that research contribute to the storehouse of knowledge, which everyone knows means basic research. There is no attempt here to go into the subtleties of the basic–applied controversy, which in any event is based on a somewhat false premise. What one calls basic or applied depends on the questions that are asked of research; basic research is supposed to answer questions of general interest and applied research is supposed to have a more limited sphere.

One must make a distinction between the common conception of applied research as being subordinate to or less valuable than basic research and the fact that application, as it is exemplified in system development, has fundamental problems also. In other words, (a) system development is a fundamental function of HFE, and (b) the most fundamental problem of system development is the translation of behavioral principles into design characteristics. Therefore, the author maintains that, as a process, application (behavioral system development) is just as important as any other so-called basic research; it contains problems crying for solution that are just as basic for HFE as those ordinarily examined in HFE research. The fact that these problems usually stem from a nonacademic and engineering context does not reduce their importance and make them less basic.

One can argue that the behavioral aspects of system development are as fundamental to HFE as a science as any other theme studied by its research, and that the system has dimensions and variables that are as important to HFE as any psychological variables. The effect of the second assumption in Table 2.2, in conjunction with preceding negatives, is to perpetuate the psychological orientation in HFE research and contribute to the gulf separating researchers and practitioners.

HFE Research Is Intended Solely to Add to Scientific Knowledge. This negative assumption is simply a clarification and extension of previous negative assumptions and needs no further amplification.

Problems Inherent in Applying HFE to System Development and Other Applications Are Not Appropriate Subject Matter for HFE Research. See the preceding items in this section for arguments and counterarguments on this point.

The Purpose of Performing Research Is Publication. Phrased in this way, probably all researchers would reject this assumption as being patently false. They would call on the precepts of science as justification for their research. Nevertheless, if one looks at the actual behavior of researchers, the conclusion is inescapable: What motivates them is publication and the resultant prestige among peers, as well as other more tangible benefits, such as academic tenure and better job opportunities. This is not peculiar to HFE. The world has witnessed many unseemly squabbles about priority of discovery (e.g., in AIDS research), which in practice is determined by priority of publication. However, much more needs to be said about publication and HFE.

Of the two major functions of HFE, research and application, one—research—is heavily dominated by the HFE publication process. It is a paradox that what would otherwise be considered a subordinate aspect of the discipline actually controls a major part of that discipline.

The publication function is a conceptual process because, although it superficially deals only with pieces of paper, it disposes of the concepts within these papers. If a research paper submitted to one of the HFE journals is not accepted, the concepts within that paper in essence die unless the paper is sent informally to individual professionals in the discipline. Those who have followed recent Russian literary history are aware that fiction and nonfiction were transmitted laboriously through an underground process of mimeographing material and handing copies over by hand to various members of the public. The product was called *Samizdat*. It is possible for a professional to do the same in the United States if there is no other publication medium available to the author (possibly the vanity press, which is, however, quite expensive), but the range of communications for technical material is limited. In a media-conscious civilization, whoever controls its communications channels in essence controls the discipline.

Publication is essential to a discipline because the primary—perhaps the only—way in which the discipline can contribute to the storehouse of knowledge about the discipline is through publication. The Russian *Samizdat* process is feeble in a potential receiving population of 5,000 (HFES membership). Moreover, what is often forgotten is that publication in a peer review Journal provides a sort of imprimatur on that material. That, rather than communication, may be its primary function. Lacking that imprimatur, the concept transmitted in a personal press (i.e., copies sent through the mail to interested respondents) cannot be formally accepted. To use an unlikely example, what would have happened to Einstein's special theory of relativity if he had been unable to publish in an *accepted* journal? What would have happened to psychoanalysis if Freud had not been able to find a publisher for his *Interpretation of Dreams*? It is not too much to say that the viability of any concept is absolutely dependent on its publication. However, this does not mean that published concepts cannot be discarded or revised (see Kuhn, 1962). The process of publication is a conceptual process of some complexity because it involves the following stages: (a) reception of a paper (or book) by a recognized publication; (b) analysis of the material in the paper or book, which means its concepts, the research findings associated with the concepts, and the conclusions; (c) evaluation of these, which presumes certain criteria and the expert knowledge of those who evaluate the paper or book; and (d) a recommendation for its publication or rejection.

This process is considerably more important than the research that preceded it. McLuhan's famous phrase, *the medium is the message*, is appropriate here. More concretely, the message will not get through unless the medium approves. The medium is a gatekeeper whose favor must be achieved; all the research in the world is powerless against the gatekeeper. One can always create research, but how many can keep the door?

The gatekeepers are a special breed. They are a product of the concepts that inform them, which is why concepts are so important to a discipline. The discipline does not exist without the concepts, but the only ones who recognize the concepts, and thereby give them respectability (and thus give life to the discipline), are the gatekeepers—an elite group of people selected by their peers for their special knowledge and expertise. There are HFE professionals who resent the power of the gatekeepers; however, if there are to be judges and evaluation, the gatekeepers are essential. It is not as if publication were a rare event; it was so in ancient times, but in a technological civilization there are more people clamoring to have their products transformed into print (or placed electronically in the Internet) than the discipline can cope with. There may come a time when electronic reproduction of concepts is so simple that everyone will communicate. Should that happen, and if there are no gatekeepers, it will be impossible to discriminate among the good, the bad, and the mediocre, and a discipline may die of a surfeit of concepts. However, that day is not yet.

The gatekeepers must make decisions based on the concepts they have developed for themselves or have had imposed on them. A gatekeeper who makes use of the negative assumptions in this section, without being aware that he or she is utilizing them, exercises a potentially great negative impact on the discipline. A professional who has a psychological orientation to research can do little damage, but a gatekeeper with such a bias is likely to eliminate certain papers that do not accord with that orientation. The gatekeeper is not alone, of course; he or she has an editorial board and other reviewers to develop a consensus. What is one to do if everyone involved has the same orientation?

The publication professionals—journal editors, paper reviewers—are an elite group of people; they are recognized for their accomplishments and are often academics (because academics seem to have greater prestige than other professionals) who have the power of life or death over the concept-generation process. This is an exaggeration, of course, but not by far. Nongatekeepers often resent these people because of the power they possess; but if they did not exist, it would be necessary to invent them. If the professional submitting his or her paper finds it accepted for publication, he or she approves of the gatekeepers; if not, the gatekeepers are disliked. Because many more papers are rejected than accepted, there must be a great store of resentment among professionals. What is so frustrating to them is that there appears to be no alternative to the publication process.

From the standpoint of the discipline as a whole, publication (the gatekeepers, really) establishes the criteria of what is adequate research, what research topics should be studied, and, by implication, the concepts around which research revolves. The criteria stem in part from general principles of science and from the predecessor discipline—psychology. These criteria are: (a) objectivity in the research methodology; (b) the experimental method as reflected in experimental design; (c) the laboratory or something as close to the laboratory situation as possible; (d) abstruse statistics (simple *t* tests need not apply), and the more interaction terms the better; and (e) originality of subject matter and method, but nothing too extravagantly new.

These are positive criteria for publication, but there are negative (rejection) ones as well. The negative criteria are improper experimental design, too few subjects, inadequacies in the conduct of the study, and failure to use proper statistics. These are the major negative criteria, all of which are entirely justifiable.

There are other negative criteria that do not represent specific inadequacies. These secondary negative criteria represent tendencies on the part of the gatekeepers to ignore negative aspects of the research. Negative criteria that are often ignored are:

1. *Nonrepresentativeness of the measurement situation.* This criterion is related to the existence of the operational environment (OE) that serves as the referent for all HFE data and concepts. The importance of the OE to the nonrepresentativeness criterion is that, if a research finding cannot be related to human performance in the real world, the research has no significance. This criterion is constantly violated in HFE publications because it is assumed (without any proof) that if a study is performed in the laboratory, it is automatically meaningful to the real world.
2. *The absence of validation.* If the existence of the OE requires that findings be validated, then almost all papers are accepted without any evidence of the validation of their conclusions. This is so common that one can infer from this an unspoken assumption—that validation is unnecessary.
3. *System development.* Few published papers deal with system development questions (see chap. 6); this means that HFE has accepted the criterion that research dealing with system development is unnecessary. This does not mean that system development papers would be rejected by the journals, but it creates a presumption that such papers do not really satisfy the prior positive criteria.
4. *Application.* There is also a presumption that application of research findings is unnecessary; most papers are published without such applications described in the paper.

The publication criteria (both positive and negative) are not clearly enunciated in any document; they can only be inferred by examining the papers that are accepted and published.

Most research papers are rejected (at a rate of 80% for *Human Factors*) because the publication channel is so narrow (500 or 600 pages a year available). Even accepting the necessity for such a high rejection rate (which is probably not out of line for HFE in comparison with other disciplines), questions arise: Are the rejected papers truly inadequate or invalid? What happens to the rejections?

The gatekeepers would probably say that the rejected papers are not, except in a few cases, bad papers; they are simply not good enough to warrant publication in a journal with limited space. Does this mean that 80% of the research submitted to *Human Factors* and rejected is mediocre research? If this is true (and the author would never maintain this), the discipline is in trouble. In actual fact, the situation is probably not so dismal. However, it does raise questions as to what concepts are accepted and others rejected and what the effects on the discipline are.

If paper acceptance is the equivalent of concept acceptance, the discipline is rejecting a great many concepts. Of course, not every paper contains an important concept, and much research does not find its way to journals dealing with general HFE topics. The government and private industry have publication resources that, although much more limited in terms of audience than the general journals, do have an audience. However, if one looks at research as a whole, a great deal of it seems to fall by the wayside and will bear no fruit.

Therefore, the competition for publication is tremendous. What is disturbing is that, if even half of the papers rejected have some value, a great many concepts have been lost. More important than their concepts, however, are the data they contain. One may be able to do without concepts because these are somewhat speculative, but can one do without data? The author does not pretend to have a solution for this problem. The answer would obviously be to establish more journals, but there is a limit to the number of journals that any profession can support. Perhaps computer communications like the Internet have a solution? Authors rejected by the journals might develop their own web page and publicize their work.

The importance of publication is recognized by the fact that, in the HFES, more attention is paid to publication than anything else except finances. However important publication is to the discipline, publication limits its responsibilities. It takes no responsibility for the validity or representativeness of what it publishes; it is unconcerned about application of results; it does not think about its role vis-à-vis the discipline as a whole. It is a gatekeeper and only a gatekeeper. Who then monitors the gatekeepers?

The discipline makes certain assumptions about its publications. It recognizes their necessity, of course. It accepts the positive and negative criteria the gatekeepers have developed. It assumes that, as far as anything like an HFE database is concerned, the act of publication adds to that database and no other special arrangements for that database need be considered. One can term this *benign neglect.* It is paradoxical (and it is presented here only as an amusing footnote to the discussion) that, although objectivity is one of the major criteria of adequate HFE research, the entire publication process is mired in subjectivity. This is only a footnote because it is difficult to conceive of provisions that would make the process more objective.

The discipline gives the publication process tremendous responsibility as the storekeeper (as well as gatekeeper) of its formal CS. The discipline takes no responsibility for the compilation of data for quantitative prediction purposes; it takes no responsibility for evaluating its research as a whole (except in terms of accepting or rejecting the individual paper) and its CS. The discipline (and this is probably true of other disciplines as well) is highly democratic (some might call it essentially anarchic); as far as research is concerned, any research can and should be performed that can be given funding and is likely to be accepted for publication. The major factor directing the research to be performed is the research that has been published in the past.

Because the publication process is inherently elitist (and there is nothing wrong with this), there is a narrow hierarchy involved in the CS. There is the small number of gatekeepers, a few society functionaries who keep a mostly negligent eye on them, and there is the rest of the HFE professionals who have limited access to the *corridors of power*. The discipline as a whole, to which it is difficult to attach a face other than that of the functionaries who volunteer their services to HFES and who work at the job only part time, is not responsible for its CS except as the CS becomes visible via a few publication channels.

If one looks at the individual professional, the publication process becomes the motive for the research he or she will conduct. Would anyone perform research if there were no way to communicate the findings of the research? Publication is the strongest manifestation of the self-interest of the professional. One sees it in full flower in the academic premise, "publish or perish." The same premise (although with fewer consequences) even applies to HFE nonacademics. One does research to expand one's curriculum vitae.

Some readers may feel that this analysis of the publication function of HFE has been too negative. Surely publication fulfills its responsibility of getting the good news out to the parishioners. However, the way it does can be a source of negative as well as positive consequences.

The Human Is So Variable That Quantitative Prediction of Human Performance Is Impossible. This belief relates to the issue of deliberately developing a quantitative prediction database. HFE employs prediction because every research conclusion is a prediction (usually quite general and certainly nonquantitive) that, given the variables and circumstances employed in the study, the results found in the research would occur again. There is no deliberate application to the OE, but that is certainly implied in the research conclusion.

What those who adhere to this assumption are really saying is that the nature of human performance reduces the precision of the prediction; that prediction in quantitative terms requires greater precision than human variability permits. The criterion that these professionals have adopted is much more stringent than that which developers of such databases have set themselves, because databases have been developed in the past (see chap. 7). Certainly objections can be raised by critics that the quantitative precision available is not great enough, but that the principle—that it is possible to develop a database for quantitative prediction—has been established each time such a database has been published. These databases have been found to be incomplete or imperfect, but the possibility of creating the database has been demonstrated.

Moreover, in countering the critics, one might say that even if a database is initially not as precise as one would wish, it should be possible to improve its precision incrementally. Perhaps the critics are responding more to the possibility that the prediction would be invalid. It should be pointed out, however, that the predictive database is based on raw experimental data or recorded observations. If the experimental data are invalid, presumably the entire research is invalid. Critics might rejoin that they have no objection to the data on which a prediction is based, but they have less confidence in the process used to transform those data into the quantitative prediction. This is a valid point, but does not negate that, with sufficient trial and error, a more efficient transformation procedure could be developed.

The effect of the previous assumption is to reduce the incentive to develop a quantitative prediction database, the foundation of which would be experimental data. As in the case of application generally, the researcher is perfectly willing to accept unvalidated research conclusions, but perfectly unwilling to apply the data producing those conclusions to more substantive uses.

Here again, the psychological orientation may have played a role. The original motivation for developing a quantitative database stemmed from the introduction of early HFE professionals to reliability engineering processes. As is seen in chapter 4, HFE professionals engaged in the aerospace industry were often located organizationally as part of the reliability or quality control division. Regardless of where they were located, they certainly met such engineers and a certain amount of cross-fertilization occurred; reliability engineers became interested in applying their techniques to human error quantification, and HFE specialists were attracted by the notion of quantitative prediction. This kind of thinking would have been (and still is) foreign to researchers who have never worked in industry.

Because Personnel Performance Is So Strongly Influenced by Contextual Factors, It Is Useless to Look for General HFE Principles. One wonders why any researchers would subscribe to this point of view because it would seem to contradict the rationale for their work. Have they no confidence in their own research? Nevertheless, there are professionals, even distinguished ones (e.g., Moray, 1994), who espouse it.

Of course, it is true that human performance is sensitive to context. That is the reason that elaborate experimental designs are developed to exclude or hold constant all contextual factors except those related to the particular experimental variables at issue in a study.

It may be that those (hopefully only a few) who maintain this assumption are concerned about the degree of generality in general principles. It often seems difficult to take a research conclusion, which is in fact a general principle, and apply it to a more specific instance. If one does so, one is making a prediction and one could be wrong. The available research may not seem to relate to the specific instance to which it is to be applied. The author has the impression that the research literature has gaps and is much more complete in perceptual research, for example, than in decision making. In any event, this point of view may be the genesis of the opposition to general HFE principles. This is a reflection on the research themes being studied, and many research gaps must be expected when research is idiosyncratic and not directed by authority.

Those who advance this negative assumption may feel that it applies only to system development and other applications, and that within the research function general principles can be found. However, of what use are general principles if they cannot be applied to something other than themselves? In research, general conclusions seem to be applied only to the generation of additional research. Some might consider this a conceptual tautology.

The author feels that the reluctance to entertain the notion of general principles is actually a reluctance to apply research because a general principle connotes application. In any event, this behalf can only have negative consequences for the adequacy of HFE research (and application).

The Only Research That Is Truly Scientific Is the Experiment When Conducted in a Laboratory. The laboratory and experiment are manifestations of a profound scientific attitude: the need for control. Because the variables with which the researcher is concerned are invisible, they can be discovered only if conditions for their expression are optimal and if all other variables and conditions are neutralized. This is the function of experimental design.

If the concern of the researcher is only the discovery of those variables affecting performance, then the experiment and laboratory should be as important as they appear to be. There are other questions in HFE research, however. The research conclusion implies a prediction that a certain performance will occur if certain variables are exercised. However, before specifying the values of the variables, it is necessary to determine which variables are relevant to which performance. This is not what the experiment does; the experiment attempts to indicate (with indifferent success, however) the relative strength of the variable. The experiment assumes that certain variables are important to a type of performance; it attempts to discover just how important.

Until now, the discovery of which variables to study in research has been largely a matter of logic and, to a lesser extent, observation of phenomena. Previous studies also help in the selection of variables. Before variables can be specified, however, it is logically necessary to describe the performance that is of interest. Description of phenomena should lead to experimentation, but HFE leaps over description directly to the experiment. This is because HFE attempts to mimic hard science disciplines, which, because of their greater longevity and for other reasons, may no longer need this initial stage of scientific investigation.

The point is that this assumption, although acceptable perhaps as a general statement, can have, when carried to excess, negative consequences for HFE. That is because there are situations in which the experiment and laboratory are not the most desirable entities with which to begin research on a topic.

A case can be made for investigations to begin in the OE, perhaps by systematic description of the phenomena and the systems one is interested in, to be followed by a possibly more rigorous investigation by the experiment, although not necessarily in the laboratory (the test site and the simulator are alternatives to the laboratory). There may be some point in looking at phenomena in their raw, undigested state in the OE before refining knowledge of these phenomena in an experiment. In other words, perhaps the scientific paradigm for HFE should be real-world observation and description first, followed by more sophisticated experimentation in more controlled circumstances. Presently, however, the procedure is for the researcher to rely on previously published papers as the equivalent for on-site examination of a phenomenon.

If the researcher is obsessed with the experiment and the laboratory, it is because he or she wants control and control is extremely difficult to impose on the OE. Control makes the researcher feel more confident of what he or she is doing; the manner in which experimentation is performed can be specified more precisely than can studies in the real world, and thus are easier on the researcher.

It is the exclusive preoccupation with the experiment, preferably in a laboratory, that represents a danger to HFE because, as has been demonstrated previously, the experiment is artificial and must therefore be validated. A strategy of investigating phenomena in the OE, followed by experiments on the same phenomena, might make it unnecessary to validate research conclusions because the experiment would serve as a validation of the initial OE investigation. That initial investigation might well include use of SMEs, observations, interviews, system tests, and so on.

That this revised strategy will ever occur is doubtful because the strength of the belief in the experiment is so great. Yet it should be considered even if the idea is rejected because the discipline must be open to all ideas. The negative consequences of negative assumptions occur primarily because they inhibit consideration of other concepts.

Research Performed Under Nonlaboratory Conditions Is Almost Always Contaminated by Its Environment. Professionals who believe this assumption also believe that outside the laboratory there can be no adequate control.

This, of course, is nonsense unless one believes that it is possible to exert absolute control over measurement absolutely. Heisenberg's paradigm suggests that complete control even in the laboratory is impossible. Would one hanker after absolute control when one considers that the subject whose performance is being measured is also analyzing the experiment, evaluating it, and possibly changing his or her performance as a result of all this analysis? One may be able to exert complete control over external conditions, but such control is impossible with a thinking human.

If one believes that there are legitimate research venues and operations other than that of the experiment and the laboratory, the effect of this assumption is to restrict opportunity for research to be performed and for research papers to be published because studies performed outside the laboratory will receive less attention from editors.

Validation of Research Results Can Be Achieved Only in the Laboratory. This supposition assumes that if one reproduces the results of one experiment in another laboratory or venue other than the OE, one has validated the original findings. If one studies the same variables in the same controlled manner in the same or a similar laboratory, the presumption fostered by this assumption is that the original research conclusion has been validated. After all, there is no other environment in which validation can occur. Unfortunately, all one has as a result of replication in the laboratory is a confirmation of data reliability.

If the laboratory and the experiment are used to validate findings first discovered in the OE, then this assumption is correct, but this assumes that research begins in the OE and unfortunately most research begins and ends in the laboratory.

It is a given in our concept of HFE that validation must occur either in the OE or that the validation is the result of findings initially observed in the OE. As a consequence of this assumption, with its emphasis on the laboratory as a venue, validation does not occur because the OE is not used either as the initial stage of a two-part research program or as a second stage of research initially performed in the laboratory.

The HFE Database Consists of the Papers It Publishes. This is an assumption generally believed by almost all professionals. The belief is inferred from the fact that it has been almost impossible to create sufficient interest in HFE in the development of a specific quantitative database. Therefore, one can think of this assumption as a parallel to and supporting a prior assumption (Table 2.2, 3[a]).

The belief that the database grows with the publication of each new paper rests on the impression that all one needs to do to make a database available is simply have the published papers accessible for reading. This is far from the truth. The database is formed from the raw data in the study rather than from the conclusions derived from those data. Those conclusions are simply the interpretation of the data and, in any event, are only general and qualitative.

It serves the self-interest of the mass of professionals to adopt this assumption because it makes it unnecessary to exert the great effort that would be required to develop a proper database. This assumption, which is not phrased in a negative fashion although it has negative effects, assumes that the individual professional, when faced with the need to make use of particular items of research, will read the individual paper (or a series of them), extract the necessary data, and transform the data on his or her own into a quantitative prediction.

This scenario assumes that the individual professional has the intellectual capability and determination that will permit him or her to perform the necessary database operations. A cynic might suggest that all he or she will do is ponder the conclusions in the paper. For the professional to make use of the data in the paper, it is necessary to transform the data (in the event that the raw data are published) into a form useful to the professional. Only a specialist in the area described by the paper will make such an effort. Hence, for the casual reader, the paper is disregarded and the database in effect does not exist for that individual.

To qualify as part of a database, the individual published papers must be acted on by database specialists (there are such people, even if database development is not a recognized speciality). The database is operationally defined by what the professional can do with it; if he or she can do little with it (as is the case with the untransformed papers), it does not exist as a database for him or her.

Any Difficulties HFE Has Had With Other Disciplines or Non-HFE Organizations Is the Product of Ignorance and Hostility of Those Who Are Not Behavioral Scientists. Chapters 4 and 5 describe the initial experience of the first generation of professionals who encountered engineers and the system development process for the first time when HFE first moved into industry shortly after World War II. Prior to that time, only a few engineers had interacted with behavioral scientists, and then only in a research capacity. Thus, they viewed the newcomers with suspicion. A great deal of hostility toward HFE was expressed, which still exists, although much reduced (one hopes).

This historical hostility has engendered a pervasive belief in HFE professionals (even those of later generations) that the negative reaction to HFE concepts and suggestions for system development (what the author calls *behavioral design*) is essentially malice. The symbiotic relationship of HFE to engineering that has developed over the years has led to feelings of quasiparanoia in professionals: Nonbehavioralists are simply by reason of engineering training (or lack of behavioral training) opposed to HFE.

It is true that ignorance and the engineer's preference for physicalistic concepts predisposes the engineer and manager to a certain resistance to

HFE, but the matter is overstated and the genesis of the negative feelings is more complex than it is ordinarily phrased by HFE people.

Much of the opposition to HFE in the past (and presently, if it still endures) is the result of the inability of HFE to live up to the promises HFE has made, both explicitly and implicitly. For example, engineers are habituated to quantitative terms. To the extent that HFE has design guidance principles (and there are fewer of these than are supposed), these are largely qualitative and therefore lack the impressiveness that quantification could lend them. If it were possible to say to the engineer that X principle when applied to a design would result in 15% reduction in operator error in operating the equipment being designed, the engineer would be much more willing to accept X principle. In honesty, however, almost no HFE design principle other than anthropometrics has quantitative values associated with it. This makes, one assumes, the engineer a little nervous about accepting X principle.

(The HFE specialist could lie to the engineer about the 15% [how would he or she ever find out about the lie?], but the truth would inevitably leak.)

The previous scenario can be repeated in almost all the interactions between HFE and engineering. In the absence of a capability demonstrated by numbers, the HFE specialist may seem no more qualified to pronounce on the human aspects of equipment design than the engineer, who is, after all, a human also and can be expected to exercise his or her innate human ingenuity to make behavioral pronouncements on engineering matters. The absence of quantification retards the applications of HFE to physical problems and makes it appear dubious. Engineers may express their uncertainty about HFE with resistance.

The obverse side of this resistance is the defensiveness of the HFE professional, which is expressed in the need to persuade nonprofessionals of the verity and value of HFE. The assumption that parallels this assumption is that verbal cajolery will induce a more positive attitude in engineers. Thus, in the early history of HFE, great emphasis was placed on providing both formal and informal training of engineers in principles of HFE. The expectation was that this instruction would lessen the manifest resistance of engineers to HFE interventions.

Such training has undoubtedly had a positive effect, but there is still much resistance. Engineers and others utilize an implicit cost–benefit analysis when it comes to evaluating HFE recommendations in system development and testing, and they find little benefit in HFE suggestions. This is because no quantitative prediction of anticipated consequences in terms of human performance are associated with the recommendation. It is not sufficient to make the general statement that the recommendation will reduce the probability of human error; such a statement comes under the heading of *motherhood and apple pie*. (A recent study [Lund, 1997] suggests that HFE specialists actually utilize such maxims as *keep it simple* as primary tools for design analysis and evaluation. If true, this negates the value of HFE research for providing design guidelines.) Engineers are smart enough to recognize and accept a HFE recommendation as it relates to human limitations (e.g., moving a 500-pound weight without mechanical aid is impossible for almost all humans), but less obvious recommendations that are not associated with human thresholds are not accepted so readily.

When these last are rejected by designers, the specialist feels much put upon and infers that the rejection is motivated by malevolence, which is in large part untrue. If one suggests to a designer that a particular aspect of the design should be modified to reduce operator stress, the designer will ask: How do you know? Can you prove it? The specialist will be unable to cite chapter and verse because no quantitative values attend his or her statement, and no research conclusion is specific enough to cover the specific design problem. Therefore, it is much easier to blame a negative reaction on an innate hostility to behavioral concepts, as if the engineer were born with anti-HFE genes.

To deal with this hostility or difficulty in interacting with engineers, the specialist would prefer to look outward and counter the negative reaction with persuasion. A recognition that the causes of negativism toward HFE may lie within the discipline is too difficult for HFE as a whole to accept. Far better to ascribe negativism to innate tendencies and embark on a propaganda campaign to check these tendencies than to seek the source of HFE difficulties in the failure to do certain things that should be done. It is easier to believe in persuasion as a mechanism for change than to make the necessary changes in HFE procedures that would enable HFE to overcome the problem. When HFE becomes more quantitative in its orientation, it will be possible to satisfy engineering expectations. When HFE is needed by engineering design and is no longer an *add-on* mandated by government, the hostility will largely disappear, but that means casting qualitative HFE statements into a more rigorous, quantitative form. It means developing a useful quantitative prediction database. All of this will require a good deal of effort and there is much doubt that HFE is willing to exert that effort. It is much easier to attack the problem by, for example, publishing advertisements exhorting designers to use HFE. One can sell beer this way because, along with the advertisement, there is the actual beer; along with the exhortation of the necessity of HFE there is . . . what?

The fault lies in ourselves, not in genetic hostility. When HFE can perform on what it implicitly and explicitly promises, the negativism will largely disappear. Until then, however, this last assumption is comforting because it requires of HFE only belief, not a genuine effort to improve the discipline.

3

The System as a Fundamental Construct

The previous chapter ended by indicating that when the human interacts with an equipment, the combination forms an entity—the system—that has properties that are different than those of either one alone. The idea of the system is a *construct*, which means that it is a concept whose characteristics are manifested concretely in physical and behavioral phenomena. The distinction between the idea and the phenomena that represent the idea must always be kept in mind because it is easy to become confused if the two are not differentiated. The system is critical to HFE theorizing because it describes the substance of the human–technology relationship. Anyone observing the *stew* of humans in real-life interactions with technology and wishing to organize it conceptually, classifying it into describable units, must inevitably see this activity in terms of systems.

The essence of the system concept in its most general form is that an entity is composed of subentities that, when arranged in a rational organization, are different from the superordinate entity. The superordinate entity is the system; the subentities of which it is composed are subsystems and individual human–machine combinations.

The reader may ask: Why is the system concept so critical to the understanding of the human–technological relationship? The reason is that, without some such organizing concept, one cannot combine the individual human–machine combinations into something more complex than their individual units. Without the system as a means of combining individual phenomena into larger units, the phenomena remain individual, although one can observe that in reality individual units and their phenomena do combine into larger entities. For example, American Airlines is much more than numbers of pilots, each in his or her aircraft. The company includes the pilot and the plane, as well as the passengers, the clerk who sells the tickets, the maintenance personnel who service the aircraft, and so on. Reality enforces their organization into higher order systems. Without conceptualizing such systems, the phenomena one experiences as part of the aviation system, for example, could not be understood. Whether one calls it the system, as HFE professionals do, or X, some way of relating disparate phenomena at various levels of complexity is required.

The system is a superordinate entity. This means that the subentities making up the system change their dimensions and characteristics when they become part of the superordinate entity. They act in accordance with the higher order purposes and requirements specified by the system; they are, in a biological sense, ingested and digested by the system in the same way that food entering the human changes its initial and external form to become proteins, fats, and so on within the body.

This metaphor is perhaps less acceptable when one considers the system as a human–technological relationship because the human is supposed to have free will, but the analogy is not very forced. If a man or woman is hired to operate a shoe-making machine and the machine is part of a higher order system known as the XYZ Shoe Company, as long as he or she is employed to operate that machine, he or she becomes subordinate to the processes, procedures, and purpose of, most immediately, the machine and, less immediately and directly but no less strongly, the company as represented perhaps by the foreman.

Because the system encompasses or absorbs its subsystems, thus becoming more important than the latter, the concept of *hierarchy* immediately emerges. In a one-celled animal, the hierarchy is a single level only. In more complex systems, there may be many levels. In the XYZ Shoe Company, there is the individual machine operator, packer, the foreman, the vice president in charge of manufacturing, the president of the company, the Board of Directors, and so on. As one goes up the hierarchy, one moves progressively away from more concrete to more abstract subsystems, but the system controls all levels and higher order subsystems control lower level ones. This is part of a natural process of specialization of function, which would exist regardless of whether we conceptualized it in system terms because it permits control. Control is critical to make large entities perform purposefully.

A fundamental dimension of the system is the organization or arrangement of its elements. To become a system, the human–technological subsystems must be organized into relationships of increasing complexity. The human and his or her machine is one relationship; the foreman who monitors and directs the performance of *N* human–machine units is in another relationship (one with higher management, another with the individual units); higher order relationships progressively develop. (One could say *naturally* develop, but management theorists may develop organizational relationships that are not natural.) That functionality is involved in putting the pieces of the organization together is quite obvious; one would not ask the shoe machine operator to be CEO of the shoe company, and the CEO could not (probably) perform CEO functions while operating the shoe-making machine.

Of course, it is possible to organize the man-made system in various ways (biological systems seem not, except in special circumstances, to have this flexibility), but efficiency of function should determine the organization that is ultimately developed (although systems can limp along with inefficient organizations). Some system functions cannot be performed except in one way; other functions can be performed in different ways, but only one of these is shown by experience to be the most efficient.

Complexity is another dimension that is characteristic of all but single-cell systems. As soon as one begins to organize the system, the system, subsystem, and subsubsystem relationships immediately display complexity, with the overall system being most complex, the subsystems somewhat less complex, and so on. The organization chart of a company displays that complexity in terms of the span of control that each level of complexity has: The CEO has near total control, the vice president for manufacturing controls all manufacturing functions but not marketing and financial, and so on. The lowest system level (the operator–machine combination) has least control. The span of control defines the amount of complexity. Control, too, is an essential dimension of the system.

The implication of organization, hierarchy, complexity, and control is that, at each level of functioning, purpose, types of functions, and processes vary. One does not expect the general of an army to do what the private does.

It is obvious that the system is a dynamic entity. It exists to perform one or more functions, which means that, to the extent that it is conscious of these functions, it has a purpose that directs the system's energy. The purpose is directed toward a goal (e.g., the production of specified numbers of types of shoes within a given time), and the goal may or may not be achieved, which requires feedback of information about goal and subgoal accomplishment (processes implementing that accomplishment; e.g., what are this week's inventory figures?). It is apparent that any complex system must be able to sense stimuli and respond, which means communication among hierarchical levels and the processing of information.

This view of the system's functioning treats the system (and its elements) as if it were human. Thus, all the functions that a human performs are performed by the system and its elements, although the system has other characteristics and dimensions that are peculiar to the system and not found in the human.

This may sound familiar to the reader because we have been describing human functions and processes. This is not merely analogy. As the system becomes more complex, it develops or is endowed with humanlike characteristics. Because humans are involved in the human–technological relationship, the system that encompasses these relationships must of necessity acquire human attributes. Humans cannot perform in nonhuman ways; hence, neither can systems. The human analogy permits us to examine system entities and phenomena in ways that make sense to humans. By virtue of our brains, we cannot think of phenomena in nonhuman ways. The following are the major points made in this chapter; some of these may replicate or overlap the concepts and assumptions of the previous chapters.

DEFINITION OF THE SYSTEM

In its most general sense, a system is the arrangement of human and machine elements organized into a whole by the need to accomplish a specified goal. There are two types of human–machine systems. The first has a single owner (i.e., the system is developed for a single customer who is the owner of that system when it is completed). Single-owner systems are likely to be those developed for the military or other governmental agencies. Examples are tanks, guns, and warships. The second kind of system is the commercial system, in which the developer of the system is the owner and the users (multiple) are customers. Examples of commercial systems are personal computers, photocopiers, and cameras. Obviously there are many more commercial than governmental systems.

The development process for these two kinds of systems is slightly different. The organization, which was referred to previously and which is the essence of the system, is achieved by establishing rules of an if–then nature. In the case of single-user systems, the rules are established by the developer and owner operating in concert; the owner establishes the overall rules in the form of a mission statement (what the system is ultimately supposed to do), which is expressed in a contract specification that lays down the parameters to which the system is designed. As part of the design of the system, the developer, again in concert with the owner, establishes the more detailed rules that apply to the individual equipments or subsystems that comprise the overall system. In the case of the commercial equipment, the developer establishes all the rules because the developer is also the owner.

Anyone who uses an equipment or performs a system function must learn these rules (the rules that apply to his or her activity). This applies to users of commercial equipment as well as military personnel. In both cases, the rules may be taught in formal training courses or (as is more likely to be the case for commercial users) through written or graphic instructions.

Rules are expressed in various ways. They may be built into the design of an individual equipment (it can be operated only by manipulating controls in a predetermined manner). Rules can be formalized in written instructions (e.g., operating procedures that must be learned). There may be rules that are expressed informally or may not even be expressed at all, such as the manner with which one brings problems to the attentions of superiors (e.g., a chain of communication). The higher one functions in a hierarchy, the less concrete the rules are because at the highest system levels one is not dealing with individual equipments but with interactions among humans.

There are many types of systems—biological, physical, mathematical, and so on. The one with which HFE is concerned is the human–machine system, which describes the interaction of humans with equipments and systems.

Tools, Equipments, and Systems

Terminology here is apt to be somewhat confusing. A distinction must be made among the tool, the equipment, and the system. Examples of a tool are a hammer, saw, or fountain pen. These are independent entities (i.e., they are used independently of any other tools, although they may be used in concert with other tools). They are independent because they are separable from any other piece of hardware, although in some instances they can be plugged into other hardware. There are, for example, battery rechargeable flashlights; lamps function only after they are plugged into the electrical system. Tools have operating rules, but these are (or are supposed to be) relatively simple. Some tools can be relatively complex and, in such cases, it may be difficult to differentiate them from equipments. Although tools are used in system operations, they are auxiliary only. The physical units of a system are equipment, subsystem, system, and tools, like equipment drawings or wind tunnels, are not a system element.

The distinction between an equipment and a system, in addition to size and complexity, may be obscured because an equipment, like a lathe, can be used independently or as a unit of a system. A lathe may be found in a homeowner's garage, in which case it is an independent equipment. The same lathe, when installed in a factory, is not independent of the production system, which sets its tolerances. In the second case, the lathe has no function or identity except as a system unit.

Equipments and systems both have operating rules. When an equipment is part of a system, its rules depend on the rules established for the overall system. The major difference between the equipment and the system is that the equipment, when it is part of a system, is dependent on that system and operates in accordance with system rules. When an equipment is used independently, it operates by its own rules.

Equipment rules (procedures) are much more specific than subsystem and system rules. When we say that the equipment operates in accordance with system rules, we mean that the equipment does so only in relationships required by the overall system. The shoes in a shoe manufacturing company are, once made, packed in a distinctive way, transported to a loading dock in a particular way, and so on as determined by overall company (system) policy. That policy may ensure that N shoes are manufactured per day, but does not directly enforce how the shoes are made using the equipment.

It hardly needs saying that power flows from the top. The CEO of the shoe manufacturing company can tell the operator what to do (although it is improbable that he or she would do so directly), but the operator has no say over the CEO.

Another difference between the equipment and the system is that the latter has a goal, whereas an equipment (and a tool, too, for that matter) has only a function. Again, there are difficulties in making a distinction between a goal and a function. A function can be independent of any human agency, although obviously humans also have functions. A thermostat, which is automatic, has the function of adjusting temperature. A circuit has the function of directing electricity in a particular direction.

A goal implies consciousness, although the goal itself is not self-conscious. The self-awareness is in the human, and that awareness includes the goal, which is incorporated in and becomes part of the human's awareness. Once included in the human's awareness, the goal is used by the human to energize his or her activities in pursuit of that goal.

The goal implies futurity—something that is desired and will occur in the future after a set of operations has been completed. The function is not a goal because there is no futurity associated with the function; it occurs immediately after it is activated unless the equipment or tool is broken.

Functions may be performed in support of a future goal, as when a module is activated to cause an equipment of which it is a part to function, and the equipment is required to achieve the system goal. For example, a pilot switches an aircraft engine on to develop speed to achieve lift-off; the switch and the engine both perform functions, but only the aircraft system has a goal because a human operates directly on the system. This may be the distinction between function and goal; when the human acts directly on the system at a system level (e.g., aircraft controls), the system acquires a goal.

Tools, equipments, and systems also vary on a continuum of complexity. At a certain level of complexity, equipments tend to assume some of the characteristics of a system and this adds to the confusion. Subsequent pages discuss equipments and systems; tools are ignored because they are auxiliary only. In previous chapters, the term *machine* was used, as in human–machine interaction. The term *machine* is a generic one (a construct) that includes tools, equipments, systems, or any hardware at all.

If readers find this confusing, they are not alone. To summarize, systems and subsystems have goals (which implies futurity) and missions (which are sequential actions taken to achieve goals). What makes the goal a goal is the awareness attached to a future state by the human; a function has no awareness. Goals and missions are always associated with systems and may be associated with equipment depending on the equipment complexity and whether the equipment is part of a system (in which case it receives some of the awareness implicit in the goal).

A system usually contains multiple equipments functioning concurrently and in interaction with each other; this requires special rules to describe how these interactions are performed. These rules organize or arrange the interactions. Organization is a system and subsystem dimension, but does not apply to individual equipments.

The Human–Technology Relationship

The human–machine or human–technology relationship is a complex one. (The relationship between technology and the human has been beautifully and profoundly explored by Hancock, 1997.) Usually only one or two personnel are directly involved in operating an equipment. When a system is composed of multiple equipments, the number of personnel can be large. One would also assume that the larger the system, the more personnel are involved and, in consequence, the number of interactions among personnel and between personnel and their machines would become equally complex. For example, if one considers an aircraft carrier as a single superordinate system composed of many subsystems (e.g., aircraft, surveillance, etc.), the number of personnel reaches 5,000. A commercial company may have hundreds or thousands of employees.

The relationship of any single individual to an equipment is highly specialized. The clerk who operates a word processor in an office has little or no contact with the other equipments that support the office, such as heating and air-conditioning equipment. The degree of specialization in relation to machines is one outstanding characteristic of the system. Outside of general maintenance personnel (e.g., electricians, plumbers), hardly anyone knows anything about an equipment except the one he or she is specialized to operate. This is much less true of less automated systems. For example, a farmer must expect to operate several machines and perform many different tasks.

One effect of this specialization is that for almost all people technology serves mainly as an environment in which they function. One operates one's own automobile and all other automobiles on the road serve only as a background, like the props on a theater stage. Of course, automobiles may interact, as in a collision, and then the individual is directly involved in someone else's technology. One would suppose that systems with which people do not interact directly, but which are in proximity to them, would have little effect on them. However, on the basis that a noninteractive system may serve as environment or part of a suprasystem (a larger, encompassing system for the one with which one is directly involved), one must suppose that it does have an effect, however slight. One premise of system theory is that no part of a system can be completely without an impact on another part. A system within a suprasystem is part of that suprasystem and must therefore be affected by it. However, the effect may be so slight that our measuring instruments are unable to detect that effect. The difficulty is that the effect may be solely attitudinal within the human (e.g., "I hate to go downtown because of all the people, cars, and noise"). Because the suprasystem may function only as background stimuli to the human in the system, its background effect may not be noticeable.

Only those systems with which one has direct contact would be expected to serve as a source of measurable stimuli to which one must respond. The systems that surround one have little effect on the individual until those systems perform in undesirable ways. This occurs only when they malfunction. If the airplane on which I have a seat for a 12:18 departure does not leave for 2 hours, I become immediately affected by the aberrant system. If a highway is closed for construction or if the mail does not come when we expect it, these formerly background systems, subsystems, or equipments immediately become foreground for us—they attract our attention (an unwelcome attention).

In contrast to the operator of a machine, the rest of us are clients of an equipment or system that performs actions intended to directly benefit us. One may be a client of an individual equipment (usually automated, like an automated bank teller machine [ATM]) or a total system (as, e.g., a recipient of a check from the Social Security Administration [SSA]). The degree of involvement as a client with a system may vary; it is direct when it involves an ATM or if one calls SSA to determine one's pension; it is minimal if one merely receives the check in the mail. One may be more directly a client of some system element if, for example, one asks instructions of a hotel clerk. However, as long as the clerk is part of the hotel (the overall system), one is also a client of the total system.

If one makes no use as either an operator or client of an equipment or system, the latter are merely stage props; if one merely observes the system, one is only a spectator. Note that it is easy to pass from one relationship to another.

However, it would be wrong to assume that the technological forest that surrounds one has no effect on the citizen of a technological civilization. At the very least, one must learn how to function in the midst of this forest (e.g., how to use the subway, how to dial a telephone or adjust a TV set, what traffic lights and signs mean, etc.). There is minimal interaction with each of these and there are countless systems below the surface of the ground we walk on and within the buildings we walk by that we are completely unaware of, but the concept of technology as a factor that may affect us, particularly when it malfunctions, represents a potential anxiety-inducing factor in the human.

There is then a learning process involved—learning that begins when we are small children but will continue all our lives. The introduction of a new type of system, such as the movies or the personal computer, can revolutionize our lives even when we have no desire to accept these new systems. It is possible, as many people do (particularly the elderly), to ignore computers or refuse to attend a movie theater or watch television, but the very act of refusing these has an effect on the lives of those who refuse. No doubt the aborigines of every nontechnological civilization resent the intrusion of technological systems into their environment. The effect of technology can be quite variable depending on whether one is an operator, a client, or merely a spectator. Obviously the effect is greatest when one is an operator, less when one is a client, and least when one is a spectator.

That technology affects everyone is evident, which is why the term *technological citizen* has been created.[1] Systems surround us from the moment

---

[1]A colleague, Richard Newman, suggests that the system approach is too concerned with formally designed systems like those employed by the military. However, there are systems like the Internet that apparently grow "like Topsy."

The point that Newman makes (the author's own interpretation, of course) is that the system concept presumes formally designed systems (i.e., developed by those who are directed to develop the systems), but there are systems that develop outside formal design processes; even worse, there are formally designed systems like the Internet, which develop additional structures to support emergent attributes.

The introduction of a computer-based technology that has a direct stimulus effect on the human through information presentation in a *game* format (which appeals to the childish part of the human), means that technology now affects more humans directly. A corollary effect is that the amplified human interaction with technology may mean that systems may be created indirectly because the human user of the system may act as a system designer, although without formal knowledge of how to design. One can contemplate the possibility that the user, given a flexible technology by designers (as an information-based technology could well be), will develop his or her own information systems leading to unknown consequences. We see this arising to some extent with Internet web sites.

There is no clear demarcation among the concepts *operator*, *client*, or *user*. In other contexts, all operators become users, even of the products they have created (e.g., the operator of a shoe-making machine may buy shoes from his or her own company).

Users may also become operators. In amusement arcades and in household privacy with the Internet, the user of an information technology product serves as the operator of the system. One can go even further than this. Information technology systems have such direct effects on the minds of its users that the user may dictate the way in which the system will perform.

Many years ago, the author read a science fiction story about a computer-based technology that enabled a client who had enough cash to be hardwired via cortical implants to the device so that the client became permanently and irrevocably introduced into the computer-developed reality. We have not reached that point yet (although this is the logical end point of virtual reality technology), but the direct effects of information technology as we see it (e.g., in the Internet) may lead to undesirable effects—information technology as a new narcotic that withdraws its users from reality? New lotus-eaters?

of birth; witness the incubator for premature babies. For that reason, the effect of familiarity dulls one to the technology. Disaster movies may in part be popular because they show us systems and technology out of control and move us into a different relationship with machines.

The development of human–machine systems of great complexity creates organizations of people devoted to keeping these systems running. The system has a human infrastructure that can be viewed as subordinate to the overall system because there would be no jobs for the humans if the system had not been created. The human infrastructure consists of system personnel, its operators, as well as its clients. The system is also dependent on its clients as much as on its operators. Suppose everyone decided to walk rather than ride automobiles or use the railroads instead of flying? The automobile and airline industries would be seriously affected even if they did not disappear. This happened to buggywhip manufacturing companies when buggies disappeared. There is then a client–operator–system symbiosis; each needs the other, although because humans create systems (meaning they were here before any systems were around), they can theoretically get along without them. (Hancock, 1997, suggested that we cannot.) The saving grace for the system and its designers is that people would rather have systems than not. Despite the presumed free will in humans to accept or reject technology, it imposes its requirements even on people who do not want that technology. During the early Industrial Revolution, people accepted factory jobs they detested because these were the only jobs available, although absenteeism was endemic (Stearns, 1993). If one lives in a city and despises motorized transport, it still becomes necessary to use buses, taxis, or subways to get to one's destination if that destination is not within walking distance. Of course, one can stay home, and some people—the poor and the elderly—do so. Concurrently, commercial technology and products utilize potential rewards to attract clients. For example, use X deodorant to make oneself more attractive to females or use cellular phones to call anyone anywhere at any time. The more technology there is around one, the less one can avoid it.

The impact of the system and the technology it represents is therefore direct and indirect. HFE is concerned primarily with direct effects, but who deals with the indirect ones? Sociologists, anthropologists? (There is an

---

In the past, technology, although it affected people, did not affect their minds directly. Now with information technology, it does. Is there a moral problem here? Technology in the context of the Industrial Revolution has always had undesirable effects on people (see Stearns, 1993), but these effects were always contextual. With information technology, the effects are now direct even if they are more subtle than child labor or industrial accidents. Should a science like HFE be concerned about moral problems? Maybe not. Should it be concerned to investigate the effects of technology on human behavior (as differentiated from human performance)? Probably yes, although there may be those who would think such investigations more appropriate for psychology or sociology.

anthropology speciality called *urban anthropology*, which is probably slightly concerned with technology as a factor in people's lives.)

System Development

What has been described is the impact of the system on the human. The obverse side of the coin is that people create systems in various forms to achieve some benefit from them. The term *benefit* is used broadly. For a military system, the benefit is in improving one's military capability while indirectly providing security for the country. Many commercial systems are built to make a profit for entrepreneurs from clients. Other systems are designed to improve worker productivity because this increases profits. Some systems like aircraft are designed to benefit people directly by flying them to their destinations. Thus, the benefit may accrue to an impersonal entity (the country), corporation, or individual. Benefits from systems may be provided to all of these at the same time. Inevitably the human impresses his or her own form on the machine. One might assume that the more the equipment and system are adapted to the human, the more beneficial the equipment and system will be to the human. In fact, this is the premise that underlies almost all HFE activity.

One way in which systems are created in the human image is by forcing the machine to take account of human limitations. One does not design equipment and systems that demand more of the operator and client than these people can provide. For example, one does not design aircraft that can fly at 40,000 feet without incorporating pressurization devices in the aircraft to maintain passenger safety. One does not use infrared displays as traffic lights because humans do not respond to infrared radiation without special devices. The determination of human thresholds and limitations is something readily discovered by psychological experimentation. The inclusion by engineering of devices to compensate for those limitations is also relatively easy.

The incorporation of human capabilities into system design is a much more difficult matter. Limitations can be removed by designing around them or by providing support devices (like providing motorized pallets to move weights exceeding 100 pounds). Capabilities are a different matter. A capability is a human characteristic, usually cognitive, that, when included in the system design in some fashion, will enhance the capability of the system to perform its mission. However, we are hampered by our inability to define what those cognitive capabilities are. It is not the ability to process vast quantities of data because equipment computers can do this much better than can humans. Is it the ability to intuit patterns from widely disparate facts or make leaps in judgment in emergency situations? As machines are increasingly automated, the capacities they demand of the operator become more cognitive because the human's role has changed to being a monitor and diagnostician. At the same time, the design of systems that incorporate cognitive capabilities has lagged behind. We have expert systems, of course, that mimic human cognitive processes, but efforts to develop artificial intelligence in machines have generally been frustrated. It may be that we do not really understand how humans think, which would of course make it difficult to transform their processes into hardware and software. Regardless of whether this is the case, designers have not been able so far to develop mechanisms that reproduce these processes. It may be that the study of how humans think in relation to machines is an area of HFE research that has not yet been given the prominence it deserves.

We have already talked about the system as imposing demands on the human—demands that must be adjusted to the human's capability to satisfy those demands. The demand–adjustment relationship is what gives HFE its rationale.

As an entity that incorporates the human, the system (e.g., a company or military unit) performs many functions that are identified with the human: monitoring and analysis of inputs to determine their meaning, processing of information, decision making, and communication among subsystem units. It is untrue that, if the system uses its personnel to perform these functions, the human is independent of the system—that it has no control over the humans. Once the human volunteers or is forced to become an element of the system, the human is required to follow the rules established by the system. It makes no difference that the system was created by humans and therefore the rules were made by humans. Once the system is in place, it develops an existence of its own; the originators of the system may be gone or dead, but the system lives on. The creation exists after its creators have disappeared.

For example, the Ford Motor Car Company exists after its founder died. The U.S. Army persists even after the first Continental Congress that created it is gone. This ability to transcend its origins is one of the attributes that gives the system a quality of uniqueness.

The fact that the system persists over long periods of time does not mean that it cannot change its characteristics; indeed, the ability to do so is one of the things that enables the system to persist over time. The humans who are part of the system management can change (upgrade) it to enhance the system's ability to produce its effects more efficiently. The larger the system, the more likely it is to persist and overcome arduous circumstances; it has resources to enable it to survive. That is one reason (among others) that systems endure. Many small "Mom and Pop" stores went bankrupt during the depression of 1929–1939; during that time, F.W. Woolworth was a feature of the urban environment, although it has since succumbed to more efficient competitors.

The use of large governmental and commercial entities as examples is deliberate; because of its simplicity, there is a tendency to think of a system in terms of the conventional one human plus one equipment combination. In their full flowering, systems are composed of multiple human–machine combinations; the larger the system, the more abstract the overall system becomes. Thus, if systems are large enough, they exist as electronic symbols on a stock market display. There are humans and machines in that abstract representation of the system; but at a certain level, they cease to exist as human and machine combinations and become abstractions themselves. For example, an army exists as individual soldiers and the guns they carry; at command level, they lose their individuality and exist only as part of divisions, corps, and army elements.

Whatever other functions a system performs, so that it can be called, for example, a judicial system, police system, library, fire fighting system, or hospital, as long as the system is dependent on machines operated by humans to perform important functions, it is also a human–machine system, whatever else it is. The use of the machine may be so minimal as not to warrant use of the term *human–machine system*; it is a question of the degree of dependency on the machine. For example, one could think of a priest in his church as a religious system (or at least a religious subsystem), but it would not be a human–machine system because, except for the telephone perhaps, the use of any machine is not really required to perform religious functions.

System Dimensions

The concept of the system has certain essential dimensions. Some of these have already been discussed or implied. The system obviously contains both personnel and machines (by definition of what a human–machine system is). It has a purpose and a goal because it is a logical contradiction to have a physical entity involving humans that functions without a purpose or goal. Because a system cannot be a system without some specialization of function, it contains subsystems, each with its own purpose and goal. These subsystems cannot function in isolation. Hence, an essential dimension of the system is interaction, in which every subsystem contacts every other subsystem with varying frequency and degree of influence. Because the system controls its subsystems (otherwise it could not produce its outputs), there must be a hierarchy, which implies levels, with the system functioning simultaneously on all its levels. Because all system elements must function in a coordinated manner, there is a degree of dependence (all elements functioning in varied degrees of dependence on each other). Each subsystem and subordinate element possesses a boundary, which is determined by the degree of specialization of its functions and its relationship to other elements. Each boundary may be more or less easy for personnel to cross. For example, a clerk has difficulty because of his or her position in contacting the president of the company personally, but the vice president has much less difficulty in doing so. Systems vary of course in degree of automation, and subsystems within any single system also vary in automation. In this respect, one need only compare the marketing division of a company with its production division.

These dimensions produce certain attributes of the system that influence how well the total system performs. The most important of these attributes is *complexity*, which the author defines as the number of informational states that the system manifests to its personnel and clients during system operations; *autonomy*, or the freedom system personnel have to select ways in which to make the system perform (clients do not have this freedom); and *transparency*, or the extent to which internal operations of the machine and system are made apparent by graphic displays, for example, to those who work with them (Meister, 1996b; see also Hendrick, 1984).

The primary function of the personnel who operate the system is to enable the system to function (perform its purpose) within specified parameters. Because it is always possible for the system to exceed parametric limits, and thus to fail, the human and the system, through its personnel, function in an environment of uncertainty and indeterminism. The former says, in effect, that anything may happen at any time; the latter says, in effect, that things are not always what they appear to be on the surface. Lipshitz and Strauss (1996) and McCloskey (1996) provided additional information about how system personnel deal with uncertainty.

Systems vary in terms of the amount of uncertainty and indeterminism they encounter in adjusting to others and which they manifest, and the human role in the system is affected by that uncertainty and indeterminism. Because the system functions in an uncertain and indeterminate environment, it (or rather the personnel part of the system) must always answer questions (solve problems). Major system-relevant questions are: (a) Is the system working within parameters? (b) If not, what is wrong? and (c) What can be done to rectify the problems? Subordinate questions that are behavioral in nature and are operator-relevant are: (a) What are the relevant stimuli (because there is a steady stream of stimuli into and from the system, and these are more or less relevant to any particular problem)? (b) What is the meaning of the relevant stimuli? (c) What hypotheses can be entertained about the problem (e.g., what went wrong)? (d) What is the range of possible operator decisions and which one should be selected as a basis for action? and (e) If personnel test the system, what information will they receive?

The concept of the system as always having to solve problems may appear somewhat exaggerated. This is because potential problems can be avoided by appropriate design and, even more so, because many problem situations are repetitive and responses to them can be learned and overlearned. It is true that the system is always in a state of problem solving, but the problems are often so routine (e.g., what does this invoice mean?) that for all practical purposes the problem is dormant.

One of the assumptions in chapter 2 was that there is a constant, low-level tension between machine and human because the machine is always exercising a demand. The same can be said about the system as a whole. In routine operations, the demand and corresponding tension is minimal. When a significant problem arises, demand increases and tension may become excessive, resulting in human stress that can lead to human discrepancies like errors.

In military systems, the demand and tension is often felt first at the individual human–machine workstation because the problem-detection function in the military is specialized. If the demand is great enough, it soon passes to the total system. A naval analogy is appropriate: When sonar detects an enemy vessel, the news is flashed first to command (the system manager) and is then transmitted to other subsystems (the command, "general quarters, battle stations" energizes the total system).

This is less true of civilian systems, which usually do not have a specialized problem-detection workstation. The problem, the demand, and the resultant tension may be felt in civilian systems in almost any subsystem, depending on the nature of the problem. If a critical machine malfunctions, the demand will occur at that machine; if the problem is that the company is losing market share, it will be felt first at management levels and will be transmitted and felt, at more or less deliberate speed, by other system levels.

The human serves as the model for the system because the human has created the system. Its designer endows it with humanlike features (e.g., communication, information processing). The designer also incorporates human threshold limitations when it becomes necessary for the human to perform certain behavioral functions during system operation. At the same time, when the human becomes part of the system as an operator, he or she voluntarily or involuntarily relinquishes certain freedoms because the system imposes certain rules (developed by the designer) to which personnel must agree while they are part of the system to make the system perform efficiently.

The outputs of system operation are both behavioral and physical. Actions taken by the operator affect the physical functioning of the system, and actions taken by the system as a whole affect the individual. This relationship is limited: The operator cannot change the essential properties of the system (although under conditions of autonomy he or she may select one of a number of modes of operation). The system cannot change the essential properties of the human, but it can train him or her to perform more effectively, and it may require him or her to perform in ways specified by operating rules.

The behavioral outputs of the system consist of: (a) performance effects that are overt (e.g., product assembly), and (b) attitudinal effects (about the system in particular and technology in general) that are usually covert. The attitudinal effects vary from highly enthusiastic to very negative. Attitudinal effects toward technology may also be shown by those who are not operators (e.g., housewives, retired); these may have to be interpreted in a sociological orientation.

The system produces physical outputs like canned foods or automobiles, although service industries produce services that are often largely behavioral. Each system produces what it was designed to produce (e.g., the judicial system produces convicts, the military system produces trained soldiers).

Factors Affecting the System

The designer/system developer imposes his or her model of how the machine should function on the machine. The entrepreneur imposes his or her model of the optimal organization on the system of which the machine is a part. In the case of the individual machine or subsystem, the model provides somewhat humanlike features. In the case of an industrial system, the organizational model imposed may be a learned one (e.g., as from the Harvard Business School) or may project the outstanding personality traits of the entrepreneur.

One can hypothesize that the ultimate goal of the system developer is (or should be) to create a symbiosis between human and machine, as a result of which the boundaries between the two are blurred. This would maximize efficiency because each could draw on the other's superior capability. This can be accomplished only as the system becomes increasingly complex because complexity offers many more possibilities for human–machine relationships than do more manual systems.

It has been suggested (Meister, 1991) that, to the extent that the operator in the human–machine combination can understand and internalize the designer's model of that combination, his or her performance and that of the system as a whole will be improved. It is also reasonable to presume that, to the extent the designer can understand the limitations of the human and can incorporate these into system design, the more effective will system performance become.

With increasing automation, the role of the operator changes from one of controlling the system to another of allowing computerization to control the system. Meanwhile the operator monitors the system so that it remains within parameter limits and diagnoses and restores the system to its prior condition when it is about to or actually fails. The new human role emphasizes perceptual analysis, information processing, and higher order cognitive functions. Because of this changing role, the emphasis in HFE research is gradually switching to emphasize cognitive variables.

The human is assumed to possess certain cognitive tendencies, such that his or her analysis of information is biased in certain ways. For example, human decision making does not agree with the formal premises of Bayesian mathematics (Lehto, 1997). The author calls these *cognitive stereotypes* and, for maximum effectiveness of system performance, they must be incorporated into system design. These cognitive stereotypes are akin to better known physiological thresholds in perceiving and motor responses. Although it is comparatively easy to include psychophysiological thresholds in design because the notion of threshold is easy for the designer to understand and can be relatively easily demonstrated, how one includes cognitive thresholds in design is unclear. Thresholds in lower level cognitive functions are easier to understand (e.g., the speed with which the human can compute, the number of information items he or she can retain at any one time), but the cognitive stereotypes referred to here are those of a more analytic nature and involve the collection, integration, and interpretation of multiple concurrent stimuli and their analysis within the framework of a problem.

The role of human motivation in affecting system performance is even more unclear. When the system is highly dependent on the operator capabilities for mission accomplishment, as in sonar or radar detection, factors influencing motivation (e.g., monotony) are comparatively easy to understand. In highly computerized cognitive systems, the influence of motivation is less clear. For example, it is possible that the stimuli presented by the system may be so complex that they tend to overload and thus reduce the operator's motivation to monitor; this then influences performance negatively. One aspect of human motivation is the attitude of the operator to the equipment and system he or she operates and, more generally, to technology as a whole. Such attitudes must be taken into consideration in designing large-scale technological entities like vast housing complexes. It is possible that the characteristics of such housing are partially responsible for the negative behavior of some inhabitants. It must be remembered that, in whatever action the human participates, he or she will analyze (consciously or unconsciously) the nature of that activity and develop an attitude toward it. The author assumes that such attitudes, because they presumably influence performance, are a legitimate topic for HFE study.

Because of its uniqueness, the system is presumed to have variables that, in some contexts, are distinctly different from either machine or human variables. The system attributes and characteristics described previously also contain variables that require investigation. It is possible to categorize system variables as *general*, *structural*, and *behavioral*.

Examples of general system variables are the nature of the requirements and constraints imposed on the system and the resources it needs to function. Table 3.1 lists these. Structural system variables relate to such dimensions as size of the system, the number of personnel it requires, its organization,

TABLE 3.1
General System Variables

(1) Requirements and constraints imposed on the system.
(2) Resources required by the system.
(3) Nature of its internal components and processes.
(4) Functions and missions performed by the system.
(5) Nature, number, and specificity of goals.
(6) Structural and organizational characteristics of the system (e.g., its size, number of subsystems and units, communication channels, hierarchical levels, and amount of feedback).
(7) Degree of automation.
(8) Nature of the environment in which the system functions.
(9) System attributes (e.g., complexity, sensitivity, flexibility, vulnerability, reliability, and determinacy).
(10) Number and type of interdependencies (human–machine interactions) within the system and type of interaction (degree of dependency).
(11) Nature of the system's terminal output(s) or mission effects.

and so on. Table 3.2 lists these. Behavioral variables relate to such things as personnel skill, experience, and motivation; these can be found in Table 3.3. So far, HFE research has emphasized behavioral variables (i.e., the human response to machine characteristics and performance as stimuli). Little or no research has been performed on general and structural system variables (although see Conant, 1996; Jamieson, 1996). This deficiency must be made good in the future or the development of such systems from a behavioral standpoint will be retarded.

The system as it has been discussed so far is obviously quite important to HFE. Consequently, one can think of HFE as a system as well as a behavioral science. As such, the nature of the system and the way in which humans and machines are organized within the system should have a compelling effect on what and how HFE research is performed. Chapter 6 indicates that, to date, HFE has almost completely ignored the system because its research is oriented toward the individual rather than the system in which the individual functions (although an exception to this statement must be made in the case of macroergonomics; see chap. 7). This pattern is thought to be derived from its historical predecessor—psychology. The kind of HFE research that needs to be performed is discussed later.

The system concept (that human–machine performance always occurs in some system context) provides a superordinate framework for explaining and understanding HFE phenomena; it increases the need to relate behavioral data to system variables. Thus, the system concept presents a requirement and an opportunity for HFE researchers. Whether they will satisfy the requirement and take advantage of the opportunity is presently quite unclear because so far it has been perfectly possible for HFE to ignore the system and persist in traditional (i.e., psychological) research practices. One can

TABLE 3.2
System Structural Variables

(1) Size (large, small, intermediate).
(2) Number of units and subsystems.
(3) Number of personnel.
(4) System organization (e.g., vertical, lateral, centralized, decentralized).
(5) Communication channels (number; internal/external to system).
(6) Attributes (complexity, transparency, autonomy, determinacy, dependency).
(7) Method of control (e.g., autocratic, democratic, oligarchic).
(8) Number of hierarchical levels.
(9) Goals (e.g., single, multiple, specific, general).
(10) Internal processes (repetitive, nonrepetitive; fixed/proceduralized, flexible/nonproceduralized, automated, semiautomated, manual).
(11) System operations (continuous, intermittent; stage in the mission when performed).
(12) Subsystem role (primary mission; system support).

TABLE 3.3
General Behavioral Variables

(1) Number of personnel involved in task performance; their arrangement (i.e., individuals or team).
(2) Functions/tasks performed by personnel.
(3) Personnel aptitude for tasks performed.
(4) Amount and appropriateness of training.
(5) Amount of personnel experience/skill.
(6) Number and type of interdependencies within the team.
(7) Motivational variables (i.e., reward and punishment).
(8) Requirements and constraints imposed on personnel.
(9) Physical environment for personnel performance.
(10) Factors leading to performance deterioration (e.g., fatigue or stress).

anticipate a great deal of reluctance to expand the limits of traditional behavioral research.

What about the effects of the system on HFE practice? The role of the practitioner is that of interpreting behavioral research results to the designer, looking at design efforts in the light of that research, and making suggestions for changing design as HFE research suggests. This will not change. What has been frustrating to the practitioner is that, although the practitioner's primary role is (or should be) that of research interpreter, he or she has little of substance to interpret because HFE research has not been geared to design needs. In fact, analysis of large numbers of HFE research reports, as in chapter 6, raises the question of how that research can be used for anything, much less for design guidance. The system concept will not change practitioner responsibilities because the design level at which he or she works is that of the equipment, not the system as a whole. However, if HFE research were modified to respond to design questions, the practitioner would have more to do especially if cognitive requirements and limitations form part of that research. This will make the job of practitioner interpretation more onerous, but even more important.

The system concept does have implications for methodology. The notion of complexity as the number of information states provides the possibility of developing a quantitative metric based on those states, and this might be used for prediction purposes. Research methodology will change if, because of the system concept, the nature of the questions asked in that research changes.

The Nature of the Human–Machine System

Viewed as an abstraction, as a concept, a system is simply the arrangement of elements organized to accomplish a purpose that the system developer had in mind. Miller (1965) suggested that there is a continuum of types of living systems ranging from the unicellular amoeba to large assemblies of people formed into governments. The human–machine system as a hybrid (part human, part mechanical) forms part of the continuum.

As pointed out previously, the aspect that makes the human–machine combination a system is that, as a combination, it is different than either of its elements. The interaction between the human and machine results in a transformation into a new entity. An analogy might be the fusion/fission of atoms. Gestalt psychology (Koffka, 1935), when talking about the system, referred to emergents that at some level (about which it was quite unclear) produced a new, unexpected entity.

Just as transformation is a process required to form the human–machine system, so is the notion of a *hierarchy of levels*. The human–machine system in the real world consists of specimens that range from a single equipment with a single operator to masses of equipments with a corresponding number of personnel.

Operating rules or procedures are required for the single operator–machine combination (henceforth referred to for brevity as the *combination*) as well as for masses of equipment and personnel. In the latter case, we call these rules an *organization*, and one cannot have a system of any complexity without an organization. An organization is not required for the combination because the relationship between the two is specified in the operating procedure. However, if one combined all the operating procedures of all machines, it would still not form an organization because the variables involved in an organization are different than those involved in operating the combination. Among these new variables are relationships between and among individuals and their communication channels and new functions such as decision making, which are rarely required at the combination level. Decision making at a subsystem or system level deals with more and different factors.

HFE concerned with the combination can be called *microergonomics*; HFE concerned with larger entities (subsystems and systems) involving many combinations is called *macroergonomics*. One other characteristic of the macrosystem is that it may have subsystems or subgroupings of humans and machines with common functions. Of course, this is not possible with the combination. However, when the combination is included along with other combinations in a larger system, it may form a subsystem of the total system.

Another peculiarity of the macrosystem is that it may have subsystems that have little involvement with machines. The interactions of humans in these subsystems are primarily those of humans with humans. The CEO of a corporation is an example of this. His or her office forms a subsystem of the total corporation, but the only equipment he or she is likely to use is a telephone.

HFE as a science is devoted to understanding and developing the system functions in relation to those two levels—micro and macro. Some system attributes like complexity can be found at both levels; others, like the need for communication, can be found only at the macrolevel. We differentiate two levels, but obviously there is a continuum of complexity that runs between micro and macro.

Another characteristic of the microlevel that is not found at the macrolevel is the need for designing the machine to accommodate human thresholds. Manifestly, visual, auditory, and strength limitations apply to the individual combination but do not apply at the higher level, although it is possible to speculate that the macrosystem does indeed have thresholds of a cognitive, information-processing nature. It is doubtful that there has been study of the question.

HFE research has presumably been guided by imperatives at each level. HFE presumably supplies information about the thresholds referred to previously. Although a good deal of HFE research does involve information processing and decision making, it is not clear how much of this can be applied to microsystem design. At the macrolevel, a good deal of research has been performed about system organization (see chap. 7), but how much of this has affected the development of actual systems is unclear because it is the author's hypothesis (admittedly without any real data) that managers do not make extensive use of organizational specialists.

Technological systems, being nonbiological constructions, are developed by engineers; if the system is large enough, it is developed by a group of engineers with the assistance of many specialists, among whom one finds the HFE professional. There is a system development process that has been described a number of times (Meister, 1987, 1991), although these descriptions are static and less interesting than the actual process. Neither the system developer nor the system development process has been researched from a behavioral standpoint, although it should be obvious that the engineer has idiosyncracies and stereotypical tendencies that affect the design of the individual machine. The personality of the engineer developer and the process of developing a system are both appropriate for HFE research, but they have only occasionally been studied (e.g., Meister & Farr, 1967).

From an anthropological or sociological standpoint, the multitude of systems with which everyone interacts on a daily basis warrants the designation of the present era as a technological civilization. That civilization has profoundly affected its citizens especially since the introduction of computer technology. The development of the Internet has had profound and not necessarily desirable effects on major segments of the population.

If the system as a concept is as important for HFE as the author believes it to be, then it is necessary to study the characteristics of the system as a system. This assumes that there are attributes and parameters of the system that cannot be revealed merely by studying one or the other instances of a system. For example, one cannot learn about systems in general by studying only automobiles, aircraft, or computers. It is possible to learn about the effect of aircraft characteristics on crew performance by studying various aircraft configurations, but this will tell us nothing about systems in general because the aircraft is only one of many types of systems. One could spend a lifetime studying the automobile as a representation of the human–machine system without learning anything more than the automobile because any one type of system has only its own characteristics, which are only partially representative of systems in general. To use an analogy, if an alien from outer space wished to determine the characteristics of the human species and decided to do this by studying 14-year-old males of the inner city, the alien would develop a skewed picture of humans in general. The preceding assumes, of course, that there are certain general attributes that describe all human–machine systems. This assumption is that starting point for all classification, and classification is the starting point of science.

The point is that the general concept of the system, like the general concept of the human, must be studied by first abstracting the qualities that make a system or a human and then testing these. In behavioral research, one cannot synthesize humans to study the desired qualities (one has to sample representatives of all humans to find these). However, because systems are artificial constructions, it is possible to develop such artificial constructions (synthetic systems) specifically to emphasize whatever system qualities one is interested in.

There are obvious limitations to the physical development of synthetic systems. One cannot readily develop a macrosystem like a corporation, for example, but one can develop analogues or simulations of the corporation in the form of, in an earlier period, the board game or presently a computer simulation. Computer technology, possibly combined with performance modeling, enables the researcher to simulate a great variety of systems. With sufficient ingenuity, it might even be possible to simulate by computer various types of humans to perform as part of these simulated systems.

There is a difference between studying the system as a general category and studying the human reaction to an individual system. If one creates a synthetic (artificial) system that varies in terms of an attribute (e.g., like complexity) and uses human subjects to determine how they deal with complexity as a system attribute, this is not the same thing as studying how well pilots utilize inside or outside attitude displays in making landings (Roscoe, 1968). The former study tells us something about the effect of complexity as a system attribute (a most important topic); the latter only tells us about two different types of displays, although they are important to aviation psychology. Hence, if aviation or computer specialists tell us that they are studying systems, they are not talking about the system as the author conceptualizes the system.

The point is that the kind of HFE research that takes certain characteristics of displays (Kelso, 1965) or tools (Konz & Streets, 1984) and determines the human responses to these characteristics is *not* human–system research, however interesting the question of how humans respond to displays or tools may be. From this standpoint, there has been almost no HFE research on systems as systems (with exception of work by Rouse and his associates; see Henneman & Rouse, 1983, 1987), but only research on specific instances of systems.

Superficially, it may seem as if the difference between the two types of research is minuscule, but the actual difference is vastly significant. If there are general principles that should guide the behavioral design of all systems (and there are those who dispute this assumption), these general principles will not be learned by studying only specific instances of systems like aircraft or computers. If one studies only aircraft or computers, one is not studying systems as a general category, but merely particular representations of the systems. There is nothing wrong with this as a research procedure as long as one realizes what one is doing. It is reasonable to study aborigines to discover how aborigines behave as long as one does not try to generalize the study results to 20th-century humans. It is equally correct to study human performance as a function of piloting and aircraft as long as one does not delude oneself that the research results describe human behavior in systems generally.

The almost always unconscious assumption that studying particular instances of specific behaviors enables one to make judgments about behaviors in general is a common delusion that functions to support the researcher's feelings of self-worth.

The performance of genuine system-oriented research will (if it is actually implemented in the future) require a long, expensive effort because it will involve many individual minicomputers, the writing of lengthy software programs to integrate their operations, and the collection of many subjects who must be trained to operate the synthetic system. Imagine writing code for a system to represent something as large as, for example, American Airlines, with modules for each of the subsystems (e.g., engineering, operations, maintenance, administration, marketing, etc.) that make up the total system. One can imagine that to house such a synthetic experimental system would require its own building with a support corps of software programmers, hardware specialists, training personnel, and so on. A single experiment with such a system might require weeks of software writing and training of subjects, as well as preliminary testing of any new system configuration before a full fledged test could be performed. The synthetic system would have to be capable of being changed in terms of all the system variables listed in Tables 3.1 and 3.3 (e.g., size, functions to be performed, arrangement of subsystem relationships, controls and displays, stimuli and symptomology, etc.). All of this would require extensive financial resources, not only for initial development of the system, but for its continuing upkeep. In the 1970s, the U.S. Air Force developed a highly sophisticated state-of-the-art aircraft simulation, known popularly as ASUPT, capable of being modified to represent any present or future aircraft configuration. It almost bankrupted the Air Force's behavioral and training allocation for 1 or more years (or so we heard); its demands on electrical power "browned out" the lighting system of a nearby town. An adequate synthetic system research facility would demand almost as many resources.

This is probably why we shall not see such a facility built until much later in the 21st century, if then. True behavioral system research is beyond the capability of almost all present laboratories. This may in part be responsible for the predominance of small, psychological, and human-centered experiments, which to a system scientist (there may be a few of them in HFE) may seem relatively trivial. Undoubtedly; however, the capability of performing actual system studies was shown in the 1960s and 1970s (see Parsons, 1972), but not on the scale and in terms of the system-descriptive purposes the author envisages.

At present, small system simulations of an experimental nature (i.e., leaving aside simulation for training purposes) are possible (e.g., see the studies by Rouse and his colleagues; Henneman & Rouse, 1987), but these are only miniature representations of true systems. What are really needed are system simulators that can be modified in configuration to incorporate major system variables.

Given that such a facility is unlikely to be developed in the near future, it might be as well to spend some resources in studying real-world systems (a difficult job, but not impossible). In the 1950s and 1960s, British scientists at universities like Aston, Loughborough, and Birmingham did study actual systems; this is what created the so-called *sociotechnical approach* (see Emery & Trist, 1960). However, one wonders if the American fascination with the experiment and the absence of interest in the behavioral aspects of systems will permit such an effort.

THE HISTORY OF THE SYSTEM CONCEPT

The system concept is not peculiar to HFE. Indeed, it was developed elsewhere and taken over by HFE. The system concept is not even behavioral in origin, having been applied to operations research and engineering before HFE professionals picked it up.

The system concept is first a philosophical point of view—a belief and a method arising out of that belief. It is possible to trace it back to concepts expressed in the philosophies of Lao-Tse, Heraclitus, Leibniz, Vico, and Marx. Van Gigch (1974) emphasized Hegel, to whom the following ideas are attributed: (a) the whole is more than the sum of its parts, (b) the whole determines the nature of its parts, (c) the parts cannot be understood if considered in isolation from the whole, and (d) the parts are dynamically interrelated or interdependent. Those with a background in psychology will recognize these ideas as fundamental tenets of the Gestalt framework (Koffka, 1935). Checkland (1981a, 1981b) dated the system movement we are familiar with from the late 1940s. In the case of human factors, the impact was felt about the mid-1950s if one works back from the article by Christensen (1962).

The apparent antithesis of the system concept is reductionism, which functions in consonance with the system concept (e.g., task analysis is reductionistic). Manifestly it is possible to break entities into their component parts, but system theory would say that the action of the parts (no matter how variables can be made to interact in an experiment) cannot adequately explain how those variables function when they are part of the whole. However, reductionism is not an antithesis to the system approach, but actually can be considered part of that approach. Depending on one's goal in analyzing system processes, it will at one time be more useful to consider them as an organized, holistic entity (i.e., as a system) and at another time as a collection of elements and processes that need to be decomposed into smaller units.

TOOLS, EQUIPMENTS, AND SYSTEMS

Technology is not only represented by the system. Many physical entities and structures (e.g., office buildings and factories) are not technology because they are not dynamic, although they are associated with technology and certainly contain dynamic systems. (All technology must be housed to protect that technology and its operators from the environment and criminals.) Moreover, tools are not systems or equipments and must be differentiated from them. There are major differences among tools, equipments, and systems, although these differences may shade into each other.

It has been pointed out that the overall system imposes rules to which the operator must submit as long as he or she is part of the system. These rules control how one operator–machine combination relates to another (e.g., communication of inputs); the overall system also serves as a context or environment in which the combination works. This technological environment is both procedural (organizational) and physical. For example, merely to walk onto a functioning factory floor is to be exposed to a tremendously noisy environment and one of large, threatening machines.

The equipment also imposes rules as to how the individual operator can control the individual machine, but these rules only apply to the one machine and not to others. The system and subsystem control the human through their organizational rules; but the machine, through its operating procedure, divides control with the operator.

The tool—an object like a saw or screwdriver—is totally under the control of the one who uses it. The nature of the tool (e.g., whether it is a hammer or a saw) determines the way in which it can be used, but the user decides when and in what context it is used. Even the language of control differs between tools and equipments. One *uses* a tool and one *operates* an equipment or machines. One does not say he or she *uses the equipment*, nor does one *operate* a system; the system *functions*.

Because the tool (e.g., a saw) functions as an appendage to the human, the latter may misuse the tool with negative consequences for tool and user. HFE professionals do help design tools especially when they are sufficiently complex that they are almost equipments.

Examples are useful here. An automobile is an equipment that is largely under the control of the operator. A highway system, which has many automobiles, roads, traffic signs, policemen, guard rails, and so on, exercises control over the individual motorist. If the motorist disobeys the rules, voluntarily or inadvertently, there is apt to be a crash—the system temporarily breaks down until the debris is removed. The automobile has in its trunk a tire repair kit. Because of its nature, it must be used in specific ways, but when and under what conditions the kit must be used depends on the user or the user's circumstances. Thus, the system controls the operator, the operator shares control with the machine, and the user is solely in control of the tool.

The tool–equipment–system relationship is on a continuum. An equipment, for example, may be so complex that it becomes a subsystem within a larger system. An airliner is such a subsystem, but it is only a unit within an airline system consisting of other equipment, air traffic control stations, terminals and runways, maintenance facilities, and so on. The baggage-handling cart is still only a tool, but there are also baggage-handling equipments, like the one at Denver airport, that may be so automated and complex that they achieve subsystem status. Therefore, the identification of a piece of technology requires individual scrutiny and there are times when one type of device shades into another. However, the general principle differentiating these categories is the locus of control, although degree of complexity also enters into the definition.

The reader may ask where HFE enters into this. It is always useful to know the names of the players in the game. Beyond that, although the HFE researcher and practitioner may be involved with each type, the majority of HFE activities are performed in relation to equipments. There is a branch of HFE called *macroergonomics* that is described in chapter 7; it deals primarily with major systems. What the HFE professional does with equipments and systems depends to some extent on whether it is one category or another and this also involves the complexity of the unit to which attention is paid.

The reader may ask how the differences among tools, equipments, and systems apply to HFE. The answer is that, with increasing number and complexity of the factors affecting each type of machine, HFE principles that apply to the design of each type probably differ. We say *probably* because there is little research evidence for this point. Comparatively little research is directed at tools—the design of hammer handles (Konz & Streets, 1984) and pen points (Kao, 1973) come to mind—but as is seen in chapter 6, there is practically no HFE research on systems as such unless one considers the macroergonomics literature to fit this bill.

THE TECHNOLOGICAL ENVIRONMENT

Just as there are different machines, so there are different human roles in the technological stage play. It has already been pointed out that someone who uses a tool is called a *user*, someone who directly operates an equipment or who functions in other aspects of the system (e.g., secretary, messenger) is called an operator, and someone who controls the system or major sections of the system is called a manager.

Because technology enters to a greater or lesser degree into every life activity, the individual also participates in technology as a client far more so than as an operator. When the individual is at home, visiting, or performing an activity like eating in a restaurant or watching a play, the individual is called a client because he or she benefits from technology while not operating that technology. The client may be in contact with technology (e.g., withdrawing money from an automated teller machine [ATM] or riding in an office elevator), but he or she is not operating a machine; he or she is receiving the benefits of technology, hence, the client status. An excellent example of the client is being a passenger in an aircraft; one is enjoying the benefits of being transported, but is not the one doing the transporting; that is the role of the aircrew during flight. Sometimes the lines are blurred a bit—someone who uses an ATM machine has to perform a few simple operations to make the machine function, but otherwise is the client of the bank, which is a system, although perhaps not a highly automated one.

The human in a technological society is surrounded by and impacted on by many systems and machines. Even in bed, the human is functioning within a habitation system that contains machines like the refrigerator, stove, telephone, heater, water closet, and so on.

The ubiquity of technology raises the question of whether HFE has arbitrarily limited itself to only a small part of the study of technology in relation to the human. The study of technological effects on the larger society could be considered a matter for HFE consideration—or anthropology or sociology. Where does the HFE researcher's concern cease? It certainly need not be limited to human responses to technological stimuli (which is what it is presently) or to design of a single equipment; it can be anything involved in the human–technology equation. For example, Hancock (1997) assumed that HFE ". . . is an endeavor central to the success of the human enterprise itself" (p. 69). This is not a matter to worry the HFE system development specialist because, in system development, one rarely does HFE research. The author is not suggesting HFE research into the larger implications of technology for humans; he raises this point only for consideration.

All this technology has an effect on the human; that effect is particularly marked when the human is an operator, but there is an effect even when he or she is only a client. There are systems with which the client interacts voluntarily—as in a recreation system like games (baseball, basketball)—or involuntarily—as in governmental systems like justice, fire, social security, and so on. Those systems in which the human is only a client also contain equipments that are controlled by operators. Hence, the interaction of the human with technology is complex, although for researcher and practitioner purposes HFE generally concentrates its attention on the immediate operator–equipment relationship and only occasionally (this is where macroergonomics enters in) on the operator–system relationship.

Whether operator, client, or spectator, the individual is always surrounded by human–machine systems. One may not be aware of this because much of the functioning of these systems is invisible to the individual. The infrastructure of modern life depends on human–machine units that function regardless of whether one is aware of them. For example, the quality of one's drinking water depends on a water processing plant; how many clients of this plant have visited it or are even aware of it?

The author has used the term *technological environment* to represent this immersion in systems. There are actually two environments. One is the larger system environment in which the individual operator–machine combination is embedded. The other environment is the greater one (the suprasystem) in which the individual is part of a society in which technology plays so large a role.

The technological environment exerts its effects not only by the physical presence of machines and their habitats, but even more by the organizational rules it imposes on operators, clients, and managers. These rules affect primarily operators and managers, but may also affect clients if machines are involved in the bestowal of benefits. For example, airline terminals are laid out to assist in maintaining aircraft; in moving through an airport, the passenger is constrained by that layout.

The effect of the machine on an individual is obviously greatest when the individual is the operator of that machine. However, technology still has a significant effect when one is the client of a machine system. One can be affected by the technological thicket even if one is not an operator, client, or manager. Technology has an effect even if one is merely a spectator. This is because technology has raw power, and power that great is intimidating and frightening to all, although individual reactions to that power will differ.

In addition, there can be no doubt that the introduction of the computer has had widespread effects, not least by creating a younger generation that thinks in terms of computer-generated graphic images. For those who engage in a great deal of television and computer display viewing, the nature of reality as the individual perceives it may change. The medium through which one perceives so-called *objective* reality may affect the viewer's physiology so that what is seen is insensibly affected by the individual's changed viewing mechanisms—in this case, the brain.

Our understanding of these mechanisms and cognitive stereotypes (i.e., preferences for ways of thinking) may eventually influence equipment design to permit the machine to accommodate these stereotypes, but not yet and not until HFE researchers have learned much more about them.

It stands to reason that an operator–equipment combination embedded in a much larger system—other equipments, operators, and managers—is affected by what goes on in other parts of the system. A comprehensive explanation of operator–system behavior would take all the effects of the overall system on the individual operator into account, but this might make the situation too complex for HFE. Some of these effects might more properly be left to the disciplines of sociology or urban anthropology.

Thus far, the total technology picture is more complex than the single operator–machine combination, but as a practical matter or one of interest HFE professionals narrow their attention span to a subunit of the whole. This may also help explain why the total HFE discipline tends to fragment into specialties. Quite apart from the interest of the individual professional, the total picture is so large that to deal with it effectively one must break it down into more manageable units. If one attempted to plot the interactions of the various system elements with each other, the resultant grid would become so complex that only a computer program could disentangle the individual units. The author once attended a meeting dealing with Naval task forces, at which a flow chart, displaying the links between individual equipments on the force's ships, was presented. The resultant grid was so complex that any individual link could be discerned only with the aid of a magnifying glass.

The single operator–equipment combination is always part of a larger system, and that system in which it is embedded provides an environment for it. The overall system establishes the rules within which the combination functions. Because of this, the overall system exercises an effect on the combination. The effect is mutual, however. The combination can affect how well the overall system performs, although only if the individual equipment is critical to performance of the whole.

It is necessary to introduce the notion of *dependency*. A high degree of dependency produces a greater effect when the unit that is dependent is affected by some phenomenon. Workers in Factory 1 produce parts for Factory 2, which assembles them into the complete automobile. When workers in Factory 1 go out on strike, workers in Factory 2 are often laid off for lack of materials on which to work. Obviously Factory 2 is highly dependent on Factory 1, but in other circumstances the reverse dependency might occur: If owners decide to cancel production of autos at Plant 2, Plant 1 will also shut down unless it can find other automobile assembly plants to which to ship its products.

Dependency is on a continuum varying from one extreme of no dependency to the other extreme of total dependency. Dependency exists at all system levels and is obviously determined by how the system is organized. The point is that one cannot begin to understand the workings of a large human–machine system unless one takes into account the various dependencies that interweave the system elements.

From the picture that has been presented, it is obvious that HFE can claim to study a great variety of activities found in technological civilization. If HFE during WWII dealt largely with *knobs and dials*, the palette which is available to the descendants of the founding fathers is much richer, much more complex, and allows much more to be done. This expansion of viewpoint can be seen more in research than practice, although even in system development there are many more things that the HFE professional can do. The conceptualizations and speculations of researchers and theorists have expanded as they have expanded their grasp of the technology palette. Many of our present concepts (including some in this book) would have been incomprehensible to HFE people working in the 1940s and 1950s. This is as good a place as any to introduce the reader to other characteristics of the system.

SYSTEM COMPONENTS

Van Gigch (1974) summarized the components one finds in systems: (a) *elements* (in the human–machine entity, these are represented by personnel, equipment, procedures, technical data, etc.); (b) *conversion processes*, changes in system state, elements combining with each other to form new elements (in the human–machine entity, these are represented by personnel outputs that combine with physical outputs to produce new system states); (c) *inputs and/or resources* (e.g., visual stimuli, personnel skills, technical data); (d) *outputs* (in a system like a factory, number of units assembled per shift); (e) the *environment* (the human–machine entity has both an internal and external, a physical and psychological environment); (f) *purpose* and *functions* (every artificial system begins its development with these); (g) *attributes* (in the human–machine entity, these may include complexity, determinacy, sensitivity, reliability, etc.); (h) *goals* and *objectives* (derived from the purpose of the entity); (i) *components, programs*, and *missions* (this is what the human–machine entity is programmed to do); (j) *managements, agents*, and *decision makers* (in artificial systems, personnel assume these roles); and (k) *structure*, or relationships binding system elements together (in the human–machine complex, this is its organization). These characteristics, which all artificial systems possess, must be considered in designing and evaluating every system.

The system concept has certain implications for behavioral analysis:

1. The system concept requires one to consider all factors that could possibly influence design and performance. Therefore, the behavioral design of systems must consider all the system elements, although those that are largely irrelevant can be ignored.
2. It is necessary to relate behavioral analysis, design, and evaluation to tasks, system goals, and outputs. This makes it mandatory not only to note system outputs but also to measure the relationship between the personnel subsystem and those outputs.
3. The notion of hierarchical levels of performance (workstation, subsystem, and system) means that it now becomes necessary to analyze and measure at all these levels and relate the processes at one level to processes at other levels.
4. Systems cannot be understood outside the larger whole (suprasystem) in which they are embedded. This forces us to consider the effect of higher system levels, which form the environment in which the system functions.
5. Because the system functions through the manner in which its elements are organized, system organization is a necessary component of design and must be treated during design as all other system elements are treated.

System Interactions

Influence. Every system element influences all other system elements. Dynamic elements in the system may influence other dynamic elements and static elements may influence dynamic elements; but noncommutatively, dynamic elements cannot influence static ones. Thus, characteristics of ordinary software programs may influence the operator's performance, but the operator's performance does not influence the software because the latter is fixed (except when an error is found in the software or a modification must be made). The amount of influence varies from slight (perhaps unmeasurable with our crude instruments) to overt (perhaps obvious).

Demand. Energizing the system automatically creates a demand on each dynamic system element, to which that element applies available resources to satisfy the demand. Demand has both positive and negative effects. If the demand is not excessive, it mobilizes the parts of the system, each in its appropriate way. If demand becomes excessive, however, it may hurt the system by burning out equipment and confusing or demotivating personnel.

Resources. The resources referred to are the mechanisms developed and built into the system to satisfy increasing demand. These mechanisms may be designed into the software or provided by selection and training of system personnel. System resources may be sufficient for the demand presented or they may not; if the latter, the human–machine unit may fail to complete its task satisfactorily.

Goal/Mission. The system is directed by its goal/mission as interpreted by system personnel. All system elements (including the human) are subordinate to that goal/mission. The system functions that are performed support the goal/mission. If they do not, they are inappropriate or the system has malfunctioned.

Information and Control. The operations leading to accomplishment of the goal or mission are or should be controlled by information-control mechanisms built into the system. These continuously compare actual system performance with required performance and, when properly designed, call attention to discrepancies through feedback.

The previous discussions can be made more specific as they relate to system personnel:

1. System personnel form a specific subsystem of the system (the personnel subsystem [PSS]). Humans who merely make use of a system (e.g., Social

Security recipients) or who enjoy the system's outputs are not a subsystem of that system.

2. Personnel performance interacts with machine functioning and, in the process, contributes to and melds with machine outputs.

3. Personnel can be found at every level of the system hierarchy. These levels are, in order of decreasing size, system, subsystems of the units, and workstations.

4. System personnel obey the rules of the system because, if they do not, they are eliminated from the system. Anyone else who makes use of the system does so also because there is no other way of using the system. This does not mean that they consciously adopt the purpose or goal of the system; using the system for one's own purposes does not mean that one necessarily strives to achieve system goals. For example, if one uses a bank because one has to, this does not mean that one consciously wishes to maximize bank profits.

5. Service systems consciously seek clients, whereas nonservice systems (e.g., military systems) have no clients and do not wish any. (Of course, military systems may have subsystems—e.g., daycare centers for military infants—that do have clients, but the military system considered at the overall system level has no clients.) Operating rules for service systems must take into account client preferences; nonservice systems do not. Systems that sell production outputs must take consumer preferences into account; for these, persuasion in the form of advertising becomes quite important.

6. Because the system is organized hierarchically, relationships among the various levels are implemented by information transmission among personnel and between personnel and equipment.

7. Control over the system is exercised by personnel (managers) acting to implement the system purpose. That control is aided by feedback information transmitted from all system levels and from equipment, software, and/or personnel.

8. Failure to achieve the system goal and accomplish the mission, when this is recognized by system personnel, elicits compensatory performance on their part, including diagnosis of what went wrong and attempts to rectify the problem. Such compensatory performance seeks to restore system functions to the designers' goal.

9. Failure of system personnel to perform in accordance with system goals and missions results in ineffective outputs and weakens system structure.

System Functioning

Figure 3.1 presents a model of system functioning in graphic form. The critical events in this figure are: (a) sensing the stimulus input; (b) analysis of the input to develop a reasonable interpretation of it; (c) generation and

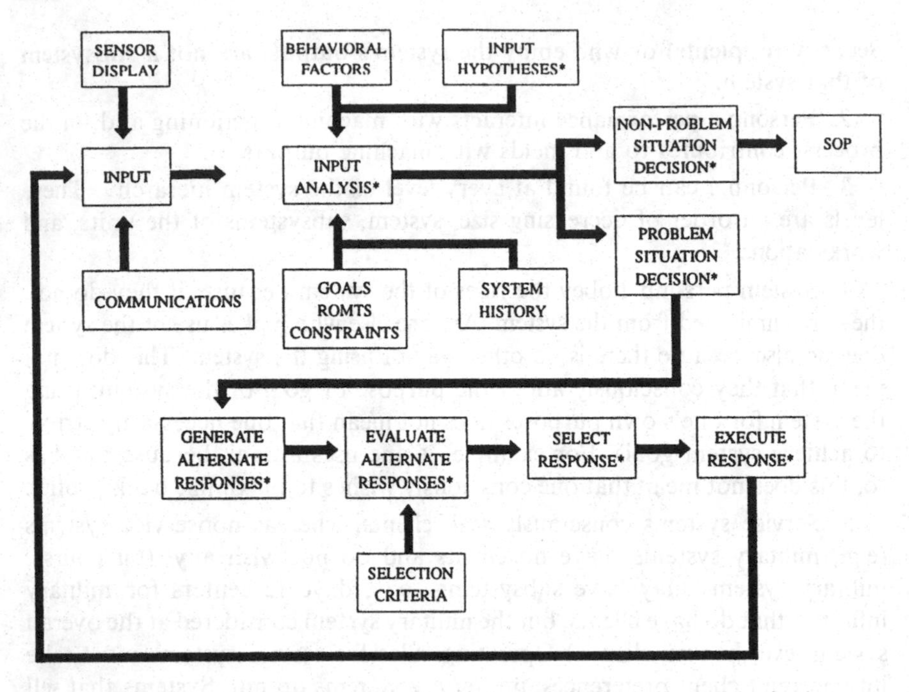

FIG. 3.1. A model of system functioning. (Asterisks indicate behavioral processes.)

evaluation of problem–solution response options; (d) selection of a single response that represents the problem–solution decision; and (e) execution of the decision and its feedback, which initiates a repetition of the process.

These are also and preeminently human functions. Although the human is only one component of the total system, he or she is as important as all other hardware/software functions, and system developers organize system functions using a human model. They have no alternative because they cannot think in other than human terms. Each input reaching the system is evaluated by the system (primarily through its personnel) to determine whether a situation requiring action by the system has arisen or is likely to arise. These situations may be routine or may present a problem requiring innovative system operations.

The system is constantly monitoring its internal and external status relative to goal accomplishment. External status describes the relationship between one's own system and other systems and/or the physical environment; internal status is monitored to determine whether system functions are being performed within specified limits. The monitoring requirement makes it necessary to examine inputs to determine the information each input contains. The interpretation of the input may be simple or difficult depending on the nature of the input and the interpretations that can be made of the input information. In the course of the interpretation, the interpreter constructs various hypotheses about what the input means and settles on one.

If the input hypothesis selected is that a problem exists that the system must resolve immediately or in the near future, it becomes necessary for the system to generate alternative solutions (potential responses) and evaluate these on the basis of standard operating procedures or more covert criteria such as risk involved in performing the response and the anticipated effectiveness of each potential solution. A response is then selected and implemented. Feedback from the response executed generates follow-on input stimuli to be evaluated in the same way as previous inputs were interpreted.

The model described is an ideal model. In actuality, system functioning often breaks down. Problems may not be recognized as such. Decisions may be postponed or made incorrectly. Problems may disappear as a function of time and changing external events, or the system may be destroyed. The whole process may be undermined by uncontrollable events external to the system. Even when the system functions as the model says it should, the process is usually iterative and may be clumsy. Another thing should be noted about this model. The system it describes functions in an uncertain, probabilistic context. Every input to the system is potentially a problem to be solved because not everything that should be known about it is known. Uncertainty is a key factor.

SYSTEM ATTRIBUTES

Concepts of Reality

Previous sections introduced a new concept—uncertainty—that requires information gathering, interpretation, and decision making by the human. It is assumed that the individual with (and even without) technology lives in a problematic world, in which the correct response is by no means immediately evident (there may indeed be no correct response); that response must be sought for by strenuous information collection and analysis efforts.

The relevance of technology to this framework is the assumption that, quite often and particularly when systems break down (and invariably they do so), the increasing technological sophistication of these systems increases the human's uncertainty. This occurs because he or she is presented with increasing amounts of ambiguous information that cannot be interpreted readily and certainly not without the gathering of additional information. When this uncertainty is allied to high-risk systems (e.g., chemical processing factories, nuclear power plants), the potential for catastrophe becomes unacceptable.

Let us examine what this uncertainty consists of. First of all, it stems from an important continuous attribute of all systems: determinacy/indeterminacy. Systems may vary from those that are highly determinate to those that are highly indeterminate. It is the latter that gives us all the trouble; hence, it is indeterminacy that is emphasized in subsequent pages. The system has other attributes also (e.g., complexity, flexibility, sensitivity, reliability, and vulnerability; see Meister, 1991, for a discussion of these), but this section focuses on indeterminacy, from which uncertainty stems.

Indeterminacy. The elements that describe indeterminacy are: (a) the ambiguity of stimuli presented to the personnel (inputs may be incorrect, irrelevant, and difficult to interpret); (b) the need for interpretation of those inputs—a need that obviously stems from the nature of the stimuli; (c) relatively greater emphasis on information and information processing, as compared with other functions (e.g., perceptual and psychomotor); (d) the amount of decision making required of the system to deal with uncertainty (decision making stems from the need to choose among response alternatives); (e) the degree of procedural variability built into the system; and (f) the number of response options available to the system and the freedom to choose among them.

Degrees of indeterminacy are possible because these elements are not completely independent. For example, one can have ambiguous inputs and relatively precise system responses. Sonar stimuli, for example, are often highly ambiguous, but once a decision is made that the stimuli indicate a submarine system, responses are highly proceduralized and precise. Moreover, stimulus ambiguity is on a continuum. The amount of decision making required will vary depending on the nature of the inputs and the system's experience in dealing with these inputs.

At the other extreme of the determinacy continuum, a very determinate system is one in which system inputs are unambiguous (i.e., almost instantaneously interpretable in only one way) and a single response for each stimulus condition is prescribed. The simplest example (although real systems are not really this simple) is to have two signal lights: one red and one green. When the red light appears, the operator must throw one switch; when the green light appears, he or she must throw a second switch—no choice in interpretation of stimuli, no flexibility of response. The procedure that determines the response is invariant. The operator's uncertainty is low. A highly indeterminate system is the reverse of all this. Stimuli produced by the system are susceptible of at least several interpretations, and personnel cannot be certain of the truth of the interpretation on which they finally decide. Responses to be made are flexible depending on the stimulus interpretation and the risk consequences of the response; thus, for any one input condition, more than one response is possible. The operator's uncertainty is high.

Highly determinate and highly indeterminate systems are easier to identify than semideterminate (or semiindeterminate) systems, which have the following characteristics: (a) stimulus characteristics may vary, but the range of variation is small; (b) some interpretation of stimuli may be required, but it is tightly controlled; (c) the system processes some information; (d) if decision making is required, the number of options is relatively few; and (e) operating procedures are invariant but they can be applied at different times, in different times, in different sequences during the operating period, at the operator's discretion, and in response to different operating conditions (which the operator must recognize). The operator functions in slight to moderate uncertainty.

The following factors may produce indeterminacy. However, not every condition is equally important, nor must all these function concurrently for indeterminacy to result.

Situational Uncertainty. Internally, system parameters are off limits: The system has malfunctioned or a malfunction is impending; or it appears evident that system goals are unlikely to be achieved; potentially there is failure to accomplish the system mission; or resources are being exhausted with little hope of finding more. Externally, an actual or potential adversary relationship exists with another system or the environment is becoming hostile. Alternatively, an opportunity exists to aggrandize (e.g., expand, increase) the system, but the opportunity also has considerable risk associated with it.

Input Interpretation. The meaning of the input is unclear, meaning that it can only be interpreted probabilistically. Conflicting interpretations of the input are possible. The information in the input is incomplete and/or inconsistent with the previous inputs, or other instances of the input are not recorded in the system's data banks, meaning that the input is a loner.

Information-Processing Characteristics. The type of information processing required in indeterminate situations is deliberate examination of cue characteristics to interpret stimuli. Multiple cues may have to be correlated; hypotheses concerning the meaning of the input must be developed and analyzed.

Decision-Making Characteristics. Standard operating procedures (SOP) fail to solve the problem presented. This means that alternative potential solution responses must be generated. A plan of action involving more than one discrete response is required for problem solution.

System Flexibility. Rules for operating the system permit alternative responses to varying problem situations. Rules for response selection are general rather than detailed. The selection of a solution response is the decision maker's choice and is not predetermined by the system.

Response Uncertainty. A choice must be made among alternative responses. No potential response is known unequivocally to be the correct one. Each potential response has some probability of risk and less than certain probability of effectiveness. The decision maker's confidence in the selected response is not necessarily absolute. Any one of these conditions may produce uncertainty; when several elements occur in the same situation, the degree of uncertainty of course increases proportionately.

Indeterminacy is mediated through both equipment and human processes as well as through their interaction. Equipment-determined indeterminacy is produced by failure of equipment to perform at all or to perform as desired (unreliability). Human-determined indeterminacy is produced by failure to understand one's system and the environment (what one is supposed to do). Indeterminacy caused by humans is far more frequent than that caused by equipment failures because equipment failure is comparatively rare, although when it occurs it is often catastrophic. Human-determined indeterminacy is a constant background to and affects the system's success in solving problems, but it is unlikely to destroy the system.

Because determinate systems are simpler to design, system developers seek to impose determinacy on the system by attempting to proceduralize all operations. Often this cannot be done because of the situational context; the external world usually determines the nature of the stimuli and hence of the system response that must be made. For example, in a military situation, if one knew precisely what a hypothetical enemy would be doing at all times, one could build a completely determinate system to deal with him, but this is of course completely unrealistic.

However, one hypothesis that has a great deal of intuitive attractiveness is the possibility that the natural tendency in systems is to become less indeterminate with increasing experience. If, as we suspect, systems learn, with increased learning it should become more possible to standardize operating procedures. Certainly when system functioning depends largely on how well personnel perform, the system will learn because personnel will learn to accomplish indeterminate tasks more efficiently.

Computerization of systems beyond mere automation—the development of expert and decision supporting systems—should have the effect of reducing indeterminacy if only because the effort to computerize forces the system designer to proceduralize ambiguous input situations and turn these procedures into software algorithms.

The implications of indeterminacy for the system are important. More human error can be expected in indeterminate systems. To compensate for this, additional error checks and balances have to be designed. The human in an indeterminate system becomes more important than he or she is in a determinate system because presently no computerization and automation can completely respond to the demands of an indeterminate system; only the human can deal even ineffectively with very indeterminate situations.

It is probable that there are qualitative and quantitative differences between determinate and indeterminate systems. We hypothesize that in determinate systems human errors are mostly what Norman (1981) called *slips*—forgetting and miskeying. In indeterminate systems, errors are more likely to be those of an inappropriate intention—choosing the wrong solution, the wrong diagnosis, or the wrong option (in Norman's terminology, *mistakes*). The failures at Three Mile Island and Chernobyl were of the latter type. Such errors are much more sophisticated than slips—much more difficult to prevent by training or to correct. In the skill–rule–knowledge-based taxonomy of Rasmussen (1983), one would expect actions and errors to be more knowledge-based in indeterminate systems and more skill-based in determinate systems. Errors of intention are much more closely connected with the terminal output of the system, and hence more catastrophic than slip-type errors, which are often trivial provided they can be recognized and quickly corrected. Errors of intention are not easily recognized because they are tied to fundamental misperceptions of the environment. Human errors presumably have greater mission consequences in indeterminate systems because they tend to occur in the context of trying to assist a system with a problem—a system that has malfunctioned, is about to malfunction, or has a high probability of failing its mission.

Systems are neither totally determinate nor totally indeterminate. The more common situation is that in which some units and subsystems are more determinate and others are more indeterminate. For example, the production line in a manufacturing facility may be very determinate—all operations are highly proceduralized and predetermined and uncertainty is minimal—whereas the marketing section of that facility is very indeterminate because the competition of competing products is great and innovative marketing solutions are prized.

A determinate system may also have occasions when the action of its environment renders its situation uncertain. Thus, one can think of farming as being very determinate. However, when there is a drought, the farm system becomes more indeterminate because the farmer has more difficulty predicting when to sow, when to harvest, and what to plant.

The amount of indeterminacy is not a constant for the indeterminate system. Determinacy is relatively constant because the essence of a deterministic system is that very little changes. The indeterminate system may have a certain degree of indeterminacy built into it because of the nature of its sensors and information flow, its procedural flexibility, and so on, and this obviously does not change. However, external situational uncertainty (e.g., a potential threat) may increase or decrease. Thus, if one could calculate the amount of situational indeterminacy on a chronological basis, the value would fluctuate.

If one were to ask which component of indeterminacy is more important, structural (system) uncertainty or situational (external or environmental) uncertainty, the answer would have to be the latter. Training can reduce structural uncertainty by making personnel more adept at using system hardware and software, thus reducing uncertainty. However, one has little control over situational uncertainty, although one can still train personnel to deal with this uncertainty more efficiently, which helps reduce its negative effect.

Uncertainty

Uncertainty is central to a problem-solving model of system functioning. Uncertainty stems from ambiguity, which results from limited knowledge of the processes that generate outcomes. It is essentially subjective; because the environment can only be known through perception and cognition, it will always be perceived as somewhat uncertain. Uncertainty describes both the state of the external real world and the state of the perceiver who lacks information about that world. Theorists define *uncertainty* in three ways: (a) the inability to assign probabilities about the likelihood of future events occurring; (b) a lack of information about cause-and-effect relationships; and (c) an inability to predict accurately decision outcomes.

Systems may deliberately create uncertainty as well as merely adapting to it. This is particularly true of systems that are adversaries, like enemy armies. Systems organize themselves in different ways to deal with uncertainty, particularly in the way in which they develop information-gathering subsystems to collect information to reduce uncertainty. Some search the environment more actively than others. Other systems adopt an active, exploratory strategy without presumptions as to the nature of the solution. Still others explore, but they focus on detecting a single correct answer. Other systems are more passive, relying on established data-collection procedures and traditional interpretations; if they assume the environment is unanalyzable, they do not rely on hard, objective data.

In a simple and stable environment, the environment produces low uncertainty for the system. In complex but stable environments, there is low to moderate uncertainty. In simple but unstable environments, there is moderate to high uncertainty. In complex and unstable environments, the environment projects great uncertainty.

Uncertainty has been a central concept in the literature of both decision-making research and organizational theory. Despite the many studies performed to explicate the meaning and function of decision making, that meaning is still unclear. The difficulties begin with the definition of *uncertainty*. Uncertainty may stem from ambiguity, which is produced by limited knowledge of the process that generates outcomes. This suggests that uncertainty is essentially subjective: the perception of a state of affairs in the environment that may or may not be uncertain, but that, because it cannot be known except through perception and cognition, must always be somewhat indefinite. Most psychological work on inference has been directed by a subjectivist view of probability.

Some of the most interesting thinking about uncertainty comes from the literature of organizational theory, which views the system not in terms of discrete human–machine interactions, as do human factors specialists, but in terms of more molar organizational relationships, although these also encompass individual human–machine interactions.

In the psychological approach, uncertainty is measured by probability, which in turn can be operationally measured by choices among gambles. This mathematical paradigm, frequently represented in experiments by the selection of varicolored balls from two or more urns, seems curiously restricted when one considers the uncertainty of situations facing complex real-world systems. (However, in the last few years, a new speciality has been developed—naturalistic decision making; Klein, 1993.) A number of factors influence uncertainty.

It appears that people are more confident and less uncertain about information that comes from a high-status source—one that is reliable, relevant, and orderly. The same problems can produce different choice responses depending on the way in which the uncertainty is expressed. Task characteristics—notably response requirements and types of responses—exercise an important influence on uncertainty. Part of the confusion about uncertainty stems from the fact that the term is used to describe both the state of the real-world environment and the state of the perceiver who examines that environment.

Reality must be thought of as on two levels. The first consists of physical phenomena, which can be pretty well apprehended. The second level is one of dynamic processes, in which decisions are being considered and made. This kind of reality may be obscure because often we do not understand the meaning of the processes. The attempt to apprehend reality can be seen as a search for the meaning of dynamic processes and events in the environment.

Given the two types of uncertainty—actual and subjectively perceived—how do they correlate? There is little reason to expect one-to-one correspondence between any objective indicator of uncertainty, however sophisticated, and a perceptual indicator because perception varies as a function of context and individual attributes. For example, when people feel some control over events, they tend to perceive these events as more certain than objective evidence justifies.

The three types of uncertainty noted at the beginning of this section—state uncertainty, cause-and-effect relationships, and decision outcome uncertainty—are differentiated by the type of information that system personnel receive. In state uncertainty, personnel lack information about the nature of the environment. In contrast, effect uncertainty does not necessarily involve a lack of information about environmental conditions. Rather, the shortage of information is in knowledge of how environmental events or changes in these will affect one's system. In decision outcome uncertainty, there is a perceived lack of information about what the system's response options are and/or the value or effectiveness of each course of action in terms of achieving desired system outcomes. Uncertainty appears to be a negative factor because it threatens system performance and integrity. Certainly people prefer situations in which uncertainty is low rather than high.

If uncertainty is as important a variable in system operations as the author believes it to be, one would expect that the system would organize itself in different ways to deal with that uncertainty, particularly by developing information-gathering subsystems to collect information with which to create more certainty. There is some evidence (but not a great deal) that people who are highly uncertain about a situation do seek more information. Certainly this seems a logical action to take.

One may ask why it is necessary for technologists to study major constructs like uncertainty—constructs that are so abstract that they may be accused of having philosophical trappings. All the abstract, quasiphilosophical constructs that have been mentioned lead in a more concrete context to problems that have been plaguing us under other guises for many years. For example, uncertainty leads to, or is a result of, task difficulty (a reciprocal relationship). System/task complexity, about which we know almost nothing, is also linked to task difficulty.

Moreover, as technology becomes sophisticated as a result of computerization, the concepts that HFE specialists create to deal with this sophistication must also become more sophisticated. The thought patterns we had when we studied knobs and dials in the early years of the discipline will not do in dealing with highly complex systems, particularly those that will face us in the next century. It would be more comfortable to return to the earlier HFE world, but as Thomas Wolfe, the American author, pointed out, "you can't go home again." Unless we want to willfully ignore what is happening to modern technology, we cannot avoid confronting the more difficult problems.

There are those who feel uncomfortable with abstract concepts. However, there is an excitement in constructing and examining these concepts and their implications for human performance. Those who do not feel this excitement are to be pitied because they lack some quality that prevents them from experiencing a great feeling of intellectual satisfaction. Those who are interested in intellectual edifices should not be discouraged by those who are not.

At the same time, it is not sufficient to merely contemplate these abstractions, however beautiful they may be. The term *contemplation* includes research as long as that research is not intended to produce concrete outputs. HFE is a pragmatic science, which means that the abstractions must eventually be translated into forms that enhance the well-being and efficiency of the humans who work with physical entities. Although we may begin by being intellectual aesthetes, we must ultimately become *journeymen of the possible*.

THE TECHNOLOGICAL ENVIRONMENT

It may seem strange to talk about the world view or inherent philosophy of HFE, but HFE does have one and it suggests what we attempt to do. However, it is well concealed because philosophy does not seem to coordinate well with so highly empirical a discipline. We begin by assuming that the human is a reactive animal—that initially he or she responds largely to external stimuli, although with increasing maturity the animal can generate stimuli on its own and respond to those stimuli. Those stimuli (even the ones that are self-generated) are often obscure and the correct response to them is unclear. The process of learning is essentially a process of assigning meanings to those stimuli. One can view all human activity as an attempt to impose order on what essentially is (or appears to be) a disorderly world.

The human is always in a state of uncertainty. (Not to have some uncertainty and consequent anxiety is to be a sociopath.) For many or even most situations, the degree of uncertainty is minimal. To well-known stimuli for which appropriate responses have been learned, the degree of uncertainty is so low that we are in effect unaware of that uncertainty. For other more unfamiliar situations, the amount of uncertainty is much greater and causes the human much anxiety. Of course, we have no anxiety for large areas of human activity, which we wall off from ourselves. In the days of the cold war, there was great uncertainty with regard to nuclear weapons, but we did not think about these matters for long periods of time. Hence, uncertainty effects exist primarily for those activities with which we allow ourselves to become involved.

The degree of uncertainty and the resultant anxiety is correlated with the number of options the human has. When the human is faced with stimuli the meaning of which is unclear, he or she has many options as to the meaning of these stimuli. In the familiar stimulus–organism–response paradigm, there may be uncertainty about all three elements, including how one feels about a particular situation.

The human's continuing effort (throughout his or her life) is to reduce the number of stimulus–response options to as few as possible—to one, if this can be done. This refers primarily to the average human. The brighter the individual, the more uncertainty he or she can endure. In fact, the most creative individuals (poets, artists, philosophers, mathematicians) may revel in uncertainty and deliberately create obscurities. However, the average person finds uncertainty difficult, unsettling, unpleasant, and to be resolved as quickly as possible.

Uncertainty is correlated with task difficulty; difficulty gives rise to error and inefficient performance, anxiety, and stress, all of which the HFE specialist is dedicated to eradicating in a technological civilization. Therefore, the ultimate goal of HFE is to reduce uncertainty to some acceptable level. What that level is is unclear and it may vary with intelligence and personality characteristics. Complete lack of uncertainty, as in monitoring a blank sonar scope for 30 minutes, may be boring, just as an excessive level of uncertainty will produce stress, anxiety, and error. Automation of technology reduces uncertainty, of course. However, as has been shown elsewhere, automation merely transfers uncertainty from the operation of systems to the maintenance of these systems when they malfunction.

The goal of our discipline, as it should be for engineering also, is to simplify as much as possible—to create as much order and regularity as one can in an uncertain and ambiguous world even when we cannot simplify to the extent we would wish. The first question we need to ask in examining a new design or proposed design solution is this: How much uncertainty will this produce? If error is a natural consequence of uncertainty, then research needs to concentrate on determining how much error the human is likely to make when he or she is faced with the many system and situational parameters that technology produces. Hence the need for the development of a human performance databank, its application to new design, and the investigation of factors that produce uncertainty and error.

In the broadest sense, the job of the HFE specialist is to make the technological world less frightening by introducing order into its potential disorder. Our counterparts, the engineers/designers, also attempt to order the world by designing machines that enable humans to gain control over that world. However, every action creates a reaction. In the design process, the engineer may inadvertently introduce a certain disorder by increasing more complexity than is needed.

Technology in large part creates the world in which we find so much that is obscure and frightening. The more sophisticated the machines engineers build for us to run our world, the more difficult it is for us to deal with this sophistication except by prolonged and continuous training. Those who must manage this technological world often plead with engineers to reduce system complexity and the uncertainty it produces. As a topic for behavioral research, the subject of complexity is a worthy one.

Every human in every working hour is engaged in problem solving: the interpretation of stimuli and the selection of an appropriate response. For familiar, well-learned situations, the need for problem solving is minimal because it is essentially unconscious at the highly skilled level. For less well-learned situations, the amount of conscious problem solving is considerably greater and we face a plethora of taxing decisions to make. This suggests that research emphasis in HFE should be focused on how people interpret complex stimuli, how they deal with uncertainty, and how they make decisions.

The human's effort to reduce uncertainty leads to an effort to gather information that will reduce the number of response options and thus reduce uncertainty. Along with many other HFE specialists, we conceptualize an information acquisition and processing model as a theoretical framework for explaining human performance, when the operator is unclear as to what he or she must do (as when the system presents a difficult problem to solve).

Hence, research on operator performance in complex systems must emphasize investigation of how and how much information the operator seeks and how he or she utilizes that information. Information as a major system dimension and the variables inherent in information become one axis of an archetypical graph, with human performance, in whatever performance units are chosen, as the other axis.

COMPLEXITY

Arguably the most significant factor influencing the behavior of systems is their complexity. Complexity is determined by the number of equipment subsystems, the manner in which these interact, and what is required of the system operator. With the exception of the last, the infrastructure of system complexity is essentially invisible to the human. What is apparent to the human and what defines complexity is the amount and type of information provided to the operator. Unless the operator also functions as a maintenance person in direct contact with equipment components (e.g., circuit boards, transistors), he or she is never aware of system complexity except as an effect of these physical components—an effect presented as information to be used in solving a problem. Information interacting with task requirements summarizes the state of system complexity, which may vary from time to time as more or fewer system functions are engaged.

It is now possible to define *complexity* more precisely as the number of different informational states presented by the machine to the operator (Meister, 1996b). Complexity as a physical attribute depends on (a) the number of internal subsystems and components within the equipment; (b) the number of their interactions; (c) the number of modes in which the system can perform; and (d) the number, type, and visibility of the malfunctions that may occur.

The analysis of the physical factors determining complexity can be performed by engineering means. The reliability of an equipment can be predicted by performing what is called a *reliability failure modes and effects analysis*, which does not need to involve any behavioral considerations. This analysis involves only considering what physical stimuli may occur, from where these come, how they interact, and their effects in terms of possible malfunctions.

From a behavioral standpoint, the measurement of complexity involves translating the purely physical analysis described earlier into information terms.

It was assumed in chapter 2 that the way to bridge the gap between the physical and behavioral domains in HFE was develop a language that would translate the special characteristics of each domain into a common means of communication (information), which is expressed in a common metric (e.g., probability of state events occurring, probability of errors occurring, etc.).

In this conception, it is possible to treat the physical manifestations of internal machine operations as stimuli (expressed physically as numerical and graphic values in displays), which, when received and interpreted by the operator, are transformed into information and then quantified.

Complexity is important to the human in the system because it largely determines how efficiently he or she and the equipment interact, which then presumably determines how well the system as a whole performs. If human efficiency is measured by such indexes as error and time (reflected by phenomena like failure to note system status, failure to recognize a system problem, or failure to organize a strategy for solving the problem), one would expect error and time difficulties as well as human stress to increase as complexity increases. Certainly this is our common experience. However, at the moment, this can be only a hypothesis because there have been few, if any, formal studies of complexity. The hypothesis seems reasonable, however; what we do not know is what factors tend to increase complexity. Complexity demands intensive HFE research.

Studies have been performed to investigate the effects of such variables as the amount and type of information presented to subjects (Schroder, Driver, & Streufert, 1967), but these variables have not been directly associated with physical systems differing in complexity as it has been defined in this book.

The importance of complexity as a system attribute is in its relation to design. It is possible to explain HFE effects in purely psychological terms because, in developing an explanation, it is unnecessary to deal with the differences between the behavioral and physical domains. When one wishes to explain human performance even in a technological environment, purely psychological processes will serve as well as any other because we look at that performance as merely a behavioral response to physical (machine) stimuli. If we wish to design into hardware or software certain characteristics that will, for example, reduce complexity, it is necessary to transcend the domain differences in the human–equipment relationship. This requires a common language (information) and a common metric.

Ideally, an HFE professional working in system development should analyze a design drawing and say to the engineer,

> This design rates as 8 on a 10-point scale of complexity, which is rather high for an operator to deal with. The problem seems to be that you have presented your information as individual items in separate displays. All the information you want to present to the operator seems reasonable, but it would be simpler for the operator to react if you combined information sources X, Y, and Z into a single predictor display and A, B, and C into another predictor display while allowing the other display sources to remain separate. This would reduce the complexity index to 5, which is much more tolerable.

Of course, this example is highly idealistic and imaginary. No HFE specialist at present would be able to make such an analysis for several reasons: There is almost no HFE research on complexity as a system attribute and an information language and metric that the nonmathematician can use have not yet been developed. The design problems and the need to transcend domain differences should be the driving forces behind HFE research, but unfortunately they are not.

The importance of complexity in practical terms is that ultimately we wish to design human–machine systems that are as simple as possible. Simplicity here equates to increased efficiency, complexity to reduced efficiency. The constraint of physical reality (i.e., what the system must accomplish and the technological means of accomplishing the goal) often does not permit us to make systems as simple as we would prefer them to be. The question then is how to include in the system design just so much complexity as is required and not a jot or tittle more.

In highly automated systems, in which the human role changes from operator to controller to energizer (to be replaced in this role by computer software), the human becomes a symptom analyzer, which means that the primary human task is now to analyze patterns of information to determine how these reflect what is going on in the innards of the system. The resemblance to a physician's diagnosis is irresistible.

It is assumed that the human possesses certain inherent methods of performing such analytic operations. To design a system effectively, we need to know what these inherent methods or stereotypes consist of. If we know this, we can construct the symptom patterns of a new system in design to match these human stereotypes so that the human will not become overwhelmed by informational stimuli.

The questions whose answers we need to know are: What level of physical system complexity produces what patterns of information (symptoms)? How does the nature of physical system interactions affect the amount and type of information presented? What inherent information analysis stereotypes does the human possess that permit him or her to deal with complexity? (This includes this secondary question: How much information and in what form can humans ingest this information without its giving them indigestion?)

It is possible to create human–machine systems that are so complex that their designers fail to understand the degree of complexity they have produced. This produces what has been called *emergents*—system-produced phenomena that were not anticipated by system designers. It is possible that such emergents may cause massive system malfunctions that may be beyond the control of humans in the system. These are practical reasons for being concerned about system complexity.

Complexity adds to the uncertainty in which systems and humans function. Lack of knowledge about the effects of complexity paradoxically adds to system complexity because it may lead the designer to construct more complexity than is actually required to perform system functions. Complexity is largely induced by the nature of the interactions among system components—an interaction that is invisible to the human operator of the system, but is more or less apparent to the designer. The designer should be aware of the effects produced by these interactions—not only in purely physical terms (e.g., a rise in a boiler temperature), but equally in the informational effects created by their interaction. To perform the latter function, the designer needs the aid of the HFE professional. In the future, it may be possible to include this feature as part of the software in computer-aided design (CAD) routines.

Complexity can be introduced into the system in two ways: inadvertently (which is ordinarily the way it is included) and deliberately. The former is produced by designer ignorance of what is being designed into the system; the latter is produced when designers ask questions about the nature of the design produced. Naturally the latter rarely occurs. However, one of the things HFE would like to do is introduce designers to the subtlety of complexity so that they are more aware of the implications of their designs.

TRANSPARENCY

Transparency is closely related to complexity. If complexity produces patterns of symptomology, transparency is what we hope to induce in the analysis of that symptomology. The function of symptomology is to allow the human to understand what is going on in the innards of the system—an area that is ordinarily terra incognita for the operator.

Transparency refers to the extent to which internal system functioning is made apparent to the human operator. Symptomology, as represented by display indications, provides a form of transparency, a *deductive* transparency, which must be interpreted from the meanings attached to the individual and patterned symptoms.

What the designer gives the human operator in the form of displays (and the ability to influence the system through controls) determines the amount of transparency provided. If an expert subsystem is provided that directly monitors internal operations and supplies diagnoses, the amount of system transparency is much enhanced.

Complexity may be and often is produced inadvertently, whereas transparency is always deliberately created because it represents the consequence of a conscious decision on the part of designers about the number and types of indexes they will provide in the system interface (e.g., the workstation computer terminal). To make this decision adequately, the designer must be aware of the kinds of internal operations possible, their potential (malfunction) effects on the system, and what information these interactions can provide if they are monitored by the operator.

It seems clear that all information about internal system operations cannot be provided because to do so would almost certainly overload the operator's capacity to absorb information (as well as being inordinately expensive). Hence, the designer must make choices (which means decisions) as to (a) the likelihood of potential effects from internal interactions, (b) the relative importance of each of these effects, and (c) the most effective way to present information about these effects (how much information, presented in what form; e.g., analogue or digitally, singly or in combination with other items of information). The first two questions are solely engineering questions; the third is largely that of the HFE professional (with the designer's assistance, of course). The answer to the third question also requires a great deal of research because, to date, display research has largely dealt with information from single sources and in automated systems we may wish to create displays integrating multiple sources of information (see Moray, 1997).

AUTONOMY

Autonomy is companion to complexity and transparency. If one reveals to the operator what is going on inside of the system, it seems reasonable that one should also give him or her the capability of doing something to select a mode of operation or correct a deviation from preset parameters. If transparency refers to displays of symptomology, *autonomy* refers to the controls available to the operator and the options he or she has in directing the system and securing a diagnosis of incipient malfunctions. Again the resemblance is to the physician who has options (which equals autonomy) to take blood samples and make tests that reveal what is wrong with the human patient.

The decision as to the amount of autonomy to give the human in the system is a deliberate choice. Testing the system by *tweaking* it (e.g., increasing or decreasing water to nuclear power plant boilers) means providing information that results from the tweaking; this adds to the information that increases system complexity. One can think of physical parallels as actions producing reactions.

Complexity, transparency, and autonomy are all interrelated. Complexity produces effects (information) that may or may not be displayed (transparency), and there is no point in providing information to the operator unless one is prepared to allow him or her to do something about the situation represented by the information. Again, conscious decision is required. How much human intervention is the designer prepared to permit? If an NPP is decaying, is the designer prepared to allow the nuclear control room team to perform surgery on the spot (intervention related to the internal component causing the trouble) or should he or she restrict the operator's autonomy to the option of shutting the system down or permitting it to continue?

A system can have autonomy only if the designers of the system are willing to allow control of the system to be exercised by the human. This requires answers to these questions:

1. What kinds of catastrophic system situations may arise?
2. Are there mechanisms built into the system to permit intervention? (The more mechanisms provided, the more complex the system becomes.)
3. What information will the system present to the operator to permit intervention?

The system attributes described assume a highly automated system in which the human's role is essentially an interpretive–diagnostic one. If the designer conceptualizes a relatively simple, proceduralized system, these system attributes are somewhat inappropriate.

SYSTEM VARIABLES

Apart from the variables associated with its human and equipment components, the system has variables of its own that function at the system level and affect how well or poorly the overall system performs. This section explores these system variables. Although the variables in Tables 3.1 and 3.2 are important in the development and operation of the system (see Meister, 1991), they have not received much attention from HFE researchers, who are much more concerned with human responses to individual stimuli. System development specialists get involved with these variables only infrequently because design usually only deals with equipments and subsystems. Nevertheless, these variables do have behavioral implications that will affect the system's ultimate design.

Although the variables in Table 3.1 describe the system as a whole, there is some overlap with behavioral variables (Table 3.3). For example, system personnel also have functions to perform, and missions assume the existence of tasks that personnel must perform. The system has goals, but so has the operator, and these should coincide (although they may not if the system is large and the operator's goal pertains only to his or her immediate equipment operations). The system goal will be more inclusive than that of any single operator. The requirements and constraints imposed on the system also have a parallel with those of the operator, but the former will be somewhat different, considering the different scope of what is involved in these. The remaining system variables are quite different from the behavioral variables listed in Table 3.3, all of which should be familiar to anyone with an HFE background.

Structural System Variables

The structural system variables listed in Table 3.2 represent an extension of Item 6 in Table 3.1. Structural variables are particularly important because many of these are internal. Hence, it is this structure that interacts most closely with complexity, transparency, and autonomy attributes. Each of the variables in these three tables is important enough to initiate an extensive research project if only because it is necessary to consider each variable in designing a system.

General System Variables

We start by examining Table 3.1. The initial step in analyzing system requirements for development is to analyze the requirements and constraints imposed on the new system in terms of their human performance implications. It is obvious that a vehicle like the Space Shuttle will have physical requirements that are not imposed on a commercial airliner, which ordinarily flies at 35,000 feet. Some of these outerspace requirements have behavioral implications, such as the design of protective clothing.

At the same time, a system may have constraints (negative requirements) that affect operator behavior. For example, the physical space constraints of early submarines meant that personnel were quite restricted in their sleeping, eating, and washing functions. The resources required by the system (e.g., certain types and amounts of fuel) may influence what operators have to do. For example, when ships were coal burners with limited capacity to store fuel, one of the tasks imposed on personnel was to load coal—a tedious and nasty job. The nature of the system's internal processes also impact operator behavior. The need to make these visible to the operator of a process control plant like nuclear power to diagnose malfunctions was previously described.

The functions performed by the system are much like those of the individual. The mission, of course, is quite distinct from the functions that implement that mission. Some missions, like sonar detection, are highly dependent on the human; others are not, and this determines the amount of attention paid to them by HFE specialists in system development. The particular system mission may emphasize certain behavioral functions like cognition and deemphasize others such as motor responses.

The system also has goals to accomplish; these are determined by the nature of the mission. These goals establish criteria (e.g., bombing accuracy) that impose requirements on personnel behavior and also serve as a means of evaluating both system and personnel performance.

The system's structural variables are described later. As has been discussed previously, the degree of automation has significant effects on operator performance: It changes the operator's role and it may, as in the glass cockpit of aircraft, lead to improved personnel performance in certain functions and degraded performance in others (see Mosier, Skitka, & Heers, 1996; Sartor & Woods, 1995; Sartor, Woods, & Billings, 1997; Wise, Guide, Abbott, & Ryan, 1993).

The nature of the environment in which the system must perform (e.g., outerspace, surface, subsurface space) affects the physical aspects of the system directly, but may also affect the operator. For example, when hard hat diving is required, the diver must wear special protective apparatus and take special precautions against overly rapid decompression. System attributes like complexity may determine the task difficulty that the operator experiences. For example, when VCRs were first put on the market, the complexity of their programming requirements dissuaded many consumers from purchasing the equipment. Internal system interdependencies have already been discussed; these may have a ripple effect on operators and may help determine the specific tasks they must perform. When the mission is completed, the system produces outputs but these do not necessarily affect personnel performance.

System Structural Variables

System structural variables (Table 3.2) expand on the more molar system variables of Table 3.1. System structural variables are more likely to have indirect rather than direct effects on human performance.

Obviously, systems vary in size and number of units and personnel. The type of system organization has implications for other variables (e.g., communication). In a highly centralized organization, one would expect more communication than if the organization is decentralized, or would one? The availability or rather the nonavailability of communications channels will obviously affect system efficiency. The major system attributes have already been described; these also affect operator performance.

The method of system control is closely linked with system organization. For example, military organizations are usually centralized and autocratic, whereas civilian organizations may or may not be. A highly autocratic style in a nonmilitary system may reflect the personality of the autocrat in control. The number of hierarchical levels is associated with organization (centralized and vertical should have more levels) and size (all other things being equal, the larger the system, the more levels it will possess). The nature of internal system processes may also influence operator performance. For example, repetitive processes such as those emphasized in manufacturing assembly lines may lead to muscular trauma. The degree of automation has been discussed.

If the system requires continous 24-hour operations, as in some military exercises, one can expect to see fatigue phenomena and performance degradation. The stage in the mission at which personnel functions are performed may or may not have an effect on the system; an error in the terminal stage may be more calamitous than one made earlier when it is possible to correct the error.

Subsystems may have a primary or supporting mission role. For example, in the Air Force, bombers have a primary mission and ground maintenance has a supporting role. The primary mission may involve danger and highly constrained time requirements, but mission requirements may be as demanding in a support role. The specific context is important; military operations show the dichotomy more obviously. The differences between a primary and supporting mission role may be very slight in a civilian system. Division of responsibility may be sharply or less well defined; this variable may have an effect on system and operator performance, but it is not clear how this occurs. In any event, the effect on human performance is likely to be indirect rather than direct.

Note that we have occasionally distinguished between system and operator performance. It may be necessary to measure operator performance apart from system performance (e.g., number of operator errors vs. mission success). To understand each type of performance more completely, it is necessary to measure both and relate each performance to the other. HFE research emphasis on behavioral performance to the exclusion of the system performance that serves as its context is understandable but undesirable.

In HFE research, most of the system variables are either ignored or considered only as context; as one would expect, the emphasis is on behavioral performance. If the research is academic, conducted in a large laboratory, there may be no mention of any system context at all because the system does not function as one of the independent or dependent variables. If the research is more applied (e.g., related to types of systems like aviation or automobiles), high-level system variables may be mentioned as explanatory factors but are rarely included in the research per se.

There is one outstanding exception to the general HFE lack of interest in system variables. This is in the speciality area of macroergonomics, which is directly concerned with systems as well as behavioral factors. Much of the most innovative thinking in HFE, as regards systems, has arisen in the context of macroergonomics.

The more traditional HFE professional will argue that in most cases it is unnecessary to examine the role of the system in human performance because the effects of the former on the latter are indirect and hence slight. Moreover, research cannot encompass everything. Thus, if one is studying a psychophysical phenomenon like visual contrast effects, one need not also consider organizational structure because the effects of structure will not extend to the visual level.

There is some limited merit in this objection, and it certainly applies when the levels of conceptualization are as far removed from each other as visual perception and organizational structure. However, when the argument is used to deliberately ignore major factors that do impact performance, the merit is slight. Moreover, the connection between levels of discourse may be closer than one thinks. In a study the author once participated in on factors causing mining accidents, a large causal factor in the initiation of these accidents did result from managerial inattention and ignorance. These were indirect factors, but they had significant effects.

At first glance, it may appear as if these variables are so global that it would be impossible to study them in a controlled manner. Examination of actual operational systems may provide a partial answer and should always be part of the research, but another answer is computer simulation. Systems of any size can be simulated, although obviously the larger, more complex systems will be more difficult to simulate, but it can be done. With larger simulated systems, it may be necessary to dedicate individual computer terminals to individual subsystems and consider as one subject a group or team.

Because of the number of system variables involved, a single experiment—even two or three—may not provide usable answers. This is because it is the interaction of these variables that is important. Statements about single variables probably do not tell us more than what can be commonly observed and provide unrealistic answers because they are simplistic. For example, it is obvious from experience that increasing system complexity creates added difficulty for the operator/manager, but the more important questions to ask are these: At what point does the complexity defeat the purpose of the system? Does complexity have differential effects depending on the type of system, type of organizational structure, and so on? What happens if, with increasing complexity, we also include increased transparency? Can one system attribute influence the effects of another? What kind of behavioral effects do we find with increased complexity, errors, procrastination, perseveration, unwillingness to make decisions? Will expert systems analyzing data in concert with the operator reduce error frequency or speed up decision time? In general, we want to know what behavioral effects are produced by system variables (e.g., what happens to excess information in system processes: does the operator store these somewhere before acting on them?) Additional behavioral effects may include the impressions that the operators and managers develop about the system with which they are interacting and their attitudes toward a particular organizational structure. It might even be possible to determine whether certain personality types are more or less affected by certain system variables.

General principles phrased in statements like *complexity creates difficulty for the operator and hence the system* are not really what we want to learn from this research. As indicated previously, general principles can usually be deduced logically or on the basis of personal experience. Moreover, there is little one can do with general principles because their generality provides no clues to specific problems. As in the case of the discipline of history, it is the details of the functioning of variables that excites the interest of the researcher and can be found useful. General principles are of little use in application because the distance between the general principle and the specific problem to be solved is so great that it is impossible to bridge that distance.

EFFECTS OF THE SYSTEM CONCEPT ON HFE

Nevertheless, general principles are important, although maybe not for solving specific problems. One general principle that has (or should have) significant impact on HFE is what is termed *the system concept*, with which the reader has been familiarized in earlier chapters. It says in effect that the elements or units with which HFE is concerned can be conceptualized as systems, or parts of systems, and that, in consequence, HFE must be concerned also with systems. This means that the system concept has important implications for the discipline.

The author makes the assumption that the system concept is the foundation of HFE. Although most professionals might accept this assumption on a conceptual level, they do not allow it to interfere with the way in which they conduct their affairs. One finds this presumably in other disciplines also. For example, it is presumed that every physicist assumes the validity of the theory of the atom. However, unless one is actually doing atomic research, it is unlikely that this theory is in the forefront of the physicist's consciousness.

Nevertheless, the system concept is important because it provides a conceptual structure that is useful for HFE professionals if they do have to think in larger conceptual units. The system concept has several effects:

1. The notion that the system in which the human is embedded imposes a demand on the human for resources is, in part, a justification for HFE because that demand can be adjusted by modifying system characteristics. Thus, by adjusting system characteristics (e.g., reducing the number of displays or positioning them more advantageously), the demand can be reduced. Reduction of demand is considered to be a primary goal and function of HFE (why else would so many researchers be so concerned about workload?), which is why the design aspect of HFE becomes so important—more important than many might be willing to admit. (Selection and training of personnel to satisfy system demands are also important, but this is a clumsy way of making the adjustment.) The author sees the role of adjusting demand by affecting design as a primary purpose of HFE. Some might say that it is the HFE role as a science, as a means of explaining human–machine interaction phenomena, that is the justification for HFE. However, unless there were a parallel attempt to influence design, no one would give a dollar to support HFE research. The author retracts the statement; maybe funding agencies would give a dollar, but not much more. (The oddity is that with HFE research doing so little for design, research funding still persists.) The major reason HFE exists on a practical level is that it is presumed to be necessary to provide behavioral design guidelines. This may seem unfair to HFE researchers who think of themselves primarily as scientists, but the author has a bias: To him, science without utility is a contradiction in terms.

2. The human–machine system as a unique phenomenon (not found otherwise in behavioral science) makes the study of that system (i.e., HFE) a unique discipline, not merely a speciality of other disciplines.

3. The system deals with totalities so that anyone who believes in the system concept must analyze every HFE problem that arises in terms of all the factors that could be inherent in the problem and that could influence design and human performance. Admittedly, this increases the amount of work for all HFE professionals, but, as some might say, "that's the way the cookie crumbles."

4. Because the human is a component of the human–machine system, in measuring human performance it becomes necessary to relate that performance to the system in which it occurs. Macroergonomists do this to a limited extent, but all other HFE professionals tend to ignore the requirement because it imposes an additional burden on the researcher. The researcher, who is all too human, would rather do things in a more simple than in a more complicated way. If the system hierarchy breaks down into the workstation, the subsystem, and the total system, the system concept requires

measurement at all these levels and the determination of the relationships among them. Again, this is a conceptual requirement that is usually ignored.

5. The individual system cannot be completely understood outside the larger whole (the suprasystem) in which it is embedded. This forces the HFE thinker to consider the effect of higher superordinate systems, which form the environment in which the individual system functions.

6. Because the system functions through the manner in which its elements are organized, system organization is a necessary component of research and design and should be treated (or at least considered) during design, as all other system elements are treated. Macroergonomists would undoubtedly agree to this because it is the foundation of their speciality. Of course, almost all design of equipments and subsystems assumes the existence of an already developed organizational structure. The system development specialist rarely gets involved in overall system (e.g., company) design. It is a weakness of macroergonomics that often it is called in to redesign a company rather than to design it *ab initio*.

7. One aspect of the relationship between the system and human is that, as we have seen, humans, particularly their functions, serve as the model for the development of systems. The human as a model may not come to the minds of the system developers as they design the system. However, because they cannot think except in terms of human functions, the resultant system product will contain these functions.

8. It is assumed that because technology is, once developed, invariant (although see Hancock, 1997), the human cannot influence that technology. However, if the system is to be maximally efficient, the design of that system must take account of human peculiarities (e.g., threshold limitations and stereotypical modes of responding). This makes these topics critical for the performance of HFE research.

9. If the importance of the system means that HFE is a system science, this requires HFE research to make the study of human–machine systems a top priority. The researcher will submit that this requirement is satisfied because the research considers variables related to operational systems (e.g., automotive, aviation, computer, etc.). However, the author would suggest that, in these cases, the study of the system is essentially the study of the human performance secured from a particular type of system and is not research of the basic dimensions of the system as a general concept. Only this last is worthy of the discipline.

4

The Formal History of HFE

The history of a discipline can be represented in several ways: as a sequence of formal events tied to chronology, in terms of participant experiences, and as intellectual and specialty history. *Formal* history describes activities in relation to known chronological events. For example, it is possible to divide HFE activities into those performed prior to and during World War I and those before, during, and after World War II. The major markers here are the two World Wars because these were tremendously influential in developing HFE. This chapter describes the formal history of HFE.

Informal history can be considered experiential (i.e., activities seen and reported by the participants in that history). Examples are the many diaries written and published by soldiers of the two wars. Chapter 5 describes how the first generation of HFE professionals saw the early years (circa 1945–1965). There is also *intellectual* history, which in this book consists primarily of an analysis of the published technical papers of the discipline. Chapter 6 describes this analysis.

The formal history described in this chapter is mostly the history of American HFE. Readers are probably aware that there is a parallel history in Great Britain and the former Soviet Union. These are also described in this chapter. The basic concepts of British HFE are largely those followed in the United States. Those of the Soviet Union, although markedly influenced by Western inputs, deviate sharply in some ways from Western concepts. Thus, some time is spent in describing Russian concepts and how these differ from Western ones.

A subspecies of the previous general histories is called *specialty* history, discussed in chapter 7. The history of the Human Factors and Ergonomics

Society (HFES), prior to 1993 the Human Factors Society (HFS), are described as part of chapter 5, the informal history. What makes it possible to consider all these in historical terms is that they were and are subject to change over time; to simplify outrageously (as professional historians would see it), that is all history is.

The formal history of American HFE has been previously reported by Meister and O'Brien (1996) and Moroney (1995). This chapter depends to a great extent on their work, which did not include the other types of historical material in this book. Additional material on the early days of the discipline can be found in the Appendix to this chapter.

AMERICAN HFE

The Premodern Period

If one wishes to go far enough back in time in considering the antecedents of HFE, it is possible to trace these through an act of imagination to our neolithic ancestors because there have always been difficulties in matching the human to his or her tool. For example, the production of swords and armor in the Middle Ages undoubtedly took into consideration the anthropometry of that era's warriors. It also illustrates the intimate relationship that has always existed between the military and HFE—a relationship that still exists today.

Meister and O'Brien (1996) pointed out that, prior to World War II, the only test of the fit of the human to the machine was one of trial and error, in which the human either functioned with the machine (and was accepted) or could not (and was rejected). This Darwinian selection process went on until a successful candidate was found.

Meister and O'Brien suggested that a slight but significant shift in concern for the human in a technological context occurred in the American Civil War. The U.S. Patent Office was concerned that mass-produced uniforms and guns fit infantrymen and that they could use the new weapons being designed at that time.

Another example comes from the southern side. During the design of the submersible *Hunley*, its developer considered certain aspects that we could call *primitive human factors* in the design of the ship: men who were shorter than the average and who possessed greater than average strength, both of which were needed in a vessel that was fashioned in part from an old boiler. Note that the efforts were made to fit the man to the machine, and that there were no efforts to design the machine to fit the adequacies or inadequacies of the human. Machine domination in the human–machine equation lasted many years. For example, there was an anecdote that in World War

II the Russians selected their tank operators by applying size criteria: Anyone who was small enough to fit the cramped quarters of the T-34 automatically became a tank operator.

Romantic studies aside, the next precursor of HFE occurred around 1900, when the American inventor Simon Lake tested operators during submarine trials for psychophysical factors, in particular their ability to withstand unpleasant and dangerous environmental conditions: lack of oxygen, toxic gases, sea sickness, and the usual cramped quarters. This type of testing considered the human as a potentially negative factor constraining the utility of the system. As the reader sees, this point of view still persists and is even expressed by HFE professionals in their concern about human error and time performance measures.

The next milestone in pre-World War I antecedents of HFE is the rise of what has been called Taylorism or the scientific study of the worker (Kanigel, 1997; Taylor, 1919). This was an attempt to increase the efficiency of humans in the workplace. In 1898, F. W. Taylor restructured an ingot loading task at Bethlehem Steel (Adams, 1989) by modifying selection, training, and work–rest schedules so that he could move 47.5 tons per day, whereas prior to the restructuring the average daily load moved per worker was 12.5 tons. Taylor also designed a series of shovels for moving different types of materials. One of the significant aspects of Taylor's work is that he employed formalized methods of data collection and statistical analysis that are not far removed from those HFE professionals use today. Moroney (1995) also suggested that Taylor's principles of work design and time and motion studies became the basis for today's task analysis methods.

Formal time and motion study was also derived from Taylor's work, primarily through the studies of Frank and Lillian Gilbreth. Frank Gilbreth was one of Taylor's students and Lillian Gilbreth became his wife and coworker. During the early 1900s, the Gilbreths studied skilled performance in bricklaying, surgical procedures, and design for the handicapped. The techniques developed by the Gilbreths became the basis of time and motion study (measurement of work elements based on molecular hand, arm, and leg movements), which is still a fundamental part of Industrial Engineering (Neibel, 1972). It should be noted that the major players prior to World War I were all engineers. It is conceivable that HFE could have turned into a branch of engineering. What made HFE a behavioral discipline was the occurrence of World War I.

The inception of World War I stimulated, as war always does, the development of more sophisticated equipment. For example, the primitive flying machines of the Bleriot type were modernized into sleek fighters of the Spad, Neuport, and Fokker type. The war also saw the development of the first tanks. The inability of personnel to make use of such systems led to an increase of interest in human capability. For example, aircraft were tested by having expert flyers (e.g., aces) evaluate these for their handling capabilities in formal flight tests (a procedure that still exists in the aircraft industry). Systematic attention was now paid to one human capability—intelligence—by the development of intelligence tests and testing procedures for soldiers (the first intelligence test for children was created by Binet in 1908). The Army alpha and beta tests were developed by psychologists such as Thorndike and Thurstone, who were attached in more or less formal ways to the American Army for the first time. Aviation psychology developed the first applied test battery used to predict flying aptitude (Taylor & Alluisi, 1993). During World War I, Yerkes headed the Committee on Psychology of the National Research Council (NRC)—an antecedent of today's Human Factors Committee of the NRC (see Nickerson, 1995). Parallel activity occurred in Germany. There the first psychological testing center for the military was established in 1915 for the selection of motor transport drivers. Tests were also used to select pilots, sound detector operators, and anti-aircraft gunners.

Koonce (1984) pointed out that the focus of aviation psychology in the early days was on the selection and training of the aviator. Only later was there a shift to the aircraft—in particular, to controls and displays and the effects of altitude, g-forces, and environmental factors on the pilot. The original selection criteria for the American World War I aviator were education and character suitable for men who would eventually become Army officers.

Because of the need to expand the number of aviators, the Council of the American Psychological Association (APA) established a Committee on Psychological Problems of Aviation and, in November 1918, that committee became a subcommittee of the National Research Council. Two members of the APA Committee developed 23 mental and physiological tests that were evaluated by trying them out on Army Aviation Cadets at MIT. A number of tests found to be promising were also tried out at Rockwell and Kelly Fields. Those that showed the greatest relationship to flight training performance were emotional stability (response to sudden excitation, such as the sound of a pistol shot), tilt perception, and mental alertness. Other research projects dealing with the selection of pilot candidates and the effects of stressors on performance were also conducted. Similar studies were conducted in Italy, France, and England. Koonce (1984) described the foreign efforts in more detail.

World War I saw the beginning of aeromedical research and the need for related test and measurement methods. As is seen, this led to the development of the first aeromedical laboratories following the war. Why did this war not create an HFE discipline? There was no critical mass of technology and personnel as there was in World War II. The aircraft had been developed 10 years before. The tank entered service only in the last days of the war. Also, American involvement in the Great War only lasted 18 months.

Between the Wars

The period between the end of World War I and the start of World War II was one of gestation, but with relatively few outstanding accomplishments. Moroney (1995) noted the incidence of studies of driver behavior because of the increasing popularity of the automobile, which Henry Ford was providing to millions of ordinary Americans. Forbes (1981) reviewed the early research, including the work at Ohio State University beginning in 1927. Texts describing behavioral factors in automotive research included such topics as studies of accidents, early driving simulators, perceptual aspects of driving such as the estimation of velocity and vision requirements, social characteristics of traffic violators, the "accident-prone" driver, and studies of traffic sign characteristics. References to this research can be found in Forbes (1939).

It is significant that the research interest in automotive systems has been maintained to the present. The number of papers published by the HFES on this topic is second only to the number of aviation psychology papers. Moreover, many of the same topics are still being investigated.

One of the few bright spots in this rather dull period was the performance of aeromedical research. By the end of World War I, two aeronautical laboratories had been established: one at Brooks Air Force Base, Texas, and the other at Wright Field outside of Dayton, Ohio.

Toward the end of the war and following the armistice, many tests were given to aviators of the AEF in Europe to attempt to determine the characteristics that differentiated successful from unsuccessful pilots. Some of these efforts were to validate tests developed on the basis of pilot trainee performance and to compare American with foreign tests. Dockeray and Isaacs (1921) concluded that what were most necessary to the pilot was intelligence, the power of quick adjustment to a new situation, and good judgment.

Much of the following information about the early days of aeromedical research is provided by Dempsey (1985) and War Department (1941). Early work in this area explored human and machine thresholds of performance at environmental extremes. Both animals and men were used to determine the effects of altitude on performance. In 1935, Captains Armstrong and Stevens set an altitude record of 72,000 feet in a balloon. Initial work on anthropometry and its effect on aircraft design and crew performance was begun. Studies were performed of acceleration forces in a 20-foot diameter centrifuge; it was recommended that, in a flight exceeding nine Gs, the pilot assume a prone position. In 1937, a primitive "G-suit" was developed.

During the early 1930s, Edwin Link developed the first flight simulator as an amusement device (Fischetti & Truxal, 1985). In 1934, the Army Air Corps purchased its first flight simulator. The trends described previously have been continued to the present and amplified with more sophisticated simulators and test equipment. The development of a flight simulator as a system test device is standard procedure for any significant aircraft development project.

During this period, some noteworthy research was performed in the civilian sector, principally at the Hawthorne plant of the Western Electric Company from 1924 to 1933 (Hanson, 1983). In this study, the effects of illumination on worker productivity were examined. The results show no significant differences between experimental and control groups because (and this was the important point) the mere knowledge that they were the subjects of an experiment induced all the workers to exert increased effort. This effect, now known as the *Hawthorne effect,* suggested that motivational factors could significantly influence human performance in various ways. The Hawthorne effect still generates efforts at explanation (Parsons, 1974).

World War II

In 1939, in anticipation of the coming war and following traditions developed in World War I, the Army established a Personnel Testing Section. The National Research Council created an Emergency Committee on Psychology, whose focus was on personnel testing and selection. This was followed in 1941 by the creation of the Army Air Force Aviation Psychology Program directed by John Flanagan (Driskell & Olmstead, 1989). The purpose of this program was to aid in the selection and training of aircrew. (For comparable efforts in Naval Aviation, see Petho, 1993.)

Any reader old enough to have been part of the 13 million personnel serving in the war will remember the various tests that were administered immediately after induction, particularly the Army General Classification Test (AGCT) and the mechanical aptitude and radio code tests. Because these tests proved useful (they directed the initial fortunes of the recruit), the Department of Defense developed the Armed Forces Qualification Test (AFQT), a test of mental ability, which was adopted by all services by 1950. In more recent years (1974), the AFQT was transformed into the Armed Services Vocational Aptitude Battery (ASVAB).

So much was reminiscent of World War I and, if it had not been accompanied by other activities, would be of little interest to HFE history. The war saw an exponential leap in technology: more highly advanced aircraft requiring complex physical and mental skills; radar and photographic systems that presented information in new ways that required special perceptual skills; sonar, which required of its watchstanders exceptional pitch discrimination; and so on.

Because this was total war, involving great masses of men and women, it was no longer possible to adopt the Tayloristic principle of selecting a few specialized individuals to match a preexisting job. The physical characteristics of the equipment now had to be designed to take advantage of human capabilities and avoid the negative effects of human limitations. Obviously such a sea change in philosophy did not occur over night; its first and logical manifestation was in research to determine the human capabilities and limitations that had to be accommodated. This work took advantage of research habits developed in aeromedical research between the wars, when the limits of aircrew tolerance to environmental extremes were established. The aeromedical laboratories of the interwar period served as a model of how the military could utilize behavioral specialists.

An outstanding example of the kind of work that was done is the now classic study by Fitts and Jones (1947), who studied the most effective configuration of control knobs for use in developing aircraft cockpits. There are two points relative to this example: The system units that were studied were at the component level, and the researchers who entered the military were experimental psychologists who adapted their laboratory techniques to applied problems. This latter aspect, which was completely logical and natural, had implications for the future development of HFE research methodology.

Early studies of signal discrimination were directed at auditory capabilities in sonar (National Research Council, 1949). Similar research was performed to determine the visual capabilities needed to detect targets on radar. The aim was to make controls and displays easier for operators to perform more efficiently.

Work at the Aeromedical Research Unit, Wright Field, involved the determination of human tolerance limits for high-altitude bailout, automatic parachute operating devices, cabin pressurization schedules, breathing equipment, G-suits, and airborne evacuation facilities (see Dempsey, 1985). In cataloging these areas of research, it is necessary to point out that they required more than *pure* research; if equipment were developed and/or evaluated as a result of this research, it forced psychologists to work closely with design engineers to make practical use of the HFE research. Slowly, but ineluctably, as a result of the enforced intimacy with engineers, applied experimental psychology (the title of the first text on the new discipline; Chapanis, Garner, & Morgan, 1949) was transitioning to HFE.

Immediately after the war, the military attempted to summarize what had been learned from research performed during the war. The Army Air Force published 19 volumes that emphasized personnel selection and testing. The ones of special interest to the historian of HFE were volume 4, *Apparatus Tests*, volume 8, *Psychological Research on Pilot Training*, and volume 19, *Psychological Research on Equipment Design*, edited by Paul Fitts (1947), which was the first significant publication on what became HFE. In 1949, a second HFE publication was *Human Factors in Undersea Warfare* (National Research Council, 1949). Other summaries of wartime aviation research include Viteles (1945a, 1945b).

The Office of Naval Research sponsored a text on Human Engineering by Chapanis, Garner, Morgan, and Sanford (1947), which was later developed into the first text for describing the new discipline (Chapanis, Garner, & Morgan, 1949) for general use. The topics of the 1947 text listed subjects familiar to any reader of an introductory HFE text (e.g., psychophysics, design of experiments, the working environment, equipment arrangement, etc.). A more recent text by Chapanis (1996) contains the same type of material, brought up to date, of course, by new research results. It was only gradually that the first generation of HFE personnel realized that they were in a new (nonpsychological) discipline. There was some debate about this, as there still is.

Post-World War II (Modern)

This period covers approximately 20 years—from 1945 to 1965. It includes the activities of the founding fathers (e.g., Fitts, Chapanis, Small; see Taylor, 1994, for their biographies) and the first generation (including the author) who followed them (see also chap. 5).

The beginning of the cold war between Soviet expansionism and Western resistance fueled a major expansion of Department of Defense (DOD)-supported research laboratories; some were devoted exclusively to human performance research, whereas others included that research as a major division in an engineering organization. One might say that the immediate postwar environment was particularly hospitable to government-supported research. The cold war was not solely responsible for this. Technological development begun in World War II continued, raising human performance questions (among others) that demanded continuing research.

This meant that laboratories established during the war expanded. For example, during the war, the University of California Division of War Research established a laboratory in San Diego. Later this laboratory became the U.S. Navy Electronics Laboratory. Through expansions and reorganizations, it evolved into the Naval Ocean Systems Center and, finally, into its present organization—the Naval Research and Development Center. In 1953, the Department of the Army established its Human Engineering Laboratory, now the Human Research Laboratory. In fact, each of the services either developed human performance research laboratories during the war or shortly thereafter. The Navy Electronics Laboratory's Human Factors Division consisted of three branches specializing in the psychophysics, human engineering, and training of sonar devices and sonar men. The first branch determined what human auditory and visual capabilities were needed, the second considered how these could be utilized in machine form, and the third examined how personnel could best be trained to operate these machines. In his excellent book, Parsons (1972) listed 43 individual laboratories and projects and described in detail the most important studies that were performed. Many of these institutions continue to evolve, providing valuable human factors research to the community.

Almost all human factors research during and immediately following the war was military sponsored. Universities were granted large sums to conduct basic and applied research (e.g., the Laboratory of Aviation Psychology at Ohio State University). Other so-called *think tanks,* like the System Development Corporation in Los Angeles and the RAND Corporation, which split off from it, were established and funded by the military. During the war, research had concentrated on smaller equipment components like individual controls and displays, whereas the new studies performed by the laboratories embraced larger equipment units such as an entire workstation or an entire system (which corresponded in point of time to acceptance by the discipline of a new concept, the "system" concept).

Some of the major psychologists in World War II continued their work. Paul Fitts remained as Chief of the Psychology Branch of the Aero Medical Laboratory until 1949. Melton and Bray built the Air Force Personnel and Training Research Center (AFPTRC), commonly called "Afpatric," into an enormous organization employing hundreds of specialists in Texas and Colorado. Human engineering research and applications were centered in the East in the Naval Electronics Laboratory at Bolling Field under Franklin V. Taylor and Henry P. Birmingham. Several university research laboratories specialized, under contract to the government, in aviation problems. The Aviation Psychology Laboratory at the University of Illinois was founded in January 1946 by another of the HFE founding fathers, Alexander C. Williams. Ohio State University's Aviation Psychology Laboratory was also opened by Fitts in 1949. Research in aviation psychology was also conducted at the University of Pennsylvania and Purdue University. Around the early 1950s, Fitts and later George Briggs began research into air traffic control at Ohio State. Later there were reorganizations of laboratories and closures. Another, although smaller, organization was the Human Resources Research Office (HumRRO). The topics that were researched included the traditional selection and training areas, as well as maintenance, continuing operations, and so on. AFPTRC published hundreds of studies that were catalogued in the Tufts University Human Engineering volumes sponsored by the Office of Naval Research. AFPTRC endured until the mid-1960s and was then closed down probably because of budgetary cutbacks and possibly because those who had funded it became disillusioned with the practical outputs of the work. (This can be only a hypothesis; those who fund research do not often reveal their private thoughts to others.) HumRRO has maintained itself to the present day.

As is seen in chapter 5, which describes the informal or experiential history of HFE, the relationship between DOD and HFE has always been a rocky one. Funders are not researchers, certainly not behavioral researchers; they seek concrete outputs for their money. Although it is possible to construct a theoretical and logical relationship between HFE research and the development of more efficient physical systems, the reality is likely to be less convincing.

At the same time, opportunities opened up in civilian industry (although the industry was in civilian hands, it was supported and dominated by the military, which was engaging in an unprecedented—except for the war years—buildup of military systems). Large organizations in aviation, such as North American, McDonnell Douglas, Martin Marietta, Boeing, and the Grumman Corporation (the number of companies would be too lengthy and boring to list), established human factors groups as part of their engineering organization, although sometimes in peripheral areas like reliability or logistics.

Electronics and communications were other areas in which HFE people found employment. An example is Bell Laboratories, which employed researchers in audition since 1925, but established a human factors group in 1946. This group advised designers on topics such as the layout of keys for telephone handsets.

The introduction of HFE to industry represented a major change in HFE. It meant HFE was no longer completely or primarily a research-oriented discipline. The interaction between HFE researchers and designers that was fostered in World War II now expanded into human factors groups that became integral elements of the system design team. The official mission of such groups was not primarily to perform research (although occasionally they did so, when necessary, or when they bid on a government contract), but to participate through advice to engineers in the design of equipments. This new mission produced a division in HFE between those who performed research and those who engaged in application work (primarily system development). This separation of functions has had significant effects on the discipline. What had formerly been the domain of those who performed basic research now had to incorporate (with some reluctance) the application of their work to the development of physical systems.

This division has produced some acrimony between researchers and appliers because the notion developed, not illogically, that research should be useful to application. Also not unnaturally, military sponsorship of behavioral research looked for concrete outputs of the work it supported. Researchers found that it is much easier to do research than to suggest how that research can be utilized.

In addition to expanded military laboratories and expanded opportunities in civilian industry, this period saw the coopting of academia into the development and testing of larger and more complex systems. For example, the Laboratory of Aviation Psychology of Ohio State University conducted studies on air traffic control under the direction of Paul Fitts. Think tanks devoted to the analysis and solution of military problems were established by DOD. Examples are the System Development Corporation and the MITRE Corporation. Even when a formal laboratory was not established, the government—through agencies like the Human Engineering Division at Wright-Patterson AFB or the Army's Behavioral Sciences Research Laboratory (BSRL)—let contracts for human performance research that were awarded to departments of psychology and industrial engineering in universities up and down the country.

To bid on these contracts, private companies were formed, like the American Institute for Research under the directorship of John Flanagan. These employed numbers of HFE professionals. This was a research activity conducted solely for money (although carried on in some instances on a nonprofit basis)—a phenomenon that could not have been imagined prior to the war. A detailed review is presented by Casey (1997). The maturation of the discipline was manifested by this contract research. During the war, there had been no time to think about the human–technology relationship in more fundamental terms. Government-sponsored research provided that opportunity. For example, task analysis (a major methodological tool) was developed by Miller under contract in 1953.

All of these activities expanded the number of HFE professionals from a small cadre during the war to at least 5,000 professionals (at present), almost all of whom have advanced degrees (M.A./M.S. and Ph.D.). The discipline drew into itself psychologists as well as those with training in industrial and other forms of engineering, those with a physiological or safety background, and so on. The connection with engineering that had only been vaguely foreseen in the war was now firm and exerted its own pressures on the discipline.

1965 to the Present (Postmodernism)

The postmodern period has seen a maturation of the discipline. The number of professionals increased to the point that the HFES now has a membership of approximately 5,000—a far cry from the initial membership of 60 in 1957. The number of universities offering graduate programs in HFE has increased significantly.

Technology has shunted the discipline in new directions. The development of the computer and the tremendous expansion of computer applications to technology have created a new speciality field for HFE. Boehm-Davis (1994) indicated that, during the 1960s, HFE contributions to computer systems were largely limited to the traditional focus on interface hardware (e.g., computer terminal keyboards). Serious empirical research in software areas did not become significant until after 1970, when the personal computer (PC) was developed. Use of the PC by the general public brought with it behavioral problems that stimulated a great deal of research, some of which was directed at the problem of making the device user-friendly (see Meister, 1991). Software designed to facilitate use of the PC by the unskilled public resulted in the development of graphics packages, icons, windows, pull-down menus, and mouse interactive systems.

HFE research in this area blossomed accordingly. Nickerson (1992) reviewed *Human Factors* for 1965, 1975, and 1985 and found that, of 58 articles published in 1965, only 1 dealt with computers. In 1975, only 2 of 62 articles dealt with the human–computer interface (HCI). However, in 1985, one third of the 55 articles dealt with this subject. International meetings of those interested in the area proliferated, as did books. Much of the attention paid to the HCI emphasized relatively molecular questions dealing with, for example, screen layout and labeling (Williges & Williges, 1984).

The effect of increasing automatization on system performance was also receiving attention, although not on the scale of HCI. This was particularly noted in the field of aviation, as represented by the term *the glass cockpit* (referring to the development of integrated computerized, graphics type displays). It was pointed out that the role of the human in interaction with automated systems was changing and that this had profound implications for the design of systems and the measurement of human performance (Meister, 1996a).

New technological areas were opened to HFE research. One of the most significant was that of nuclear power plant (NPP) operations. This occurred in large part because of the highly publicized NPP failures in 1979 at the Three Mile Island facility in the United States and the Chernobyl facility in 1986 in the Soviet Union. In both situations, operator error played a significant and perhaps primary role (Kemeny, 1979; Medvedev, 1991).

The effect of these failures on the nuclear power industry in almost all countries was devastating. In the United States, there has been a halt to new construction of NPP; attention has shifted to retrofitting and upgrading present facilities, with particular focus on supporting the operator adequately. In this HFE has received much attention in the form of commissions to examine the HFE programs of the Nuclear Regulatory Commission and the establishment of research facilities (Electric Power Research Institute, Palo Alto, California) for the application of HFE principles to NPP design.

Within the American Department of Defense (DOD), greater attention has been paid to the contribution of HFE to the development of weapon systems. A critical report by the General Accounting Office (General Accounting Office, 1981) indicated that these systems often "cannot be adequately operated, maintained, or supported" (p. i). Perhaps as a result of criticisms such as these, during the 1980s, each of the services developed programs to ensure that adequate consideration is given during system development to the capabilities and limitations of the human. The Army developed MANPRINT (Manpower-Personnel Integration; see Booher, 1990), the Navy developed HSI (Human System Integration), the Air Force developed IMPACTS (Integrated Manpower, Personnel, and Comprehensive Training and Safety). These programs attempted to develop systematic, step-by-step procedures for the inclusion of HFE in the development process. How effective these programs have been in accomplishing their goals is not known. Because each of the programs required more knowledge of how humans interact with systems, additional research to answer the questions posed by the programs was spawned.

During this period, the number of books, published articles, and meeting reports proliferated. Specific numbers in relation to articles in *Human Factors* and *Proceedings* papers are presented in chapter 6. The liveliness of a discipline is reflected in its publishing activities and, by that criterion, HFE was (and is) burgeoning. What began as an application of experimental psychology in a war situation had been transformed into a new discipline—HFE. From a conceptual standpoint, the metamorphosis has not been an easy one. In chapter 5, the reader sees that even at present many HFE specialists of the highest rank question what they are and where HFE is going. Nevertheless, although half of the HFES membership report that they consider themselves psychologists (Hendrick, 1996a), the break between the two disciplines is final.

The change from experimental psychology to HFE has been relatively smooth despite conceptual angst. What HFE did, at least as far as America is concerned and one suspects elsewhere, was to slide almost imperceptibly from psychology to HFE. The change has occurred because psychology has never been particularly interested in practical problems so there were no turf difficulties. As is seen in chapter 5, few knew that a new discipline was emerging. Few thought whether there were special problems involved in the new discipline at the foundation of the Human Factors Society in 1957; the impetus of developments in the field overwhelmed serious discussion of these issues. These new issues (who are we, what are we trying to accomplish, what are the implications of the foregoing) require a degree of self-consciousness that many HFE professionals apparently do not have.

One would think that a new discipline would require a new theoretical structure, but the old one (based on the stimulus, organism, and response), which we all learned as part of our elementary psychology classes, proved durable. This made the slide from psychology to HFE easier. As far as the author is concerned, the one significant and new (to HFE) concept was the system concept (Christensen, 1962), but that has not produced major new theory or any major modifications in the old experimental psychology methodology. The imperceptible slide made it easier for HFE specialists while masking serious problems that are still not being considered because the transformation is still not complete. Many HFE specialists have still not worked out in their own minds the implications of the new discipline.

Sheridan (1985) looked at the changes in HFE over time from a somewhat different standpoint. He approached HFE history in terms of what has happened over the years to the concept of the human–machine system and particularly the manual control of that system.

HISTORY OF HFE IN THE UNITED KINGDOM

There are no outstanding differences between HFE in the United Kingdom and the United States, certainly not from a conceptual standpoint, although there are obvious differences in scale between the two. The following description is as of 1969 taken from a paper by Cumming and Corkindale (1969).

HFE in the United Kingdom started in 1917 in World War I when the Department of Scientific and Industrial Research and the Medical Research Council were asked to investigate industrial conditions, particularly of munitions workers. As a result, the Industrial Fatigue Research Board was established to perform research on the topic. In 1929, the name was changed in accordance with an expanded scope of work (hours of work, training, accidents, lighting and ventilation, and design of machinery).

There was a parallel activity in aviation psychology. The main interest in aviation was, as in the case of the United States, in pilot selection and training. In 1921, the National Institute of Industrial Psychology was created with the director of the Cambridge Psychological Laboratory, C. S. Myers, as its head. The purpose of the agency was to apply research in psychology and physiology to industrial firms. During the 1920s and 1930s, interest in the study of industrial psychology declined perhaps because of the depression. At the beginning of 1939, the Flying Personnel Research Committee (FPRC) was formed, which indicated a revived interest in human factors. One of the first actions taken by the FPRC was to establish the Royal Air Force (RAF) Physiological Laboratory, which later became the RAF Institute of Aviation Medicine.

As in the United States, there was a relatively large group of scientists in the life sciences who were recruited to work on both military and civilian problems. Examples are to be found in studies by Mackworth (1950) on vigilance and Craik (1947, 1948) on manual control of systems. During the war, the FPRC issued some 600 reports covering every aspect of medical, psychological, and physiological aviation problems.

The Ergonomics Research Society was established in July 1949 at a meeting held by a small group of research workers at the Admiralty. Thus, the society was the earliest to be formed in the world. The history of the society has been described at some length by Edholm and Murrell (1973). As in the case of the United States, the field has always been regarded as interdisciplinary, with personnel with medicine, physiology, psychology, and engineering backgrounds.

Each of the three military groups has its own human factors group. The oldest is the RAF Institute of Aviation Medicine at Farnborough, which in 1969 had about 50 HFE professionals out of a staff of 200. The Army Personnel Research Establishment is also quartered at Farnborough; the Royal Navy has a small number of HFE personnel, but these are divided among various research groups.

The Ergonomics Research Society (ERS) publishes a flagship journal, *Ergonomics*, and a number of more specialized journals, *Applied Ergonomics* and *Work and Stress*. A review of *Ergonomics* would not suggest any significant differences in topic areas for HFE research in the United Kingdom. There has obviously been a great deal of reciprocal cross-fertilization between the United States and the United Kingdom, with American researchers presenting papers at ERS conferences and publishing in the United Kingdom and vice versa. A great deal of HFE research is proceeding at British universities, the most well known to American researchers being the Applied Psychology Research Unit at Cambridge, where a small group of researchers tackles both basic and applied problems over a wide field. Other outstanding universities for HFE training and research are the University of Technology at Loughborough and the University of Birmingham.

As of 1969, most of the nationalized industries such as the Electricity Generating Board, the National Coal Bureau, and the iron and steel industries had HFE research groups. Quite a number of HFE professionals in the United Kingdom are well known to Americans (e.g., Mackworth, Shackel, Singleton, Murrell, Corlett, and Wilson). Some of these are now retired, of course.

Many government departments have extramural university contracts, as well as their own inhouse programs. Some industrial firms sponsor university research in HFE, but this is less frequent than in the United States. One important difference in the United Kingdom as compared with the United States is the relative lack of consulting firms that support government and industry. In industry, the present centralized organization is thought to be adequate. This creates a problem for the student who wishes to intern at an appropriate research establishment. The usual UK method for entering HFE as a professional is to qualify in one of the basic sciences—physiology, psychology, medicine, or engineering—and then acquire either specialized training on the job or take postgraduate training in a university specializing in HFE, such as Loughborough, London, Liverpool, and Aston. According to Cumming and Corkindale, the content of the training provided is similar to that given in American universities, with perhaps less concentration on the military and somewhat more on industrial problems.

Although again the United States leads in the number of specialized textbooks published, the catalogue offered by Taylor & Francis, the premier HFE publishing house in the United Kingdom, is quite respectable. As of

1969, the demand for qualified HFE professionals exceeded the supply from British universities. As in the United States, there has been an attempt to widen the scope of HFE from a concentration on design factors to those in which the social environment plays a part. For example, the concept of *sociotechnical* systems originated in Britain (Emery & Trist, 1960).

RUSSIAN HFE

The reader may ask why it is important to know something about the history and concepts of Russian HFE. The reason for including this material is that the Western professional should be aware of other traditions. In addition, Russian HFE provides a contrast with Western concepts and thereby illustrates more sharply the salient characteristics of Western thinking. However, if the history and concepts of German, Japanese, French, or South American HFE are not included in this chapter, it is because Anglo-American concepts and methodology have generalized to these other countries.

The Development of Russian HFE

Predecessors. The conceptual predecessors of Russian ergonomics include Mendeleev, who discussed in 1880 the general notion of adapting machines to man, and Arendt, who in 1888 discussed the adaptation concept with relation to the development of aeronautics. In 1915, Rudnev raised questions about the development of a standard cockpit for aircraft; in 1928, Rosenberg, using anthropometric data, determined the requirements of a cockpit layout. It is significant that, in discussing the development of Russian ergonomics, Lomov and Bertone (1969), from whose article some of this material is taken and who unfortunately supplied no references in their article, also said nothing about the link between human factors and the two World Wars that were important in both Britain and the United States. These authors provide several early examples of what we today would call HFE: Gellerstein and Ittin in 1924 with regard to redesign of the Russian type face to enhance the movements of the type setter; Bernstein in 1929 with regard to redesign of the tram drivers' workstation; and Platonov and Mikhailovskii in 1934 with regard to adjustable chairs for auto workers. In 1937, Zimkin and Aeple used a tachistoscope to study information reception from aviation instruments.

Political Factors. Russian ergonomics has been significantly affected by the political changes that followed the October Revolution of 1917. In the 1920s and 1930s, there were many psychologists, some of whom specialized in work psychology and physiology. Following the Revolution, there was great pressure to introduce Marxist philosophy into psychology. One only has to read a translation of an ergonomics textbook by Russian writers to realize that all interpretations of behavioral phenomena were phrased with reference to Marxist philosophy. However, this did not deter the criticisms of the ideologues who considered psychology to be a bourgeois deviation from the Truth. In 1936, all work ceased, psychology and psychophysiology laboratories were closed, and many psychologists were imprisoned and even executed.

The need for industrial recovery following World War II and the initiation of the cold war between the West and the Soviet Union led to a revival of research in work psychology. From 1950 to 1958, Platonov taught the subject in the Department of Psychology at Moscow University. In 1962, the first national conference of philosophers, physiologists, and psychologists was convened, at which issues of work psychology were extensively discussed.

The first engineering psychology laboratory was established in 1959 at Leningrad University. The first Conference on Engineering Psychology in 1967 is considered by Russians to be the official beginning of Soviet ergonomics. In a curious parallel to the American experience in the two World Wars, because of the militarization of science and technology, ergonomics and engineering psychology initially evolved for military applications.

Efforts were made to substantiate the assumption that ergonomics is based on science. These efforts were critical because of the precarious political position of the discipline. Ergonomists justified their existence on the ground that military failures were caused by human errors—a justification not unknown in the West. They gathered examples of poor design of controls and displays, prepared normative data to establish standards for military equipment, and conducted experimental studies of military personnel work activity.

All of these efforts were profoundly influenced by Western European and American literature. For example, American standards such as MIL-STD 1472 were adopted. Many textbooks were translated (sometimes pirated) from the West during the period of 1970 to 1980, including books by this author, Chapanis, Woodson, Siegel and Wolf, and Sheridan and Ferrell. Some information on Russian concepts has become available to Westerners (e.g., Cole & Maltzman, 1969).

Although Russian ergonomics has been profoundly influenced by the West, it is not a carbon copy of that found in the West. Because mathematics and physics were quite sophisticated in the USSR, they had extensive influence on the development of ergonomics, with Soviet ergonomists devoting much time to developing mathematical methods to formalize ergonomics theory. However, the practical effect of all this effort was quite limited.

The Western reader of ergonomics theory will find much that is familiar in Russian writers, but also distinct differences between Russian and Western terminology. The concepts are much the same, but the terminology was changed to divorce the concepts from their Western predecessors (again, it can be surmised, from political motives, to protect the infant science from Marxist idealogues).

If one were to summarize the major differences between American and Russian ergonomics (at least in terms of theory), the differences are more in terms of degree than content. As is seen, the Western reader can recognize all the major Western concepts in Russian writing (however changed they are in terminology), but the emphasis given these concepts is different than in the West.

One important characteristic of Russian ergonomics thinking is that it is much more molecular and detailed than that of Western HFE. Russian engineering psychology (to be differentiated from ergonomics practice, as it is also in the West) attempts to describe behavior and performance exhaustively, in all its sensorimotor, perceptual, cognitive, motivational, and emotional aspects. Western ergonomists tend to emphasize the critical aspects of performance, whereas Russian engineering psychologists incorporate all aspects concurrently. For example, if they were asked to describe the behavior involved in throwing a switch, they would be concerned about the goal of the action, the operator's motivation, his or her perception of the switch, the image of the switch in the operator's consciousness, his or her strategy in throwing the switch, the response of the arm muscles, the feedback received by the operator, and whether the goal (throwing the switch) had been achieved. The Western phrase "the operator threw the switch" would not be an adequate description for Russian psychologists (at least in their writings; what they do in actual ergonomics analysis might be somewhat different).

Because Russian ergonomists are concerned about the totality of work activity, they are also more concerned than Westerners with context (which they term *situation* or *situational conditions*).

Russian ergonomists have been much influenced by cognitive psychology, philosophy, physiology, and cybernetics. As a consequence, they are much more concerned about the operator's inner state than Westerners are; the tie to philosophy can be shown in their use of the concept *will,* which in American thinking was discarded in the first two decades of the 20th century with the advent of Watson's behaviorism. Russian ergonomists accuse the West of being excessively behavioristic, but this is because they are still mired in the thinking of the 1920s, when behaviorism first appeared. The impact of physiology is suggested by the Russian tendency to refer overt behavior to the neurological and cerebral conditions that presumably underlie that behavior. The impact of cybernetics is shown by the manner in which they describe functions, as if these were written in software language.

Basic Concepts. The following material is derived from Bedny and Meister (1997). The theory of activity, which is the conceptual basis for Russian ergonomics, has an extensive history dating back to Vigotsky (1962) and his followers. Pavlov, whose influence in the 1920s throughout the world was tremendous, considered the basic unit of behavior to be the conditioned reflex. Vigotsky was the first Russian scientist to consider the analysis of human behavior from the activity point of view.

The concept of activity (*deytolnost*) plays a key role in Russian psychology and ergonomics. Although it translates to the Western concept of *behavior,* there are significant differences. The concept is defined (Bedny & Meister, 1997) as a coherent system of internal mental processes and external behavior and motivation that are combined and directed to achieve conscious goals.

This introduces the extremely important further concept of the goal, which is the image of desired results in the future. Goals must be differentiated from needs. The content of the work activity is not determined by needs, but by the goal to make a certain product or perform a certain task. The needs and motives that energize activity diverge in most cases from the concrete task goals that direct activity. The nature of the goal accepted by the operator determines other activity aspects.

The goal is both sensory-perceptual and cognitive. In the first instance, the goal is the image of the desired result in the future. To Russian theorists, imagery plays an important role in conceptual processes. Imagery as such does not have a comparable importance in Western thinking about the goal, or, indeed, about performance in general. The goal includes both sensory and conceptual components of future results and, in fact, represents to the operator a form of information.

The parameters of work activity are: the result (output) of the work; the method used to perform the work; the individual's style of work, which is to a certain extent idiosyncratic; the work method prescribed by instructions; and the attributes permitting organization of the activity to achieve a goal. Goals and outcomes differ. In any event, all of these are organized in the form of information, and work activity as a whole is viewed, as is the goal, as a form of information.

This view of information in the Russian theory of activity suggests that information is the major parameter of *self-regulation* (another major concept), which permits the operator during task performance to compare the desired goal with actual outcomes and adjust behavior when there is a discrepancy. The resemblance of this process to Western concepts of feedback is notable; the two concepts appear to be identical.

The first to develop the principle of self-regulation were Anokhin (1935) and Bernshtein (1935). The basis of their principle is the concept of self-regulating units of activity (Anokhin, 1955). According to Anokhin, the neurophysiological basis of an activity is a functional system that can be considered a dynamic organization. The organization selectively integrates different central and peripheral neural nechanisms; their interaction permits the achievement of the desired result. A functional system is considered a closed loop, providing continuous feedback information about the success of performance.

In the Russian conceptual system, one can identify two aspects: (a) a drive toward a continuing effort to adapt or modify the course of the activity to the requirements imposed by the goal; and (b) continuing information reception, processing, and decision making. The latter is, of course, equivalent to the Western orientation toward information in behavior. Self-regulation involves updating information during the performance of a task, deciding whether to attempt an action (part of the task), deciding on a program of action, and then performing it.

Any unit of behavior can be considered as a self-regulative subsystem. The basic unit of activity is an action directed to achieve a conscious intermediate goal. All actions have a loop structure. The starting point of any action is the moment when the goal is formulated or accepted. The end of the action occurs when the results of the action are evaluated. This permits a continuing flow of activity divided into individual units delimited by the goals of the activity (intermediate and terminal) and the evaluation of the action's outcomes.

As in Western HFE, the task is the basic component of activity. All work activity is a problem-solving process in which the task is a situation requiring achievement of a goal. The process of achieving that goal represents a problem to the operator that must be solved.

Activity is performed by motor and mental actions. Mental processes involve direct-connection action (e.g., sensory perception) and information-transforming actions. The latter involves diagnosis of a situation, hypothesis formulation, development of decision alternatives, selection of a decision, and evaluation of the effects of implementing the decision. None of this will be unfamiliar to Western readers.

There are two types of motives: sense-formative and situational. The former are relatively stable, persistent, related to personality, and determine the individual's general motivational level. Situational motives are related to ongoing task solution. Sense-formative motives are intimately connected with needs and emotions.

The question of goal formation is a fundamental problem of motivation. A distinction is made between meaning and sense; the former is cognitive and general and the latter more personal and emotional. The goal of activity must be differentiated from the Western concept of goal. Goals can be overall or partial.

All psychic phenomena, including task-determined behavior, which appear to the individual as instantaneous, actually occur in short but measurable time periods and can be represented as a series of subprocesses.

All psychic phenomena are implemented through a series of *function blocks.* The function block is not something physical or an observable process. It is a construct that is inferred as a result of certain chronological (time measurement) experiments and the qualitative analysis that forms a significant part of Russian analysis of work activity. Precisely what goes on within the block is unclear to the Russians. At the same time, although the function block is a theoretical contruct, Russians view it as an actual functional system—as a set of mechanisms that perform individual functions.

All the function blocks are interconnected by feedforward and feedback loops; they serve as self-regulative mechanisms. Context is extremely important in influencing the function blocks, their degree of development, and how they perform. The content of a function block is not always the same and how it is utilized varies according to the individual situation.

Although Russian theorists may not know precisely what is going on inside the block, they feel it is important to know how much time is spent processing information within the block. This accounts for the extraordinary emphasis placed on time measurement in Russian psychology. Presumably duration of functional activity provides clues as to what is being done within the block. Although they consider the function block as a construct, they also tend to think of it as having physical characteristics, such as an entrance and an exit. Knowing what has happened at the entrance and exit and the duration of what has gone on within the block, they can then hypothesize about the supposed mechanisms within the function block. This methodology (a *black box* approach) is acceptable to them because it is used in other sciences, such as physics. The duration and character of the interconnections among function blocks is felt to have both theoretical and practical value.

Units of activity are not cast in concrete. They change over time and transform into other units. For example, in the process of mastering a skill, the importance of some operations may change. During the automatization of a skill, the importance of feedback will decrease because the operator needs to pay less conscious attention to the regulative effect of feedback. From that standpoint, regulation becomes less of a conscious process. An element of activity that is directed to achieve a partial goal is an *action.* The image of a future outcome becomes a goal only when linked to motivation. The relationship of motive to goal provides direction to self-regulation.

All psychic phenomena, including task-determined behavior, which appear to the individual as instantaneous, actually occur in short but measurable time periods and can be represented as a series of subprocesses. Russian psychologists feel that the Western stimulus–response paradigm can be employed only for analysis of involuntary reactions—that it cannot be used to explicate conscious goals and directed actions. At the same time, they recognize that Western ergonomists have not ignored internal mental processes. Information processing and decision making do enter into the Western analysis of work behavior. Russian theorists trace these concepts back to Tolman (1932), who was the first to introduce intervening cognitive and inducing variables between the stimulus and the reaction.

The principal distinction between the Russian activity approach and the behavioral approach is the existence of the conscious goal, which determines the specificity of the selection of information and influences the strategy of its attainment. A person does not react to the stimulus or simply process information, but actively performs in a given situation based on the goal and existing motives.

The Russian goal is not necessarily the objective goal, as presumably specified by the work situation or the researcher. It is the goal as modified by the subject's analysis of the task (in which the goal is embedded). In complex situations, particularly those requiring creative activity, the goal may appear to the individual as being obscure or ambiguous, and therefore must be formulated at least in part by the operator.

Russian HFE in Practice. The preceding section dealt with the fundamental behavioral theory underlying Russian practice. If Russian practice (e.g., analysis of system requirements, design of the system, comparative evaluation of alternative system configurations) is anything like Western practice, the influence of fundamental behavioral theory on practice should be relatively small. That is, practical methodology recognizes that there is a fundamental theory, but that methodology is largely independent of theory, and this is what one also finds in Russian practice.

One of the peculiarities of Russian ergonomics is that its professionals are much more concerned about temporal factors in relation to operator performance than are Western ergonomists. Russians use time as a means of evaluating the system as well as studying information processing.

The detailed analysis by Russian psychologists of microactions leads to an interest in predetermined time systems such as MTM-1 (Barnes, 1980), although not all Russian theorists utilize MTM-1. In work activity, the index of time emerges as one of the most important criteria of work productivity and efficiency. Failure to function within time limits is viewed as the failure of the person in the human–machine system. Russian ergonomists view time as one of the more objective and easiest ways to measure a person's performance. Time not only reflects the distinguishing features of external behavior, but also the specifics of the internal psychic process.

Design analysis begins with the development of algorithms to describe the system to be designed. An algorithm is a general principle, usually in equation form, that contains variables whose values, when changed, allow the algorithm to be applied to a number of situations involving those variables. Russian algorithms are highly proceduralized methods (based on logical rules) that can be used to describe how system components interrelate. An algorithm can also be presented as a table, symbolic description, or graph.

Although Western behavioral algorithms are usually based on generalizations from human performance, some Russian algorithms derive from logic; whether based on logic or human performance, Russian algorithms function at a more microlevel than do Western ones because major units of analysis are actions and operations. However, some Russian algorithms tend to summarize complex phenomena like stress in a short-hand and, hence, overly simplistic manner. Russian ergonomists have used algorithms for descriptive, analytical, evaluational purposes and as a design aid.

Before creating the algorithm, the analyst develops a verbal description of the task and its goal, stimuli for task performance, task conditions, and temporal relationships of subtasks. Western readers will note that this verbal description of the task is the conventional Western task analysis, although as performed by Russian ergonomists one would expect their verbal description to be much more detailed because they use different units of activity analysis.

The task description is then divided into elementary actions and logical conditions and are designated by special symbols (e.g., O_x, where O means operator and x means perceiving information). The symbol O_x is associated with the execution of components of activity, such as moving a gear or rotating a wheel. Logical conditions, which are designated by using 1, have two values—0 or 1. The symbol for a logical condition must include an associated vertical arrow with a number on top. The syntax of this system is based on a system of arrows and superscripted numbers.

The algorithm for even simple tasks can become quite complex because of the need to decompose the task into highly molecular elements. For example, the helmsman's task of keeping a ship on a specific course is broken down to the man's perception of a compass reading, turning the wheel either to the left or right; even the physical action of the rudder is described. The algorithm then is a symbolic description of the task at a molecular level comparable to the MTM elements used in industrial engineering (e.g., moving the hand upward, moving it downward, flexing the wrist, etc). The algorithm is then transformed into a formula.

Where the task includes many actions and the business of developing the algorithm becomes quite difficult, the algorithm can be replaced by a block scheme, which is similar to the flow charts developed by computer programmers. Zarakovsky (1966) used the algorithm to make a quantitative evaluation of the complexity of task performance. This involves the calculation of different numbers and types of elements in the algorithm.

In a somewhat similar manner, Galaktionov (1978) used graph theory as an analytical method to determine the information to be displayed in the control room and which controls are required by the operator. Graph theory studies the interrelationships among objects. Objects may be represented abstractly as a set of points, and the relationship between the various objects may be represented by lines connecting them.

The graph is a geometric figure consisting of points (nodes) and lines (arcs) that connect these points. The arcs indicate the frequency of interconnections. Based on the graph description, one can evaluate the probabilistic characteristics of activity structure. The graph theory procedure, which has some similarities to Western *fishbone* or cause-and-effect diagrams, is quite cumbersome as applied to behavioral design.

Distinguishing Characteristics of Russian HFE. The Russian concept of work activity is largely a motivational one, in which goals play a large part. Western concepts of the goal suggest that in most work activity the goal is quite evident to the operator, whereas in Russian thinking, which is determined by a problem-solving orientation to work, the goal may or may not be apparent; even when it is, the operator examines that goal and then develops a sense of what the goal is and a strategy about how to accomplish it. This applies even when a task is a psychophysical one.

The operations performed to accomplish that goal are implemented by 19 function blocks, which are roughly similar to Western functions. These function blocks describe stages in the performance of an action. What the Russians have done is to take each action stage and translate it into a function. For example, if there is a perceptual action to be performed as part of a task, there must also be a perceptual function block; if information has to be integrated, there is an information integration block; and so on.

Another important aspect of Russian conceptualization is the notion of self-regulation, which is a sort of homeostatic process focusing heavily on feedback that constantly makes adjustments to the work activity to bring it into accord with the known, sensed, or analyzed goal. The Russian conceptual structure is complex and so is their design methodology. The emphasis is on decomposition of task performance into extremely molecular elements and the analysis of these elements in terms of symbolic algorithms.

From a conceptual standpoint, they insist on analyzing phenomena that Westerners would accept on a heuristic basis. For example, the goal is fundamental to their explanation of work activity. Westerners would accept a clearly defined and explicit goal for what it is, whereas Russians require a goal-formulation and acceptance process even for psychophysiological functions. The Russian concept structure is intensively subjective and has strong links to philosophy, as in the case of the will. Everything is seen through the eyes of the subject so that the so-called *objective situation* is not really objective; it is changed by the subject's cognition and motivation. The Western ergonomist can find this acceptable in ambiguous situations, but might think that Russians carry this to a much greater extent than is warranted.

Russians are preoccupied with time presumably because it can be measured precisely, whereas error cannot. Russians use their algorithmic methodology and temporal units to evaluate the complexity of alternative system configurations. The molecular nature of their methodology could appear to make it difficult for them to deal with complex cognitive behaviors. Indeed, their ergonomics experience has largely been applied to industrial processes for which MTM analysis (repetitive, simple perceptual-motor actions) is quite appropriate. However, this can lead to what Western ergonomists might consider absurdities, such as the measurement of cognitive activity by counting eye movements.

Russian ergonomists have been much impressed by mathematics, and there is a tendency to mathematicize concepts. This leads to a situation in which a complex concept like workload, for example, is expressed in a mathematical (but somewhat simplistic) equation. Unfortunately, many such equations are not operationally defined. If challenged on this, Russians would consider operational definition as behaviorism, which they vehemently reject because they consider it to be based on animal, not human, phenomena. Motivation plays a much greater role in the Russian conceptual structure than in the Western one, although Western theorists are quite aware that motivation affects perception, which affects action.

It is also difficult to see how the Russians can incorporate concepts such as goal and motivation in the design of equipment, although they would probably say that both possess information that can be provided for in the appropriate design of machine displays. It must also be recognized that most Western theoretical concepts cannot be directly included in machine design, although the task analytic processes that precede design directly reflect that conceptual structure. Russian ergonomists probably consider their Western colleagues somewhat *slapdash* because the latter are prone to the use of heuristics to speed up the analytic and design process.

Finally, if one were to attempt to summarize the differences between Russian and Western ergonomics, one might say that there are some peculiarities in the former (such as emphasis on time, molecular analysis), but that many of the Russian concepts are the same as Western ones but phrased in a distinctive terminology. The differences then are in many cases only those of degree and emphasis, rather than substantive. However, the information we have is rather sparse, and the picture may change if Russian ergonomists make more of their material available to us.

APPENDIX: ROOTS AND ROOTERS[1]

Military Psychology in World War I, 1917–1921

The roots seem to have begun on April 6, 1917—the day the United States entered World War I. That date can also be taken to mark the beginning of U.S. military psychology. As Uhlaner (1968) summarized the events crucial to the involvement of psychology in the war effort,

> Military psychology, in fact, got its start on April 6, 1917. On that day, a meeting of experimental psychologists under the auspices of the American Psychological Association was being held in Emerson Hall of Harvard University. In the midst of this meeting, a messenger burst into the session chambers with the grave announcement that our country had just entered the conflict. Then and there, Dr. (Robert M.) Yerkes and a small group of forward looking psychologists put in motion most energetically a series of actions, including letters that same day to members of the Council of the American Psychological Association and to the National Research Council outlining what psychology could do for the national defense effort.
>
> In the spring of 1917, the first United States military psychological effort had its inception under Captain Yerkes of the Sanitary Corps, starting with the problem of enlisted classifications.

Psychology mobilized quite rapidly. On April 21–22, 1917, the APA Council voted that the APA president "be instructed to appoint committees from the membership of the American Psychological Association to render to the Government of the United States all possible assistance in connection with psychological problems arising in the military emergency" (Yerkes, 1921, p. 8).

Between this beginning in the spring of 1917 and the end of January 1919, American psychologists working with and in the U.S. Army's psychological services:

> Developed two group mental tests based partly on the earlier work of Otis—the Army Alpha (for the classification of English-language literates) and the Army Beta (for illiterates)—and determined their reliabilities (r's = .95); their validities (e.g., with the Stanford-Binet, Alpha r = .80 to .90, Beta r = .73); and their intercorrelations (Alpha with Beta, r = .80; see Anon., 1919, p. 225);
>
> Organized mental testing . . . for personnel classification in 35 army training camps;
>
> Trained 100 officers and more than 300 enlisted men as examiners;

[1]Written by Earl A. Alluisi, Science and Technology Division, Institute for Defense Analysis, Alexandria, VA. Presented at the annual meeting of the American Psychological Association, Washington, DC, August 1992. Reprinted with permission of Division 21, APA.

> . . . Tested a total of 1,726,966 men, including more than 42,000 officers, and
>
> Assisted in the work of the development battalion's work that resulted in the qualification for military assignment of more than 50% of the nearly 230,000 men who were assigned to those units and provided treatment, physical training, or instruction for varying periods of time in them.

There were other notable psychological activities and contributions as well. For example, research on psychological problems of aviation officially came under the direction of the Aviation Medical Research Board—a branch of the Air Medical Service that took charge of all medical, physiological, and psychological problems relating to the behavior of flyers (Henderson, 1918).

Yerkes (1919) gave an extended report on the range of activities. He identified 13 committees that were "active for varying periods during the military emergency" as well as 4 areas of special work of individual members of the Psychology Committee or its subcommittees. The 13 committees examined psychological literature examining aviation; special aptitudes; recreation; vision; military training and discipline; incapacity; emotional stability, fear, and self-control; propaganda; acoustic problems; tests for deception; and adaptation of psychological instruction to military education needs. The four areas of special work were scouts and observers; the gas mask; the Students' Army Training Corps; and special problems in learning, methods of instruction, and selecting for special tasks.

The work of the Committee on Psychological Problems of Aviation, including examination of aviation recruits, is especially relevant to engineering psychology's early roots. The committee was authorized by the APA Council and also became a subcommittee of the National Research Council. In the words of Yerkes (1919), it progressed as follows:

> In the summer of 1917 the committee was reorganized . . . George M. Stratton who had been working independently on tests for aviators at Rockwell Field, San Diego, was appointed chairman. Edward L. Thorndike was chosen as executive secretary, and John B. Watson, Warner Brown, Francis Maxfield, and H.D. McComas were added to the membership.
>
> From August 4, 1918, Thorndike served as chairman of the subcommittee on aviation. From the records of over two thousand flyers, Thorndike determined the relation between actual success in the work of a military aviator over the lines[,] and age, social status, intellectual ability, business achievement, athletic ability, and many other characteristics. (pp. 96, 99)

Work was also accomplished on topics other than selection and classification. For example, Yerkes (1919) reported that:

> To Major John B. Watson was assigned, in the summer of 1917, the task of organizing methods, other than medical, to be used by the examining boards for the selection of personnel. Watson also assisted in organizing a group of

> research psychologists to collaborate with physiologists and medical officers in the study of aviational problems at the Bureau of Mines, Washington.
>
> Special mention should be made of the Psychology Section of the Medical Research Laboratory at Hazelhurst Field, Mineola, Long Island, which developed from the work inaugurated in Washington by Watson and his associates. At this station, Major Knight Dunlap was primarily responsible for the development of a series of psychological tests to assist in determining the ability of candidates for the aviation service to withstand the effects of high altitudes. (pp. 97–98)

In general, the method employed called for the performance by the subject of a group of continuous tasks involving coordinated reactions during the gradual decrease of oxygen supply.

Among the early reports of results on the "various ways in which the effects of coordination and attention are manifested in different reactors," Dunlap (1918, p. 1393) suggested that a general methodology involving continuous tasks like those he used in his oxygen-deprivation experiments could be used in studies "of the problems of drug and fatigue action" (p. 1392)—clearly a prediction of the utility of performance measures for the psychopharmaceutical dose–effects curves that were to follow decades later.

Among the committee's other research efforts in which one can sense the early roots of engineering psychology are some examples that Yerkes (1919) reported:

> H.L. Eno and O.V. Fry developed apparatus for measuring the aviator's ability to point his plane quickly and accurately in a desired direction, as at an enemy plane. Major Watson was sent to Europe to gather statistics on the qualities essential to success as a military aviator. Dr. Parsons of the Navy received help from the committee in giving tests to every candidate for flying status in the naval air service. Parsons' study of the relation of the duration of nystagmus after rotation to flying ability yielded negative results, which are corroborative of Thorndike's findings, and supported by Dodge's analysis of nystagmus reactions. (p. 99)

Early in the war, Raymond Dodge, later Lieutenant Commander Dodge, was assigned the problem of devising a test for the selection of naval gun pointers (Yerkes, 1919, pp. 106–124). In solving the selection problem, he constructed and demonstrated an instrument to test gun pointers. The device gave a series of graphic records of the basic processes involved in training a gun on a moving target. When it was tested aboard two battleships, the best gun pointers gave the best records and the untrained recruits the worst. However, although the testees had only five trials each, Dodge had observed learning curves in many cases. This led him to devise a suitable training device. In the process of doing so, Dodge constructed what must be counted among the earliest and most successful of comprehensive part-task trainers (man-in-the-loop) simulation. According to Thorndike (1919):

> He (Dodge) studied the task of the gun-trainer and pointer, the situations and responses involved, the methods of testing their ability then in use, the men from whom selections would be made, and the practical conditions which any system of selection for this work must meet. He had the problem of imitating the apparent movements of the target which are caused by the rolling and pitching of the gun-platforms as a distant object would appear to a gun-pointer on a destroyer, a battleship or an armed merchantman. He solved this by moving the imitation target through an 84-phase series of combined sine curves at variable speeds by a simple set of eccentrics, motor-run. He had the problem of imitating the essentials of the control of the gun by the gun-pointer and of recording in a fuller and more convenient form the exact nature of the gunner's reactions in picking up the target, in getting on the bullseye, in keeping on, in firing when he was on, and in following through. He solved these by a simple graphic record showing all these reactions on a single line that could be accurately measured, or roughly estimated.
>
> Subsequently he made an apparatus that could be used not only to test a prospective gun-pointer's ability, but also to train both gun-trainers and firing gun-pointers four at a time. The demand for these instruments has been so great that sixty have been built by the Navy for use at shore training stations. The success of this led to further similar work, especially on the problem of the listener, the lookout and the fire control party. (pp. 57–58)

Dodge also began experimentation on the effects of gas-mask tenancy (or duration of wearing) of the standard-production mask. He served as the sole subject, but with multiple psychophysical types of performance and physiological measures. He reported:

> The most consistent and largest effect of gas-mask tenancy was decrease of visual acuity, an average of 20 per cent. Addition was slowed 7 per cent. Eye-reactions were longer by 9 per cent. Eye-movements were 7 per cent slower. In lesser degree the finger reactions, finger movements, and dynamometer strength tests were adversely affected, three, two, and one per cent, respectively.
>
> Of vastly greater importance than the fractional falling off in efficiency of the various processes was the effect of improperly made or improperly fitted head gear. Within one hour I had reached a degree of discomfort from an ill-fitting head gear, where in spite of experimental interest in the task, in spite of patriotic sentiment, and all the scientific pride I could muster, I could stand the punishment no longer and simply took the mask off. The extreme military importance of such a condition of mind seems clear. A properly constructed and properly fitted mask can be worn almost indefinitely, after adaptation.
>
> *It was officially reported that our study and the recommendations that grew directly out of it were of substantial help in developing the modern mask.* (Yerkes, 1919, pp. 122–123; italics added)

Later in the same article, Yerkes reported that Major Knight Dunlap was assigned "to continue and extend the investigations on tenancy of the gas mask initiated by Dodge" (p. 143) and that just before the end of the war

"Dunlap had perfected a procedure for determining the effects of different types of masks on the efficiency of the wearer" (p. 143).

The preceding is certainly sufficient evidence to support the conclusion that the early roots of engineering psychology in the United States go back at least as far as World War I. These successes joined those of mental testing for personnel classification and assignment and the selection of candidates for flying training and combat flying. Thorndike (1919) touched on what is probably the major contribution of the psychologists in World War I and a primary early root of U.S. engineering psychology: the application of the scientific method to the practical problems of the Army and Navy during time of war. These successes fostered the further development and growth of all branches of applied psychology. As Thorndike described it,

> The sciences dealing with human nature were brought to bear upon the problems forced upon America by the world war. Anthropology and psychology, economics and statistics, history, sociology and education, were put in service to improve our use of manpower, just as the physical and biological sciences were put in service to increase, economize and mobilize the nation's physical resources. (p. 53)

United States Military Hibernation, 1921–1939

After World War I, the Army reverted to its prewar personnel procedures, reinstituting "what was in effect an apprenticeship system of selection and assignment" (Uhlaner, 1968, p. 9). However, military psychology was not entirely eliminated from the Army and aviation psychology was continued under the aegis of the medics.

The Army appointed the Aviation Medical Research Board in 1917, and the board established the Air Service Medical Research Laboratory in 1918 and the School for Flight Surgeons in 1919. In November 1919, both were moved from Hazelhurst Field in Mineola, Long Island, to larger quarters at Mitchel Field, New York. Then, in November 1922, both the school and laboratory were combined into a single organization named the School of Aviation Medicine (SAM). In 1926, SAM was relocated from Mitchel Field to Brooks Air Force Base, Texas, to be near the large cadet training center in San Antonio. Then, in 1931, SAM was moved along with the Air Corps Training Center to Randolph Air Force Base, Texas. Aviation psychology played a supporting role at SAM—a role that emphasized the collection and analysis of performance data from pilots who were exposed (either in aircraft or the laboratory) to the physical and physiological stresses studied.

In the meantime, the early roots of engineering psychology were spreading in the aviation psychology research of other medical or life sciences research and development groups. For example, the Physiological Research Unit of the Air Corps Materiel Division was established at Wright Field near Dayton,

Ohio, in 1934, to provide medical engineering for the design and testing of equipment to protect pilots from the hazards of their flight environments. The unit tended to focus on determining the natural limitations of flyers as well as developing support equipment to extend those capabilities or adapting the design of the aircraft being developed to conform to human limitations. It became known as the Aero Medical Research Unit during the late 1930s, and its studies of cabin pressurization, rapid decompression, oxygen supply systems, and related life support equipment greatly influenced the design and operation of many military aircraft used throughout World War II.

In his report on "Military Psychology in the United States of America" before the First International Symposium on Military Psychology, held during July 1957 in Brussels, Belgium, Arthur W. Melton (1957) summarized the hibernation era:

> Between World War I and World War II, there was almost no interest of American psychologists in military problems, perhaps because there was almost no interest of the military in gaining the assistance of psychologists. Exceptions were the development of a new and improved Army General Classification Test by the Army Adjutant General and some research on psychomotor selection tests for aircraft pilots which was done under the aegis of the Surgeon General of the Army Air Corps. Then, with the clouds of World War II on the horizon, psychologists were recruited rapidly and in large numbers to do a great variety of research studies and technical applications for the benefit of military operations. (p. 741)

Military Psychology in World War II, 1939–1945

In the Army Air Forces (AAF), military aviation psychologists applied experimental psychology to aviation problems and developed entirely new fields of technology applications and research studies: human factors engineering and engineering psychology. Melton (1957) stated that the largest single program in World War II, that of the Army Air Forces, "was organized under medical auspices" (p. 741). That program was headed by John C. Flanagan.

According to Flanagan (1948), it was partly because of a shortage of flight surgeons that the chief of the Army Air Corps Medical Division recommended, in May 1941, that a Psychological Research Agency be established in the AAC Medical Division. The recommendation was approved on June 14, 1941; a month later, on July 15, 1941, John C. Flanagan (then associate director of the Cooperative Test Service, American Council on Education) was commissioned a major in the Specialist Reserve branch of the Officers' Reserve Corps. He went to work the next day.

Four years later, at the end of June 1945, the AAF Aviation Psychology Program included Colonels John Flanagan, Frank Geldard, J.P. Guilford, and Arthur W. Melton; about 200 other officers; 750 enlisted men; and 500 civilians whose accomplishments are recorded in the 19 volumes of the Army Air Forces, Aviation Psychology Program Research Reports published in 1947 and 1948.

Their research and development on aircrew selection and classification had blossomed, especially after early 1942, when the Army shifted that responsibility from the surgeon general to the newly created AAF Flying Training Command. Their work led to the development of two examining procedures, both of which were outstanding successes from scientific as well as applied viewpoints: (a) a 150-item screening examination called the AAF Qualifying Examination, the use of which rejected from between a quarter and a half of the more than 1 million aircrew applicants; and (b) the Air-Crew Classification Test Battery consisting of 20 tests—6 apparatus tests of coordination and speed of decision and 14 printed tests that measured intellectual aptitude and abilities, perception and visualization, and temperament and motivation.

Performance was recorded in terms of a nine-category standard score with a mean of 5.0 and standard deviation of about 2.0, called a *stanine*. Testees were assigned stanine scores for the various aircrew specialties based on weighted combinations of the results from various tests in the battery. More than 600,000 men took this comprehensive battery of tests. The stanines were found to have high predictive value, especially for pilot and navigator training success, and to be correlated significantly with measures of success in operational training and in combat.

The AAF Aviation Psychology Program was probably the taproot of engineering psychology or human factors engineering. Paul M. Fitts (1947a; 1947b), generally regarded as a founder of the field, edited *Psychological Research on Equipment Design*, number 19 in the series "Army Air Forces, Aviation Psychology Program Research Reports." In addition to that publication, the reports covered all aspects of the field, including programs on aviation psychology; classification programs; research problems and techniques; apparatus tests; printed classification tests; the AAF qualifying examination; motion picture testing; training for pilots, bombardiers, navigators, radar observers, flight engineers, and for flexible gunnery; AAF convalescent hospitals; operational training in the Continental Air Forces; research in the theaters of war; and records, analysis, and test procedures.

Thus, the progress of engineering psychology during World War II can be summarized as follows. During the early years, 1941–1944, numerous eminent psychologists were brought into the AAF to devise effective methods of aircrew selection and training. They succeeded in developing and implementing stanine scores, which were based on a battery of psychomotor and paper-and-pencil tests and used principally to select and classify candidates for pilot, navigator, and bombardier training.

During the later war years, 1944–1945, AAF psychologists emphasized criterion development, test validation, and training, with projects at Randolph, Mather, and Lowry Army air bases. Some psychologists from 1945 onward were engaged in the new area of engineering psychology or human engineering that Paul Fitts and his colleagues established in the Aero Medical Laboratory at Wright Field, near Dayton (now Wright–Patterson Air Force Base).

However, the AAF psychologists were not alone in the creation of the new discipline. Rather, they appear to have been part of a zeitgeist that led military psychologists to address issues regarding the design and operation of equipment. Essentially similar efforts had been taken from 1940 onward at the Applied Psychology Unit of Cambridge University in the United Kingdom under the direction of Sir Frederic Bartlett. Similar efforts took place in Germany, too, but on a small scale (Fitts, 1946). For the United States, Kappauf (1947) stated matters quite clearly:

> When the Army and Navy recruited psychologists early in the war, assistance was sought primarily in the areas of selection and training of personnel.
>
> At the same time a few research programs sought the services of psychologists to insure the more satisfactory design of some items of military equipment, to insure design which would take account of particular psychological and physiological characteristics of human operators. Typical of this work was that which was initiated in the design of dark adaptation goggles, sun scanning devices, and communications equipment. As the war progressed this phase of research in applied psychology assumed greater and greater importance and involved more and more types of equipment. The field developed as much *through the initiative of individual psychologists* as it did through specific service requests. (p. 83; italics added)

Most of those who were engaged in military psychology during World War II left the Armed Services after the war, but not without having made their impact both on the conduct of the war and the future directions of psychology through their contributions to testing, selection, and classification; training and training devices; and the design and operation of equipment.

Postwar Engineering Psychology, 1946–1949

The growth of engineering psychology in the United States during the years immediately following World War II—from 1946 to 1949—was phenomenal, at least in comparison with its progress during the war. There was reason for its slower growth during the war. As Grether (1968) observed,

> Although engineering psychology had its birth during World War II, the level of research effort in the United States was on a modest scale until after the

> end of hostilities in August 1945. This low level of effort was apparently deliberate, because it was recognized that only during a prolonged conflict could the benefits of such research be realized. Obviously, the time lag between initiation of engineering psychology research and the design, manufacture, and deployment of new or redesigned equipment is relatively long, often 5 years or more. Thus, during wartime it was more profitable for psychologists to concentrate on other types of research, such as selection and training, with faster payoff. (p. 774)

After the war, when the Armed Services were demobilized rapidly, inhouse engineering psychology groups actually grew. It was late in August 1945 when Fitts and a small group of psychologists were transferred to the AAF Aero Medical Laboratory. At about the same time, the Navy set up two engineering psychology units: one at the Naval Research Laboratory (Anacostia) under Franklin V. Taylor and the second at the Navy Special Devices Center (Port Washington, New York) under Leonard C. Mead. The next year, the Navy established a third unit at the Navy Electronics Laboratory (San Diego) under Arnold M. Small. Both services also provided contract support for engineering psychology research at universities (e.g., the University of California at Berkeley, Harvard University, Johns Hopkins University, and the University of Maryland, among others) and at the Psychological Corporation, where Jack Dunlap had set up a Biomechanics Division for engineering psychology research and development. That division later became Dunlap and Associates.

Publications that were important stimuli to the growth of engineering psychology also appeared during this period. The publications were in journals; for example, Kappauf's paper was the first of three invited papers in the March 1947 issue of the *American Psychologist*; the other two were by Taylor and Fitts on psychology at the National Research Laboratory (NRL) and in the AAF, respectively. Books were also published—books that served both as texts for students in colleges and universities and as reference works for researchers in laboratories and research and development organizations.

It was also during this period that engineering psychology achieved recognition—not only in the academic community, but also in the areas to which its findings were being applied. Successful applications were apparent in aircraft cockpits, combat information centers, fire-control systems, and sensor systems. Engineers in the airframe industry accepted, and often sought, the input of engineering psychologists for their designs.

There were still only a few engineering psychology laboratories in the Armed Services, but they were productive. The relatively small number of universities offering training degrees in the field was growing (this is the period in which Paul Fitts joined the faculty of The Ohio State University). The success of Dunlap and Associates also portended growth in engineering psychology contractor firms. Thus, the decade ended.

The Early Growth of Engineering Psychology, 1950–1960

Grether (1968) characterized the growth of engineering psychology during the decade from 1950 to 1960 as *explosive*, and his characterization is accurate. Much of the growth took place within the existing engineering psychology units of the services. To the existing units, the Army now added its own organization, which later developed into the U.S. Army Human Engineering Laboratories at Aberdeen Proving Ground, Maryland.

The greatest increase was the establishment, within the defense industries, of human factors engineering groups, usually consisting of a majority or plurality of psychologists. Kraft (1961) reported that although there were only two industrial human factors activities in 1951, there were 24 in 1956 and 157 in 1961.

Among the earliest of the independent contractors to seek and conduct human factors or engineering psychology research and development during the early 1950s, besides Dunlap and Associates, were Bob Sleight's Applied Psychology Corporation, Harry Older's Institute of Human Relations, and John Flanagan's American Institutes for Research.

Not surprisingly, toward the end of the decade, the growth pattern increased exponentially. The airframe and aerospace industries entered the picture by rapidly expanding their human factors activities. The reason for this is that the U.S. Air Force promulgated the personnel-subsystem concept, which required early attention to the manpower, personnel, training, and human factors domains—the personnel-subsystem aspects of the system being designed and developed. The Air Force contracts of the day contained formal requirements for reports on qualitative and quantitative personnel requirements information (QQPRI), the types of training and simulators or training equipment that would be needed, and the human factors engineering and tests that were planned. Thus, the Air Force began requiring contractors to prepare data and analyses based on system concepts and design that would help increase the efficiency and effectiveness of the personnel subsystem—the manning and training of the personnel who would be called on to operate and maintain the system.

Important, too, were the new books that reached the market and the chapters that appeared during the 1950s. Among the most influential in terms of its impact on both the nature and growth of the discipline was Paul Fitts' (1951) chapter in Steven's *Handbook of Experimental Psychology*. Several generations of engineering psychology graduate students and fledgling professionals studied that chapter, nearly to the point of rote memory, which is to say to a point just a little more intense than they studied all the other chapters of that monumental text.

The first edition of Ross McFarland's *Human Factors in Air Transportation* appeared in 1953. During the next year, Wesley Woodson's (1954)

Human Engineering Guide for Equipment Designers was published—the first of the several well-designed and readable guides or handbooks for human factors engineering practitioners that have appeared since then. Ernest McCormick's *Human Engineering* was published in 1957; it was the first textbook on the topic since Chapanis, Garner, and Morgan's (1949) *Applied Experimental Psychology*.

The decade saw the publication of many chapters and reports on the topic of, or topics in, engineering psychology (e.g., see the papers included in Wally Sinaiko's *Selected Papers of Human Factors in the Design and Use of Control Systems*, which appeared in 1961 and was widely used as an adjunct to the primary texts of courses in engineering psychology). The decade also saw the beginnings of growth in the publication of specialized texts and reports, such as Floyd and Welford's (1953) *Symposium of Fatigue*; Finch and Cameron's (1958) *Air Force Human Engineering, Personnel, and Training Research*; and Ray, Martin, and Alluisi's (1961) *Human Performance as a Function of the Work-Rest Cycle*.

The third indication of growth during the 1950s was the formation of two societies to represent the profession and science of human factors engineering and engineering psychology. In the Far West, especially in Southern California in the area from Los Angeles south to San Diego—the area in which many of the airframe and aerospace industries were located—persons interested in the new emphasis on human factors began meeting for technical discussions. These meetings led to the formation of the Human Factors Society (HFS) in 1957, which in turn produced the journal *Human Factors*, intended as an archival journal for the publication of research findings, and the *HFS Bulletin*, a newsletter that also carries papers of a substantive nature pertaining principally to professional affairs.

From the beginning, the HFS was a multidisciplinary organization that accepted as members anyone who worked or even expressed interest in any of the multiple areas of human factors—areas dealing with considerations of human factors as they influence or should influence the design and operation of systems, including aspects such as human–machine interfaces, product and workplace designs, and safety. Although at its beginning between a third and a half of its members were psychologists, the HFS has never been viewed as a psychological society, nor has it indicated any desire to be so perceived. However, the situation was different on the East Coast. There, Franklin Taylor, Karl Kryter, and Harry Older organized Division 21, the Society of Engineering Psychologists—A Division of the American Psychological Association.

References

Anonymous. (1919). The measurement and utilization of brainpower in the Army. *Science, 49*, n.s., 221–226.

Chapanis, A., Garner, W. R. & Morgan, C. T. (1949). *Applied experimental psychology*. New York: Wiley.

Dunlap, K. (1918). Psychological observations and methods. *Journal of the American Medical Association, 71*, 1382–1400.

Finch, G., & Cameron, F. (Eds). (1958). *Air Force human engineering, personnel, and training research* (Publ. No. 516). Washington, DC: NAS-NRC.

Fitts, P. M. (1946). German applied psychology during World War II. *American Psychologist, 1*, 151–161.

Fitts, P. M. (1947a). Psychological research on equipment design in the AAF. *American Psychologist, 2*, 93–98.

Fitts, P. M. (Ed.). (1947b). *Psychological research on equipment design* (Report No. 19 in Army Air Forces, Aviation Psychology Program, Research Reports). Washington, DC: U.S. Government Printing Office.

Fitts, P. M. (1951). Engineering psychology and equipment design. In S. S. Stevens (Ed.), *Handbook of experimental psychology* (pp. 1287–1340). New York: Wiley.

Flanagan, J. C. (Ed.). (1948). The aviation psychology program in the Army Air Forces (Report No. 1 in *Army Air Forces, Aviation Psychology Program, Research Reports*). Washington, DC: U.S. Government Printing Office.

Floyd, W. F., & Welford, A. T. (Eds.). (1953). *Symposium on fatigue*. London: H. K. Lewis.

Grether, W. F. (1968). Engineering psychology in the United States. *American Psychologist, 23*, 743–751.

Henderson, Y. (Ed.). (1918). Medical studies in aviation. *Journal of the American Medical Association, 71*, 1382–1400.

Kappauf, W. E. (1947). History of psychological studies of the design and operation of equipment. *American Psychologist, 2*, 83–86.

Kraft, J. A. (1961). A 1961 compilation and brief history of human factors research in business and industry. *Human Factors, 4*, 253–283.

McCormick, E. (1957). *Human engineering*. New York: McGraw-Hill.

McFarland, R. (1953). *Human factors in air transportation*. New York: McGraw-Hill.

Melton, A. W. (1957). Military psychology in the United States of America. *American Psychologist, 12*, 740–746.

Panel on Psychology and Physiology, Committee on Undersea Warfare. (1949). *A survey report on human factors in undersea warfare*. Washington DC: National Research Council.

Ray, J. T., Martin, O. E., Jr., & Alluisi, E. A. (1961). *Human performance as a function of the work-rest cycle* (Publ. No. 882). Washington, DC: NAS-NRC.

Sinaiko, H. W. (1961). *Selected papers on human factors in the design and use of control systems*. New York: Dover.

Taylor, F. V. (1947). Psychology at the Naval Research Laboratory. *American Psychologist, 2*, 87–92.

Thorndike, E. L. (1919). Scientific personnel work in the Army. *Science, 49*, 53–61.

Tufts College & U.S. Naval Training Devices Center. (1949). *Handbook of human engineering data*. Medford: Author.

Uhlaner, J. E. (1968). *The research psychologist in the army—1917 to 1967* (Tech. Res. Rep. 1155). Arlington: U.S. Army Behavioral Science Research Laboratory.

Woodson, W. (1954). *Human engineering guide for equipment designers*. Berkeley: University of California Press.

Yerkes, R. M. (1919). Report of the psychology committee of the National Research Council. *Psychological Review, 26*, 83–149.

Yerkes, R. M. (Ed). (1921). *Psychological examining in the United States Army (Vol. 15, Memoirs of the National Academy of Sciences)*. Washington, DC: U.S. Government Printing Office.

5

The Informal History of HFE

The informal history consists of the memories and impressions of the people participating in the events involving HFE, the problems they recognized, and their opinions of various subjects. This chapter relies heavily on the memoirs of those who responded to questions posed by the author. The 130 individuals solicited were those selected from the HFES society's 1995–1996 Directory who had received their final degree on or before 1965 (usually the PhD in psychology, occasionally the MA or MS, and in a few cases the BA). This would include the first generation of professionals. Forty-six responses, including those of the author, were received—a 35% return, which is quite respectable for mailed questionnaire surveys.

What is involved in these recollections is not only the history of a period but also something that can be termed the *concept structure* of the HFE professional. This concept structure, previously discussed in chapter 2, is composed of a more or less organized set of positive, negative, and neutral beliefs about elements of the discipline. This chapter describes history reflected through a prism of attitudes developed by experience over many years. Therefore, one should expect (and the reader will find) variations in the way in which these professionals, who are part of the first-generation cohort, view their discipline. Despite this, the reader may note certain common conceptual trends.

To provide supporting information, a review was also made of the early issues of the Society's monthly bulletins, which contained such materials as letters to the editor, reports of impending conferences, demographic data, occasional editorials, and short biographies. Most of this material provided context for the memoirs.

A number of conclusions can be drawn from the responses, despite inevitable variations:

1. World War II was responsible for the initiation of the discipline. The crucial factors impelling the first generation into HFE after WWII were in many cases sheer chance: after receiving an advanced degree in psychology, the need to secure a job, even if it was not in a university; the influence of classmates and faculty mentors; an interest in application rather than research; and the exposure of those who had served in the military to new equipment and operational situations.

2. Prior academic training was not relevant to the new discipline. Psychological training often provided an opportunity to secure entrance to the new field, but, except for scientific methodology, was of little further value and then mostly to researchers.

3. Awareness that a new discipline was developing occurred only progressively. The fact that the new discipline was not completely defined retarded this consciousness. One of the noteworthy features of the study—one that needs to be emphasized—is that the respondents felt that the discipline is *still* not completely defined.

4. The influence of government and the military in sponsoring the new discipline immediately after the war cannot be overestimated. Without this sponsorship, the discipline would have been still-born.

5. The initial relationship between HFE and engineering was often negative. Managers resented the need to consider behavioral factors in design because it imposed an additional burden on them.

6. HFE research did not make novel demands on HFE researchers, but application (system development) did. The often negative reactions to HFE noted in Conclusion 5—the need to justify and prove HFE recommendations, the need to develop a quasi-engineering viewpoint, and the requirement to learn about physical systems—all produced a change in attitude among HFE system development professionals that was different from the scholarly tradition to which most had been exposed. They developed a much more pragmatic and possibly an anti-intellectual attitude toward HFE.

7. The special HFE training that was gradually developed had, as its distinctive elements, acquaintance with the system development process, the problems involved in that process, and analytic methods employed in design. Measurement methodology was largely taken over from traditional psychological teaching.

8. Respondents felt that the scope of HFE had changed significantly from controls and displays to equipments and systems, and that it involved more complex functions like cognition, which initially had not been considered important. Acceptance of HFE by engineering and the general public has also increased.

9. HFE effectiveness in the early days depended on acceptance by engineering of HFE recommendations in design. This has led to an emphasis in HFE on publicizing successes (Hendrick, 1997b). Many of the changes in HFE, conceptually and professionally, have resulted from the interaction of HFE with engineering and system development. This applies primarily to application; the research aspect of HFE has remained under the conceptual control of its predecessor—psychology. This last has led to a disconnect between HFE research, which is primarily human-centered, and HFE application, which is much more equipment-centered.

10. The changes that the first generation sees are those of increasing complexity: systems rather than individual controls and displays; multidimensional performance rather than that involving single variables; and cognitive and emotional functions that now have to be considered along with simpler perceptual and motor functions.

11. Among the respondents are those who consider HFE as a distinctive discipline, those who see it as a form of either psychology or engineering, and those who see it as a multidisciplinary approach, which in effect rejects the concept of a discipline and sees HFE as only a way of attacking certain problems.

12. Those in the sample who were/are practitioners see research published in HFE journals as largely irrelevant to practitioner problems.

13. The first generation sees a tendency toward greater acceptance of HFE by other disciplines, but there is also anxiety that HFE will be assimilated (disappear) into disciplines like industrial engineering.

The general conclusion one comes to from a review of these informal recollections is that the interaction of behavioral science with engineering has profoundly changed the nature of the discipline and has given it its distinctive character. This change has been recognized primarily by practitioners; researchers are still, to an extraordinary extent, functioning in accordance with a psychological model. This disconnect has reduced HFE effectiveness. The new discipline is still incompletely defined. HFE specialists are still not certain of what they are dealing with in HFE.

METHODOLOGY

The 19 questions in Table 5.1 were developed to stimulate respondents' thinking. They were told in an introductory letter that they could write anything they wished in response to all, some, or none of the questions in as much detail as they wished. As it turned out, most respondents wrote to the suggested topics; a few ignored the questions completely. A list of those who participated in the project is provided in Appendix 5.1 to this chapter.

TABLE 5.1
Suggested Topic Questions for Consideration

(1) What impelled you to go into Human Factors as opposed to psychological teaching or research or some other activity?
(2) How relevant do you feel your graduate studies were to the new work? Was your training primarily in psychology, engineering, or human factors?
(3) Did you feel that you were going into a new discipline or was Human Factors just another branch of psychology or engineering? When did you feel that a new discipline was being created?
(4) What kind of work did you do in your early years in Human Factors: research, system development, teaching? Were you in government, industry, or at a university?
(5) Were you satisfied in the early days with the status of Human Factors vis-à-vis engineering? With the capabilities of the discipline to answer the research or system development questions that arose?
(6) Did you have any distinctive experiences as a Human Factors specialist in your first years in the discipline? Or later?
(7) If your primary work in Human Factors was in teaching, was there anything distinctive about what you taught in Human Factors—different, that is, from traditional psychological material?
(8) If you think about *then* as compared with *now*, are there any major differences in theory, research, methodology, or practice that come to your mind?
(9) If you began your work in industry, what was the reaction of engineers to the introduction of Human Factors? Have human engineering practices in system development changed significantly since you began? In what way?
(10) Leaving aside computerization, what major changes in theory, methodology, and practice do you find have occurred over the years?
(11) If you worked in industry in the early years, how productive do you think human engineering efforts were in affecting design and, if they were not, what were the causes? How does this aspect compare with the present?
(12) Are you satisfied with what Human Factors has accomplished and where we are at the present time?
(13) When you started your professional career, did you have any special hopes and/or expectations as to what Human Factors would become? Have these been achieved?
(14) How would you evaluate the adequacy of present training programs for Human Factors specialists? How would you evaluate current Human Factors theories, methodology, and so on?
(15) Are there any significant gaps in Human Factors theory, methodology, or practice that need to be filled? What are they?
(16) Do you think that the research published in the Human Factors journals has changed over the years and, if so, in what ways? Were you satisfied with the research published earlier, and are you satisfied with the research being published today?
(17) In your opinion, what makes Human Factors as a discipline different than applied psychology, or do you think of it in that way?
(18) How do you see the relationship between Human Factors and engineering?
(19) What changes do you anticipate in Human Factors/Ergonomics in the 21st century?

No information has been included about their background, but many of them were/are, in their time (to use baseball parlance), the *heavy hitters* of their profession; all of them are known personally or by reputation to the author. If the author was not aware of them, they were not solicited. In addition to this source of bias, there is no pretense that all the most important professionals of that time were canvassed because some of the most important, like Paul Fitts and Arnold Small, had died and others had broken all connection with the Society upon their retirement. With the exception of Dora Dougherty Strother, the sample is male; this reflects that, in the 1950 to 1965 period, the number of female HFE professionals was minuscule.

Another more subjective criterion was also applied: Respondents had the same kind of background as that of the author (i.e., in most cases, a behavioral or engineering degree, involvement in system development and/or system development research, with or without teaching experience). Many of the older respondents had served in World War II or the Korean War.

Almost everyone selected and responding was what one might call a *developmental ergonomist*, meaning that the major focus of their work experience related to system development, either directly in industry, government research, or teaching. The kinds of people who are involved in industrial ergonomics, with its emphasis on production processes and muscular trauma/rehabilitation, are not found in great number in the sample because they are not members of the Society or were too young to satisfy the chronological criterion. Probably both factors were responsible for their noninclusion. Obviously the sample is limited and it is limited to Americans, although there were at that time some distinguished foreign HFE personnel, such as Singleton of England (see chap. 4). If the reader asks whether this sample is truly representative of the early pioneers of the discipline, the answer is that they are probably fairly representative, based to a large extent on the author's own experience of those days.

As one would expect, with even a relatively small group (or perhaps because the sample *is* small), there is some variation in the conclusions they drew and their attitudes about events and factors that influenced early HFE. Moreover, those who are still practicing their profession cannot but be influenced by their more recent experiences. The variations are entirely understandable because the sample, although uniformly intelligent and observant, varied, as one would expect, in terms of personality, training, work background, and experiences. Regardless of this, there was much homogeneity in their responses, and that is quite satisfying.

Is 46 a large enough sample? No one can answer that question, but it should be pointed out that in the very early days the number of HFE professionals was much smaller than it is today (hundreds rather than thousands).

The answers and anecdotes that the sample provided have been analyzed in accordance with the 19 questions in Table 5.1. Specific illustrative responses have been extracted and noted by the respondent's surname ("X says thus and so . . ."). Comments were selected to illustrate a particular point made in the body of the chapter and have been edited, but only to summarize the response.

RESULTS

Entrance to HFE

There were several factors that impelled the first generation of HFE professionals to enter the profession. (It is necessary to make a distinction between *profession* and *discipline*. The discipline began, quite without the people engaged in it knowing this, in World War II. The profession represented the formalization of the discipline; its beginnings can be traced to the establishment of a formal society [The Human Factors Society of America[1]] in 1957.)

The major factor responsible for the development of the discipline was the war. Eminent academics in experimental psychology—people like Fitts, Christensen, Small, and Chapanis—were drawn from their classrooms to work at laboratories like the Aeromedical Laboratory at Wright Field in Dayton, Ohio. They considered that what they were doing in the war effort was the application of their experimental psychology research; it is almost certain that, without their knowledge, they were initiating a new discipline.

It was significant for those who came after them that they were psychologists by training. It would be natural for those who were psychologists to enter HFE in the 5 or 10 years after the war ended. This was because HFE appeared to all (including employers) to be a variant form of psychology. If a psychologist could not find an academic or research position, it would be natural for him (not women; there were almost no women in the first generation of HFE professionals) to look to the new industries like aerospace, which were employing psychologists, and the new research laboratories that had been established by the military.

Just as important as the entrance into war work of a handful of academics was the fact that many who were to follow them were now exposed to systems they had never encountered before. For example, Strother's strong interest in aviation created an auxiliary interest in cockpit design. Harris had a degree in industrial psychology; a dissertation on creativity in engineering was a motivating force. A significant number had been fighter pilots (Nicklas) and naval officers (M. Parsons)—not only in World War II, but immediately after. (For example, Wherry took psychological courses in school, but was

[1]Later simply the Human Factors Society and, in 1993, The Human Factors and Ergonomics Society. Henceforth referred to as the Society.

not particularly interested in psychology. He feels that his operational experience in the Navy was more influential in directing him toward HFE.) As a minor illustration, the author's first airplane ride (as a passenger) was in a B-17 bomber en route to a new station. Others who took a more active part in combat became sensitized or predisposed to work with equipment in a behavioral sense in some place other than a laboratory. In addition to these experiences, if they also had degrees in psychology or ambitions to work in some aspect of behavioral science, it was more rather than less probable that they would end up working in a more applied part of that field. (For example, S. Parsons reports that he was not particularly interested in academia and academic research.)

Others drifted into HFE because it provided opportunities that academia did not possess after the war. Economics was important. Woodson was without a job and knew Arnold Small who, as head of the Human Factors division at the Navy Electronics Laboratory in San Diego, was in a position to give him a job. Kurke wanted to get a PhD in experimental psychology, but needed a job. The author, who received his PhD in 1951, encountered the Korean War, which temporarily dried up slots on university faculties. The new research laboratories and the aircraft and electronic companies that had been developed during the war now provided an opportunity for people who could not make a living in academia.

In some cases, a genuine interest in HFE suddenly appeared. For example, Askren was teaching industrial psychology and became excited by the material on engineering psychology. Sometimes it was a mentor or a group of students with common interests who stimulated the entrance into the discipline. Both Revesman and Roebuck credit the influence of mentors and associates. This was perhaps more characteristic of a slightly younger generation, when HFE was somewhat more organized and better known. Entrance into HFE was multidetermined and anything could have shut it off. If the government and aircraft industries had not been hiring, the profession would have been still-born; if it existed at all, it would have been as a minor aspect of applied psychology. The original cohort of HFE thought of themselves as psychologists, albeit applied researchers. The field was so young (10 years old when the Society was established) that it was more readily possible to transition from an academic degree in psychology to the new work. This is no longer possible; with accredited courses of instruction, one must now have *correct* academic qualifications.

Many of the first generation had begun in an area related to Human Factors (it is perhaps inappropriate to use the term HFE in relation to this early period because only relatively recently has the term *ergonomics* been joined to that of *Human Factors*). The closer the relationship between past academic training and the new work, the easier the transition was accomplished. In a few cases, the outstanding example of which is Woodson, there was no relationship at all. In other cases, pure chance determined the choice of profession. The author, with a new PhD and a new wife, had to scramble to take what he could, which fortunately was in Small's Human Factors division. Champney, married with children, had to get a job and found himself at Eastman Kodak, where fortunately he was sponsored by the head of ergonomics, Harry Davis. In still other cases, bizarre circumstances had an influence. Jones, as a regular artillery officer, was given four choices by the Army for graduate training; the only one he did not select (psychology) was the one to which he was assigned; the psychology in his records influenced the army assignments he subsequently received (personnel related) and led to his becoming a HFE professional when he retired from the Army.

The ideal qualifications for a HFE professional are, as the late Paul Fitts put it, an engineering degree followed by intensive training in experimental psychology and HFE. However, such a regimen was too strenuous to be adopted so that the majority of new HFE people were primarily psychologists by training (although not of the clinical variety). This was a source of some concern in those early days (HFS Bulletin, November, 1964) and still is because the received HFE wisdom at the time was (and may still be) that the discipline required inputs from many disciplines. Where did these come from if the majority of HFE professionals were essentially psychologists? Because of its somewhat amorphous nature, unconstrained by formal job descriptions, the field tended to pick up people with various backgrounds. Today some professionals still think of it as a multidisciplinary approach rather than a unitary discipline. The field also attracted *tinkerers*—people with creative mechanical ability, like M. (Mac) Parsons.

The war, one's prior training, chance, and the availability of a sponsor (new opportunities that had not existed before the war) were all catalysts, which in combination led to the new profession. The fact that many of its practitioners did not think of it as a new discipline paradoxically helped the profession develop because it avoided *turf wars* and excessive formalization.

Those who came later on sometimes had a career goal already developed. In other cases, the interest began with animal research and switched to people in part because of contacts at psychological meetings. An interest in applied work or instrumentation was important for some. The intersection of engineering and aviation was most important for people like Vreuls and Strother. Then, when an opportunity arose, they snatched at it.

Advanced Training

Because advanced training in HFE did not exist until there was at least some recognition that a new discipline had been developed, it seems reasonable to ask whether prior training was relevant to the new work field and to what extent.

The relevance of advanced training (in the university) to later work in HFE falls into several categories in the sample:

1. Those who, like Woodson, had no graduate study at all; there are few of these.
2. Those professionals (e.g., Swain, Uhlaner, Askren, Van Cott, Teel, Nicklas, Irwin) who felt that their psychological training was relevant to their work and helpful.
3. Those who were not particularly interested in general psychological training because they were pursing other more specific interests like aviation (e.g., Strother, Vreuls).
4. Those who found nothing in psychological training and learned on the job. For example, Erickson had had formal training in engineering and physics, some of which had a little relevance, but most of what he learned was on the job.
5. Those who utilized their training in scientific method and statistics, but who found other psychological material relatively useless. For example, Chapanis had already completed his PhD in psychology when World War II broke out. He considers that the education he had received to that time was not really relevant to his wartime research, but the methodology was.
6. Those who, like Wherry, felt that general scientific training was most important.
7. A number, like Wolbers, had had undergraduate engineering training followed by psychological training. As indicated previously, Fitts thought this was the ideal training progression, as does the author, but the regime is demanding. However, it does recognize that HFE crosses domain boundaries (physical and behavioral).
8. Those whose careers drove their educational goals. Kurke notes that his psychology courses were useful, but that for him education did not determine his career. Rather his career determined the education he obtained.

There were generally two routes into HFE: through psychology (soft science) or engineering and physics (hard science). The fact that most HFE specialists in the sample had psychological training may be because engineers, lacking an interest in behavioral considerations, would not be attracted to HFE. This may be changing, albeit slowly.

To the extent that psychological training provided a research methodology, and one went into HFE research, the training was relevant. If one went into more applied work, like system development, which did not usually involve research, the relevance of the training seemed less.

The time frame in which one studied was important. In the later 1950s, formal coursework in HFE was being developed; this work would be considered more relevant than that which the professional received previously. Earlier, as in the author's own training, which did not include any HFE, only research methodology would be important, the rest inconsequential, except as background. The training one received may also have been more important in terms of introducing one to a role model who suggested further training and career choices. It is also possible that the more specific the interest in any particular field, as in aviation, the less the general training mattered.

In summary, the relevance of training to one's future career was largely idiosyncratic. For some it was important; for others, much less so. Of course, the longer one is in the profession, the less relevant previous formal training becomes because one's career becomes essentially on-the-job training.

It is also possible that the particular training one received influenced one's view of HFE as a discipline. Previous psychological training did not necessarily lead one to think of HFE as applied psychology (although it helped). There are various attitudes on this point because almost all professionals had some psychological training. Whether this led to an advanced degree, however, the relevance or lack of relevance of that training to attitudes toward HFE is ambiguous. Because any training (whether in engineering or psychology) provides mostly background information for career activities (one thinks here of the standard psychological courses in attention, perception, or learning), it probably does not have an exceptional effect on attitudes. Some training, when it is specific (e.g., statistics, experimental design, computer software), is useful if one goes into work that demands that training. The nature of the career determines the utility of the training; if one goes into perception research, for example, then courses in visual and auditory perception and psychophysics are obviously relevant. However, it is the author's impression that this only occurred in a few cases. Training for the later generation of HFE professionals is more tightly focused toward the discipline than it was for the earlier generation; these days it is impossible to get a doctorate in HFE without receiving training in software or human engineering design.

The Concept of a New Discipline

In the development of a new discipline descendant from a previous one, it is relevant to ask when (roughly) new HFE professionals felt that they were engaged in something different. This recognition, if it occurred at all, would obviously influence attitudes about what HFE actually was.

HFE was only slowly recognized as such. Roebuck reports that the question of whether a new discipline was developing was debated in the 1950s, until the organization of the Society in 1957 seemed to settle the matter.

There were different degrees of consciousness. For example, Erickson was aware that HFE was a new endeavor (so he reports), but this consciousness was mitigated by the discipline's continuing original roots in engineering and psychology. References in the literature, and particularly the organization of the Society, made people more aware of the discipline. The importance of advertising even in scientific areas cannot be overestimated; giving an activity a distinctive name can make all the difference in its subsequent development.

For some individuals like the author, there was a gradual process of sensitization, manifested by dropping membership in the American Psychological Association (APA) and joining the Society. There were others who questioned the organization of the Society because the discipline's overlap with psychology made it seem reasonable to them that the Society should amalgamate with the APA (HFS Bulletin, August, 1962). In fact, this was what happened with the East Coast contingent (i.e., people like Taylor and Birmingham). One gets the impression that the Society arose largely because of the West Coast contingent (mostly people working in aerospace; only Grumman was located in New York). The author admits to being prejudiced, but feels that the progress of both the discipline and profession would have been retarded if a distinctive Society had not been organized.

The organization of the Society must be emphasized; it is the most overt manifestation of the profession. Those who organized the Society knew what they were doing, but it was a tremendous learning process for younger people. Originally Seminara thought he was doing applied psychology. Attendance at the Society's 1957 organizational meeting at Tulsa sensitized him to HFE, at which time he became more interested in the Society and less in the APA. In 1952, Champney felt that HFE was a branch of psychology; O'Hare felt that he was no longer solely in psychology. Neal indicates that, until the mid-1970s, he considered himself a psychologist and that what he was doing was interdisciplinary; then he dropped the APA. The Society was the culmination of what was originally an underground process and at the same time the start of a new process, which makes the history of the Society so important (see Appendix 5.2).

The relationship of the predecessor disciplines to the new discipline was clear to many who became HFE professionals, but not to others. Some have never accepted HFE as a separate discipline (Kurke). For others (e.g., Chapanis), HFE is a branch of engineering, definitely not psychology. For others, psychology had no significance (e.g., those who were impelled by an interest in aviation). The shift from psychology or another predecessor discipline to HFE was often a matter of what the task was (Obermayer). Some (S. Parsons, Strother) immediately saw that they were in a new discipline. Initially some were psychologists (O'Hare, Van Cott, Champney); then, because of their work, they lost the psychological influence.

Psychology still has a continuing influence, as seen in the 50% of the Society who think that what they do is applied psychology (Hendrick, 1996a). For these, HFE and psychology co-exist and one or the other may become more important as the nature of the task or the variables they study change. The influence of psychology is perhaps greater for researchers than it is for applied practitioners.

One need only look at the literature over the years to see that the discipline has expanded its scope—going from a special concern for controls and displays to equipments, systems, and software; from motor functions to cognition and artificial intelligence. As it has expanded its scope, the nature of the discipline has changed to encompass more things. One sees a progression: from applied psychology, to engineering psychology, from Human Engineering to Human Factors, to HFE. The field redefines itself, which is tremendously important, because it means that HFE in 2050 may well be significantly different from what it is now. This also points up the importance of debating what HFE is and what professionals do because, without such debate, there is no systematized expansion of the discipline—no change of scope. When the discipline was only half formed, this debate did occur, but it is the author's impression that the debate has since died down (out of weariness? because all the answers have been discovered?). However, the debate needs to be renewed as a continuing aspect of the discipline. Some people want an unchanging discipline; others feel that lack of change is dangerous.

The point is, HFE can and should change, just as in reviewing the history of other sciences one sees that their fields changed radically. A principle may be involved: A discipline that is not dynamic and does not change becomes obsolete. Of course, one cannot separate HFE from technology and the changes that technological advances bring. However, today HFE can be seen as the product of additional forces—economic, social, and cultural—that need to be taken into account when redefining what HFE is (see Hancock, 1997).

One may ask whether different conceptions of what HFE is make any real differences to the work being performed. On the applied level, probably not. However, it may be significant on research and theoretical levels. Because HFE has such a wide scope, it draws in people who are not psychologists (e.g., Krendel) and not interested in psychology (e.g., Strother), and perhaps it is the wide scope that causes questions of identity to be raised.

Early Work

This question asks what work the respondents engaged in during these early years. What strikes one immediately on reading the responses is that, even with the variety of backgrounds the sample has, the influence of government (directly or in government-supported contract research) was paramount. Whatever activities these professionals evolved into doing, their initial and often critical work experience was in something related to government.

Of course, they moved around a bit—from government to contract research, to industry, to government once again (the author's experience).

Often, as in the cases of Teel and Ritchie, the work began with government, moved onto industrial system development and/or contract research, and wound up in academia. Wolbers originally worked in Ruch's applied contract research as an extension of his student training at USC. During this time, a contract involving Douglas Aircraft was let. He then joined Douglas and remained there for 35 years. HFE professionals often taught as an ancillary activity. However, except in a few cases, such as Chapanis, Brown, and Pearson, academia did not seem to play a tremendous role in their careers. There was also civilian work in contract research (Harris); others, like Zeidner, started in government and remained there.

This early period was a major learning experience for the early professionals because, at this time, there were few if any academic courses in HFE. For many of them, whatever was available (such as industrial psychology) did not quite accord with what they eventually went on to do. In a period of great flux, attitudes toward the discipline and the work were developed as part of their work. The nature of their tasks (both research and system development) was often pivotal in determining their later careers (this is probably true of other disciplines as well). One gets the impression of new opportunities, uncertainties, and unknowns that were somehow conquered.

The Status of HFE

This question dealt with the status of HFE relative to engineering and the capability of the new discipline to answer the system development questions posed to it. The importance of engineering relative to HFE cannot be overemphasized. HFE practice (as differentiated from research) meant that many, if not most, early HFE professionals worked in an environment vastly different from that in which they had received their training. This could not help but determine points of view. If HFE is in many respects highly pragmatic, less scholarly than it should be, and more other-directed than internally oriented, it is largely because of the engineering influence. Those who were psychologists by training had to become more like the engineers they worked with if they were to be successful.

At the same time, they encountered problems they had never faced before: The application of behavioral principles to hardware design meant that they had to cross domain lines—an unsettling process for many of them. They had neither been trained to application nor were there any guidelines. They also had to (and still do) face the scrutiny of professionals vastly different from them in training and interests. Because of this, there was the need to rationalize and justify their work, which, in an earlier, more academic and scholarly environment, had not been necessary. If justification of research in the university was necessary, it was on different grounds than those of the engineering environment.

One can look at the status of HFE in two ways: (a) its ability to contribute to engineering development and provide solutions to design problems, and (b) its relationship with the engineering community. The answer to the first aspect depends on the nature of the questions asked of the discipline. If these questions related to the capability of the human to withstand, for example, cold climates, HFE was able to provide answers because there was no need to translate behavioral principles into hardware equivalents. If the questions asked of the discipline dealt with the behavioral consequences of a certain design configuration, HFE's capability to answer these questions was much more doubtful.

The effectiveness of HFE in system development was therefore fairly weak. Wherry reports that, in 1966, for example, HFE was poorly defined (and he believes it still is). There was a lack of HFE in many systems. The Navy, for which he worked, considered each human engineering discrepancy individually and so minimized them. It cost too much, they said, to change these deficiencies. There was a lack of objective data. Despite government standards, the impact of HFE on many systems was minimal. The effectiveness of the discipline often depended on the willingness of project managers to fund HFE. HFE professionals were able to evaluate design but had difficulty in transferring HFE requirements into design specifications.

If the question asked of HFE depended on the translation of behavioral knowledge into physical systems, the HFE lack of capability became quite visible and distressed many professionals working in system development. Pearson feels that much of the HFE literature could not be translated into design. The inability to cross domain lines (to turn research results into solutions of development problems) was noted by several respondents in the author's survey.

There were also problems involved in having enough funding and time to do research that applied to a development problem. This brings us to the second aspect—that in many cases there was a poor relationship between HFE and engineering. S. Parsons reports a distinct lack of respect for HFE from engineers and managers. The government mandated HFE, but engineers felt that this was an intrusion on their turf. Ritchie reports much prejudice against HFE, particularly in engineering. He puts this down to HFE specialists being pioneers. However, he feels that the early professionals could easily move into areas with significant problems.

Part of the difficulty was the sheer ignorance of the engineer regarding human factors and its impact on system operations, which required a great deal of indoctrination of engineers (both formal and informal training courses). Difficulties of the first type (inability to translate behavior into systems) contributed to the poor relationship between HFE and engineering because it was impossible to prove the value of HFE in terms the engineer (who valued precision, quantification, etc.) could conceptualize. For example,

Irwin reports that there was a constant battle with engineering. At that time, HFE lacked measurement and quantification methods. He believes this has been solved with simulations. Because they were human, engineers felt that they could solve HFE problems (Heglin).

It should not be assumed that everywhere there was continuing acrimony with engineering. Some respondents (e.g., Swain, Seminara, Kurke, Teel) saw no problems, whereas others, such as Askren and Van Cott, definitely did. HFE professionals resented that HFE took second place to engineering. HFE people initially had difficulty fitting into an engineering context because that context was vastly different from the cloistered, scientific research orientation they had inherited from their university days. Van Cott reports that it was difficult to explain the science/research orientation to engineers. HFE professionals had difficulty understanding that what HFE was supposed to be doing in system development was not so much science and research as utilizing professional judgment. The engineering environment demanded quick answers, although these answers were partial and relied on subjective processes. Many HFE people did not interact sufficiently with the engineers, preferring to seclude themselves so that they would not be challenged by problems they felt they could not solve.

The research literature did not provide the answers that were applicable to design questions. O'Hare reports that HFE professionals were more inclined to ask questions than provide answers, and hence to recommend studies that the engineers rejected because there was no money or time for these.

It is apparent that no single answer can be supplied to Question 5 without considering the conditions under which one worked. Many of the sample had poor experiences with engineers and/or their feeling about the ability of HFE to contribute to development. If one had a supportive management, the problems were eased. Woodson indicates that the relationship between HFE and engineering depended on the attitude of top-level management, which, if neutral or negative, would be quick to eliminate what they considered nonessential development aspects (of which HFE was a prime type for them). This is an example of how nontechnical factors (Meister, 1993c) can significantly influence technical aspects. At the highest level, government supported HFE by promulgating MIL-STD 1472 (in its latest revision, 1994) and MIL-H 46855 (1979). However, in actual practice, at the lower working levels much of the support disappeared. Some of the early problems have continued to the present day—most notably, HFE's inability to translate behavioral principles into hardware requirements.

It should not be thought that these were days of only gloom and anxiety. For example, Wolbers reports that the early days were exciting. There were only a few HFE professionals, so people like Fitts and Small knew each other well. There were, so Wolbers indicates, certain focal points: Fitts at Wright Field, Williams at Illinois, Small in San Diego. Fifteen or twenty people in Los Angeles would get together monthly and, with ONR funding and the help of Douglas Aircraft's Tulsa Division, they started the Society in 1957. Wolbers feels that HFE made significant contributions to the new advanced aircraft then being developed. The biggest problem, he says, was the lack of qualified people; in consequence, they relied for support on consulting companies, the Air Force and Navy laboratories, and universities.

Distinctive Experiences

The rationale behind Question 6 was that any special experiences—*critical incidents* in Flanagan's (1954) terminology—might illustrate the conditions under which professionals worked in the early days. The reader of these memoirs cannot help but be impressed by the variety of experiences the early HFE specialists had—quite different from those of the research laboratory or classroom. The need to develop sufficient system expertise to deal with development problems often made it necessary for them to undergo military training (for his first job, the author trained in a school for Navy sonarmen) and participate in operational exercises (he sailed on a destroyer escort in naval operations), all of which placed heavy demands on one's flexibility and adaptability.

One wonders what the effect on the personality of HFE people experiencing this became. The ability to adapt to many different situations was definitely an asset. Those who could work outside a highly controlled situation were likely to be more successful; adaptability was a prerequisite for a highly fluid environment in which one had to establish his own procedures. One had to transcend one's academic training. Without being able to prove it, it is a reasonable hypothesis that success in the applied area depended to a great extent on the personality of the HFE specialist. McGrath reports that he once challenged an entire group of engineers on a prediction he made and triumphantly demonstrated that he was right. S. Parsons, in the absence of more qualified missile engineers, once had to explain the workings of the Polaris missile to Werner von Braun and some visiting German scientists and did it successfully.

If one were able to find a job that enabled him to seclude himself, the most introverted individual could make a success of the job, but this happened infrequently. Work in applied research and more especially in system development required a more outgoing personality—the ability to interact with people with interests different from one's own and the ability to communicate abstract concepts simply to engineers who lacked the necessary background. Some of those with psychological training could not do this and were unsuccessful in the sense that they were encouraged to seek other employment. In this connection, Harris points out that much of the data needed by the HFE specialist could be secured from other disciplines, but what was unique to HFE was its human-centered orientation.

This raises the question of what success is in an applied (as contrasted with a purely research) environment. In the former, acceptance by engineers was a primary criterion: acceptance of self, acceptance of HFE. Neal reports that, because he was in a highly respected Human Factors organization, he did not have the problems younger HFE people had. Knowledge gathering in a research mode was essentially irrelevant because, in most cases, one did no research. One had to be able to endure many small defeats without many tangible rewards so that feedback was lacking in many, if not most, cases. Indeed it was commonplace to wonder whether one's contributions actually had an effect when the product was a design drawing, in which no single individual's contributions, including those of design engineers, was acknowledged. (Champney, however, says that in industry as opposed to research one could quickly see tangible [i.e., hardware] results of one's efforts.)

Where one had to engage in military operations (e.g., data taking on naval vessels or on helicopter flights), a sort of adventurous spirit (or a fatalistic one, perhaps) was essential for a civilian. What is striking is that so many different types of professionals could adapt to situations that were often inherently highly stressful. Of course, this does not suggest that the HFE specialist was a sort of superman. However, the work required the development of special qualities or skills to withstand the applied/operational environment, qualities, or skills that could be learned.

One wonders about the effect of such unusual experiences on people, who were in most cases trained in a different (academic) environment. Those who could *stand the gaff* were undoubtedly changed by their experiences, but it is not clear how. It may have made practitioners more pragmatic—less willing to accept abstractions and inadequacies of research and the slower pace of the research environment. This may in part be responsible for the tensions between practitioners and researchers, which has played a part in the development of HFE.

The Distinctiveness of HFE Teaching

Question 7 also attempted to get at the differences between the predecessor, psychology, and the new emerging discipline. Early on, training was given in HFE. For example, the System Development Corporation (a military think tank) announced (HFS Bulletin, March, 1959) special training in the system concept, but how did this new training vary from previous academic courses? It is significant that we do not ask this question about engineering training for HFE specialists; only one of the respondents (Ritchie) even mentioned training in engineering.

There were not many responses to this topic perhaps because not many of the sample engaged in teaching, except possibly as an ancillary part of their activity. Wolbers indicates that what distinguished HFE training from traditional psychological material was the inclusion of the system development process as the context in which one applied behavioral principles. Obviously, system development completely re-oriented HFE. HFE poses a problem in which one must make use of accepted behavioral principles or find new ones. In the teaching of psychology, the problem orientation is not so evident. In responding to stimuli, the human is behaving mostly in relation to him or herself, but in HFE, one must think of the human performing a task. The task orientation was the one thing that required a new way of conceptualizing the human. If there is one aspect that distinguished HFE from psychology, it was that the human was conceptualized as behaving in a work context, and that context plays a much lesser role in psychology.

Perhaps the question was poorly expressed. It was phrased in terms of formal academic training, although, as Vreuls suggests and certainly this was true of the author's own experience, there was a great deal of informal training—not of HFE specialists, but of design engineers. This was performed in the engineering facility in classrooms and, just as frequently, at the engineer's desk. This was practical teaching of human engineering principles directed at the design of control panels, for example.

For those who taught HFE in an academic setting, what was distinctive about their teaching was that they put the individual they were describing into an action orientation—a problem context (Howell) in which the following were most important: the system development process (Pearson) or a mission/task that had to be accomplished (in other words, a goal orientation in which one could succeed or fail and, in relation to the latter, error was extremely important; Van Cott). In traditional psychological teaching, although there is consideration of individual differences, variations in performance are not particularly significant. In HFE, performance variations (e.g., errors) were very early tied to causal factors in the design of equipment and the task (see Chapanis et al., 1949, in which there is a discussion of system-induced vs. chance-induced error). The need for the individual to analyze a situation and the need to evaluate and measure performance in relation to a standard (which implies a goal) is critical. Traditional psychological teaching is inner-centered (it is about how the individual behaves), but not in relation to specific task-induced goals, whereas performance in HFE is externally centered in relation to an overt goal. In HFE, the methods that are taught are those needed to understand the individual in his or her performance toward that goal, whereas in psychology the methods are in relation to the individual within himself: psychophysical methods, in particular, although learning measurement and processes are related to a goal. The need to apply psychological or behavioral processes to system design does not exist in standard psychological texts. Or so it was 50 years ago; the author has only his general psychology texts he used as a student with which to reference these points.

In HFE, teaching multiple causality was emphasized (O'Hare taught a course in games and simulations, which introduced this topic to undergraduates). Causal or performance shaping factors (the term was introduced by HFE) are much more *molar* in HFE than in psychology. In psychology, the stimulus may be a light with a given wavelength; in HFE, it is the task as a whole, in which a light, if one exists, is only one part of the total task stimuli. In HFE, the *scale* of things—of events, of behavioral phenomena—is much larger than in that of psychology. For example, the time scale is different: Psychology deals with milliseconds, seconds, and minutes, whereas HFE deals with behavior that is manifested over hours and days, if not longer.

Then and Now: Differences in Theory, Research, and Practice

This topic proved difficult for respondents because it required a much broader time-based outlook than most people are accustomed to or feel comfortable with. Moreover, it requires making concrete (at least verbally) vague feelings, at which most people are not competent.

The answer to the question cannot be a single, total one, but must be subdivided into changes occurring in theory, research, methodology, and practice. Before going into each of these, several respondents pointed out that the scope of HFE has changed tremendously—from emphasis on controls and displays to systems, from simpler function variables like motor responses to variables that had not been considered before (e.g., cognitive, social, motivational, emotional). However, others like Harris felt that, despite new technology like the computer, there have not been that many significant changes. Chapanis notes that methodology has progressed from the formal experiment to other ways of collecting data (e.g., activity analysis and critical incidents). These additional methods, he feels, add a major strength to the discipline.

The extension of the discipline's scope has given us new problems. Consequently, the question arises as to whether it can deal with this enlarged scope. A simple answer to the question (which could be utterly wrong) is that we now have the computer. Indeed, this answer was common among respondents, but it ignores the now classic formulation for computers: garbage in, garbage out. It may be that the reliance on the computer as a potential solution of HFE problems is a way for professionals to avoid dealing with their problems directly.

One can look at the changes that have occurred by distinguishing between a technical, knowledge-based discipline and the profession as distinct from the discipline. (The author has made this distinction repeatedly; to him, the discipline is content-oriented and the profession is method-oriented. However, many HFE professionals might not think of the difference, if there is one, in this manner.) Also, advances can be physical (based on technological changes) and intellectual (based on concept shifts). There have been significant changes in the profession as distinct from those of the discipline and changes in the physical aspects of the discipline as distinct from its conceptual aspects. The major change observed by the respondents is the introduction of the computer, which to them has revolutionized simulation. Whether one is an optimist or pessimist may determine how one views the changes in the discipline. A few respondents saw little change in the discipline as distinct from the profession.

There are more changes in the profession. HFE is more accepted and is now more involved in earlier planning stages of development. Military and other system standards have been developed to systematize the work. O'Hare suggests that, in theory, the change has been from servomechanisms and feedback loops to the concept of the entire human, using the computer software development process and human anatomy and physiology as a model. If true, this has implications for the type of research we perform, with particular emphasis on modeling and information processing. Research earlier involved small, tightly controlled studies of single variables; now we are willing to tackle larger problems. In HFE practice, there has been a shift to issues of standardization.

Whether there have been significant changes in HFE and in what ways depends on which aspect of the discipline one is talking about. The scope has certainly changed, as have the physical modalities available to us. Changes in conceptual issues have been far fewer—the same questions often reemerging with new names, a characteristic termed *reinventing the wheel.* The profession has obviously changed in terms of employment, but has our knowledge base and our capability to answer questions changed as much? If our scope has expanded, so have the problems associated with multiple interactive variables. Are simple univariate behavioral principles adequate for the new challenges?

What is clear is that if the sample of respondents is representative of most HFE professionals, few of them think seriously about any but the most obvious problems—those of an individual speciality type. They do not do as well on broader-based, general disciplinary problems.

The Engineering Reaction to HFE

This question sought to determine the reaction of engineers to the introduction of HFE and HFE professionals in system development. This topic provoked a more vigorous response than did many others probably because it involved the respondents personally. The following is a sample of the responses received (summarized):

Erickson: There was good support in those days at the local level, least support came from Washington. When HFE was included as a requirement in system development projects, program managers fought it as another hurdle.

O'Hare: Engineers working at the same level as the HFE professional were generally supportive. Higher level management types were less so. The requirement to include HFE in a system project was resented by managers. In later years, as HFE standards were incorporated into contracts, there was greater appreciation of their validity and more attention was given to these. Curiously, as engineers took short courses in HFE, there appeared to them to be less need to rely on qualified HFE people. If there was a HFE problem requiring empirical research, management would *out source* to a consulting organization.

Wolbers: Because engineers are human, they believed they could deal with HFE issues on their own; they did what they considered best, but did not understand HFE nuances. Engineers have become more accepting of HFE as information about it spread, with the issuance of standards and Woodson's (1954) book. Legislating HFE had a negative effect on program managers because of tight time and budget constraints. Many so-called human engineers were not capable of dealing with engineers directly. In recent years, there has been much more support; this may be because younger people were better taught than older ones. There is a more general awareness in the population of the need for user-friendliness.

Teel: Initially engineers were skeptical, but after we showed them we had something to offer, they became more receptive.

Obermayer: The early difficulties of making an input to design have vanished. It was difficult to get issues considered unless they were noted in the system specification. Where the need for HFE is clearly identified, HFE does get in. Now, in addition to engineers, we have computer programmers to worry about.

Seminara: Engineers were not universally happy to see human engineers. Acceptance had to be earned by improving their systems. This could be done only if one participated as part of the initial design team, not as a critic after the fact.

Champney: Engineers were and still are skeptical of the value of HFE except for the converted engineer who often became a zealot.

Van Cott: In the early years, engineers mistrusted HFE people, but trust and communication increased as they discovered we had solutions.

Nicklas: If system specifications lacked HFE requirements, management designated certain engineers to take care of the human aspect. Opportunities for HFE professionals to "crash" the development project were difficult to obtain. When the military included HFE people on the military monitoring teams, they demanded HFE participation. HFE procedures and documentation requirements were hard for the engineers to accept.

S. Parsons: Engineering reaction to HFE varied. A few were hostile, most neutral, a few supportive. Today almost all engineers are supportive (relatively). Program managers were cautious because they had to budget for HFE. The system development process and HFE support are much more standardized today. However, HFE inputs must be tailored to the project, type of human performance, budget, and schedule.

Strother: The reaction of engineers to HFE was negative. Budgets did not include HFE, engineers thought they could handle human problems themselves. In aviation, if there was a human problem, engineers went to the pilot for advice; they still do. Today HFE is accepted in many companies, but only because the government demands it and provides a budget for it.

The previous comments speak for themselves, but if they were summarized, they would read like this: Engineers felt that because they were human it was within their capability to handle HFE issues. Managers did not like to conform to HFE demands, which also required budgets and time. If the military provided the money, fine; otherwise, no. Even then, however, project managers felt that they were being forced to accede to demands they would not otherwise have to satisfy.

Many HFE professionals were not trained to deal with engineers in their development work. HFE concerns were initially research and not development; engineers felt that HFE should make a positive contribution to design, and this, even in optimal circumstances, was hard to demonstrate because design is a team effort. If HFE was not specifically addressed in system specifications, it was difficult to get adequate consideration of it. Even when required, there were no quantitative criteria associated with it so it became part of *motherhood and apple pie* requirements (which meant that it was of secondary importance).

Managers sometimes designated incompetent engineers to handle HFE issues, and they did a poor job, which then backfired on the HFE profession as a whole (Ritchie). Engineers lacked understanding of the relationship between human performance (expressed in the form of error) and subsequent system performance. The machine, being so powerful and sophisticated, would not deviate because of operator performance (or so they felt; it was more a feeling than a consciously expressed thought). Also, in the best American tradition, they felt that humans could compensate if inadvertently there were some design inadequacies; in fact, they could not conceptualize inadequate design from a behavioral standpoint because they felt they had done their best in the design.

The contrast between the machine and the human was all in favor of the machine: the machine was strong, powerful, and invariant (all characteristics of the male) and the human was weak and variable (characteristics of the woman), thus the relative importance of HFE seemed quite small. The inability to attach quantitative values to HFE variables was also a serious difficulty and still is because one cannot quantitatively and precisely prove the importance of HFE in terms of its impact on performance.

The situation has changed for the better, although it is not completely satisfactory. There has been a gradual diffusion of knowledge about HFE in the general population, which has generalized to the engineering profession. The budgetary problem exists today as it did before. If there is no budget for HFE in design, it will not be included. HFE is only considered an ancillary discipline because it does not deal directly with performance of machine functions.

Because the engineer lacked behavioral knowledge and had a negative reaction to HFE, strenuous efforts were made by government and the individual HFE professional to propagandize engineers and put HFE into step-by-step proceduralized guidelines. There was almost a cottage industry in writing materials for engineers. Researchers like Meister and Farr (1967) investigated how well engineers could make use of HFE information.

Changes That Have Occurred

This question, like some of the preceding ones, was an attempt to define what HFE is now, as contrasted with what it was 40 or more years ago. This is a difficult question because it asked respondents to exclude computerization, and this obviously has been the primary change factor that HFE professionals recognize. It is not easy to think of changes occurring over many years, and a few in the sample even felt that no substantive change has occurred. However, it stands to reason that changes have occurred even if these are not visible.

In terms of theory, cognitive ergonomics (however that is defined) was mentioned most frequently (Askren, O'Hare). O'Hare mentioned the tendency toward the development of specialities, the expansion of HFE into areas like office furniture (a phenomenon that has been mentioned previously), and because of computers the entrance into HFE (by the back door) of people who have only a peripheral relationship with HFE.

Outside of computerization, HFE professionals (if the sample is representative of the whole) are aware that changes in the profession have occurred over the years, but they are unable to indicate clearly what these changes are. This may be a common human failing, or the pace of the change in

HFE may have been so slight as to be only vaguely perceptible. As far as system development is concerned, the conditions for HFE people working in this area have improved because specialists are introduced into the design process earlier (some in the sample indicated this) than in previous years. Changes in practice may be more evident than changes in theory and methodology, although the emergence of cognitive ergonomics (Rasmussen, Pettersen & Goodstein, 1994), situation analysis (Endsley, 1995), and naturalistic decision making (Klein, 1993) and virtual reality (Kalawsky, 1993) belies this. The inability of respondents to specify changes in theory and methodology suggests that, despite much beating of the drum about theory, only researchers pay much attention to it. Some, like Chapanis, are not certain that theory is important or, as in the case of Debons, that there is a distinctive HFE theory other than that provided by psychology.

O'Hare's point about specialization as a continuing trend in HFE is quite valid. It may be that the trend to specialization is the most obvious change that one can discern. Whether there was ever a generic HFE may be a misconception because, in the early days, we were mostly working in a few specialty areas like aviation, sonar/radar, and electronics and primarily with controls and displays. As a consequence, there were only a few specialty languages in which to communicate with each other (hence the myth, perhaps, of a relatively homogeneous HFE). Now that HFE has expanded into a much larger number of specialties (the latest HFES Directory lists 20 speciality groups), it may appear that there is no common body of theory and methodology—or, more likely, that the commonality resides in the psychological background knowledge that most HFE professionals bring to the discipline. All specialties are based on the S–O–R paradigm and common measurement methodologies, of which the experiment is preeminent.

One change that was not mentioned by the sample, because it is not inherently a part of the discipline, is the increasing awareness by the general public of ergonomics in the exploitation of commercial products like office furniture. When the respondents began their careers, no one literally knew of HFE; this was certainly true of engineering. Now many more people (and hopefully many more engineers) recognize the name, although whether this has enabled HFE to be better understood is doubtful.

The inability to specify significant HFE changes over time may be a function of the fact (and the author assumes it is a fact) that many of the same problems that were first encountered in 1950, say, still persist. The fact is that we cannot predict the effect of behavioral variables on design and performance (although we can model them, more or less successfully; see Laughery & Corker, 1997) and we have no quantitative database (although we have increasingly sophisticated statistical techniques). We cannot predict human performance except in vague generalities of research conclusions, and we absolutely cannot predict quantitatively. We cannot relate behavioral principles to equipment design characteristics, although we have some heuristic design guidelines. We are aware of computerization as a change factor, but curiously that has not enabled HFE to perform its basic functions (analysis, design, evaluation, prediction) much more effectively than in the past.

What we do is useful; however, in terms of what we should be able to do, our progress seems pathetically slow. Of course, that may be the fault of the youth of the discipline. An awareness of progress or lack of it requires an ability to conceptualize activity in terms of lengthy intervals, and whether HFE professionals possess this capability (more than the general public) is not clear.

From a professional standpoint, there has been change: There has been expansion into more specialties, something the first generation definitely desired (as indicated by the number of pages given in early HFS meetings that dealt with the application of human factors to unlikely specialties like dentistry), and an increased number of HFE professionals, as evidenced by the increased membership in the Society—aspects that are most obviously evident. The fact that after 50 years HFE as a profession is still around is perhaps cause for minor rejoicing. Theory change (if one thinks of a generic HFE theory) has been almost nonexistent, although the specialties have specific theories. Our original concept of the human in the work situation has changed; originally it could be summarized in the basic psychological paradigms of S–R and S–O–R, but we are now aware of many more performance-shaping factors. With the advent of the computer, it is suggested (perhaps mostly by computer specialists) that the human is a sort of biological computer processing information as a computer does.

The Effectiveness of HFE

The question of how effective HFE was in the early days is a question relative only to HFE practice in system development. HFE was effective when recommendations made by the specialist were accepted by the design engineer. There is then an intimate relationship between acceptance and effectiveness, which is in part a factor in the great sensitivity HFE professionals felt (and still feel) about the acceptance accorded the profession.

That effectiveness varied depending on idiosyncratic factors (and to a certain extent does so even now; Harris, Wolbers). There was a general feeling among respondents that HFE was less effective than it should have been, but a few respondents felt that their inputs did make a difference (e.g., Woodson, because of his reputation, and Teel, who said that 90% of his recommendations were accepted). If one were a talented HFE specialist, one would be more successful than others who were less talented. What constituted talent is not at all clear (prior training as an engineer? a feeling for hardware? an engaging personality?). S. Parsons points out that, in a relatively negative climate, the ability to sell engineers on HFE, just as one would sell shoes, was most important.

Much depended on higher management. If management was supportive, this was a positive factor because lower level engineers took their cue from the management. However, management was felt by many HFE professionals to be ignorant of behavioral factors and uninterested (much like engineers in general). If the HFE specialist was invited into a project early enough in the development sequence, his imputs were more successful than if he were called in toward the end (Erickson). If one could solve a problem that arose because there had been no HFE on the project, the HFE input was accepted (Kurke).

Unfortunately, there was no objective way to demonstrate effectiveness to the engineer. Even if operational system testing of the completed equipment or system showed that operator performance with it was satisfactory, this was not proof that what the HFE specialist had recommended was in fact responsible for the satisfactory personnel performance because the design configuration was the product of many inputs. Paradoxically, if the final tests revealed no human engineering discrepancies (that needed fixing), the design engineer might also feel that HFE had made no contribution at all because obviously engineering design was responsible for the satisfactory state. It was a no-win situation.

A number of respondents (Swain, Obermayer) felt that the absence of data to support HFE recommendations was a primary factor that negated HFE effort. The author sat in meetings at which the primary question was whether HFE could demonstrate that its activities would be effective: Does human error really affect system performance? Where are the data? The lack of data for evidentiary purposes was very damaging.

Of the three functions performed in system development—analysis, design, and evaluation—evaluation was easiest for HFE because its academic origins and expertise were measurement-oriented. Analysis and design were and are the most difficult for the HFE practitioner because he lacked guidelines for his judgments and also because entree to participation in the project was not easy to obtain. Because evaluation was easiest for the specialist to perform, this is what he wound up doing most; engineers resented this because they considered it criticism after the fact.

The effectiveness of HFE in industry depended not only on idiosyncratic factors like talent and salesmanship, but also on contextual factors—management knowledge and confidence in HFE (low), engineering knowledge of HFE (almost nonexistent), biases (I can do what the HFE specialist does because I am a human), turf wars (constantly), funding nonavailability (constantly), and governmental supervision and support (occasional). The introduction of military standards like MIL-H 46855 (1979) gave a rationale and justification for HFE efforts (what the customer—the military—wants, the customer gets), but military standards were often ignored (who could find out about this?) because engineers (thinking of themselves as creative free spirits) resented requirements imposed from above. However, the fact remains that without governmental sponsorship, weak as this was, HFE would never have gained a foothold because in industry the bottom line was performance and money. One question that frequently arose was, what does HFE contribute to these?

The picture then is uneven. Depending on the concatenation of circumstances, good HFE people, supportive management, a system heavily dependent on the human operator, such as the aircraft (and it is no coincidence that HFE practice began in aerospace and much of its success was in aerospace), adequate funding, and so on, HFE would be successful. If any of these was lacking, the probability of HFE effectiveness was much reduced.

The author's own experience during the years from 1956 to 1964, while working on the Atlas ICBM project for Convair, was that, in introducing HFE to a company that had had no previous acquaintance with the discipline, one did anything one could to bring oneself positively to the attention of management, one actively scoured engineering departments to discover what could be done, and one wrote reports that magnified HFE accomplishments (although everyone in engineering wrote reports that dramatized achievements). Parsons' point about selling engineering reproduces the author's experiences. HFE accomplishment was something like public relations—one had to give the impression that one was doing good things, and it was necessary to demonstrate this to inflate accomplishments in whatever way one could.

If the HFE practitioner was the type of individual who could deal with engineers (and personality was important here), the chances improved that he would be invited into design early and consequently be productive. The hierarchical structure of importance in industry strongly affected HFE effectiveness: first the customer, then company management, then the design engineer, followed by support engineering like reliability and logistics, and then anything else (including HFE).

The HFE specialist always worked against the odds so that anything that was accomplished was almost unexpected. The remarkable thing is that, with all the negative factors, the HFE professional managed to accomplish things of value. What is the situation today? Does anyone know? There is a real need to determine whether the initial development situation has changed.

Satisfaction with HFE

One way to determine how far HFE has progressed (a reasonable question to ask in studying the history of a discipline) is to determine how satisfied professionals are with the present state of the discipline compared with that of the past. Whether one is satisfied with HFE progress depends on a number of factors:

1. Whether one is aware of the distinction between the discipline and the profession. The first is an abstraction that describes its conceptual structure. The second represents the external trappings of the discipline, the job situation, recognition by other professionals, the awards one has received, and so on. A personality factor may direct these varying attitudes. In the 1940s, a sociologist—Reisman—made a distinction that is relevant here between inner- and outer-directed people. The former march to their own drum beat (are internally directed), whereas the latter march in accordance with everyone else's drumbeat, which means that they are directed by commonalities, objective goals, impressions, and so on. Inner-directed HFE people are more attuned to HFE as a discipline, whereas outer-directed people are concerned with the *here and now* and progress in application. The latter are much more concerned about recognition for their efforts. (Of course, recognition is also important for inner-directed professionals, but they distinguish between objective professional activities and *the life of the mind.*)

2. Whether one equates the adequacy of the discipline with personal satisfaction. If one has had a satisfactory career, one is more likely to feel that the discipline/profession has also had a satisfactory career. For example, S. Parsons is quite satisfied with his career. However, there are those who are personally satisfied, but not on a disciplinary level, because there are still problems that need to be solved and questions that have not yet been answered. For example, O'Hare says he is satisfied, but then goes on to criticize the state of HFE as not having produced outputs of a novel kind. In his opinion, we do not produce unique additions to knowledge and may eventually be folded into industrial engineering. (The fear that HFE may eventually be assimilated by a more dominant discipline was not an inconsiderable one among respondents.)

3. It is also necessary to distinguish between the progress that has been made (which may be considerable), the present status of HFE, and what we could do to satisfy HFE goals. Such distinctions require a recognition of those goals and whether and to what degree we have accomplished them. For example, a goal might be to quantify and predict human performance—an area in which we are grossly deficient (certainly Swain, who has worked all his professional life in support of this goal, feels that).

4. Whether one is thinking of progress in an applied context (i.e., in system development) and progress as a research science, as a branch of knowledge. It would appear from comments made by the respondents that major improvements have occurred in system development. However, these may not be matched by our accomplishments as a branch of knowledge. For example, Wherry feels that HFE has accomplished a good deal, but needs

improvement as a discipline. He feels that HFE professionals may not understand the tremendous complexity of their tasks; for him, HFE is one of the most complex fields imaginable.

There are then several ways of looking at satisfaction (if satisfaction equals progress, which is a question that needs discussion): what we have accomplished so far (which may be a great deal when one thinks of the early days) and what more we have to do (which may be a great deal also). What is astonishing (or should one consider it surprising?) is the large proportion of the sample who are relatively well satisfied. The author is inner-directed and, concerned about the discipline as a branch of knowledge, had expected much more dissatisfaction. Obviously a considerable number of HFE professionals are concerned primarily about what happens to them in terms of recognition, status, ability to contribute to development, and whether we have university training programs, standards, laws regulating ergonomics in the workplace, and so on (all of which are worthy considerations, but not necessarily for HFE as a branch of knowledge). One suspects that only a minority are concerned about HFE as a science. It may be that the proportion of those interested in HFE as a discipline is smaller among the younger cohorts (which may be an unjustified denigration of them). If this is true, this is bad news for the discipline because, although the profession may flourish, the discipline may not progress.

Hopes and Expectations

This is another effort to discover the inner feelings of HFE professionals, but with less than maximum success. It may be that after a certain age one no longer has hopes and expectations except of a personal type. There are two ways of answering Question 13: in terms of what one would like to see in system development and in terms of research. For example, Woodson says that he began by hoping that business and government would recognize the need for user-related research and design. However, he feels that decisions are still being made on the basis of opinion, and industry has redefined HFE as anything it takes to sell a product.

For some respondents, the lack of definition of HFE in the early days prevented them from developing hopes and expectations. As the discipline gradually evolved, they developed a point of view about it, which enabled them to have hopes and expectations. For example, Roebuck says that in the early days HFE had not been invented yet and thus he had no expectations or goals. Gradually he saw changes in the discipline so he developed hopes of what it might become. He is generally satisfied with HFE progress, but looks forward to additional changes.

Another factor that must be considered is that some respondents, like Strother and Vreuls, are so specialized in their interests that they hardly think of HFE as a whole. What has happened in aviation, for example, has determined their expectations and the feeling that these expectations have or have not been achieved. Still others think of HFE in a more comprehensive manner. Almost 50 years on, there is still confusion as to what is included in HFE and what it is. For example, Van Cott has no hopes for HFE because he is not sure what it actually is. Others (Klemmer) feel that the field is better defined now than it was at the beginning, but this does not mean that it is fully defined even now.

Some are satisfied with the progress that has been made and others are not. Satisfaction may be defined by personal experience. For example, Seminara had hoped that HFE would become more diversified, but he is pleased with the variety of projects he has worked on. Others have a more abstract concept of the discipline. More important than all this is the fact that (apparently) we have not yet adequately defined the discipline in which we work. There is perhaps a need to return to basics in thinking about HFE. This seems extraordinary, but it is the author's feeling that HFE people have not generally attempted to probe deeply into their work.

Adequacy of Present Training; Evaluation of Current HFE Theories and Methodology

Professionals who began their careers in the early days do not have any clear ideas about what is going on now in training, but they seem fairly certain that present training is much better than what they had had (or did not have). There is a feeling that HFE training should involve hands-on experience, as well as a knowledge of the system development process. For example, Chapanis suggests that a weakness of present training is a reliance on general guidelines. There is also a failure to understand the development process and the use of trade-offs in design. Those who are into applied work more fully do not see the utility of advanced training (the PhD) except where research or teaching is to be performed. Seminara feels that his early training was inadequate. He feels, however, that much advanced training is essentially irrelevant to applied work. A master's degree with some engineering, industrial design, and computer training would be sufficient for those working in applied HFE. He feels that doctoral work should be reserved for those going into research, teaching, consulting, or management.

One can look at theory in terms of the expansion of interest into processes like work load or stress, which were not originally areas of interest (Askren). From that standpoint, theory has much advanced over the early years. Present theorizing is considered to be somewhat shallow, as essentially descriptive only. O'Hare generally denigrates the effectiveness of theory and methodology. He says description and the naming of phenomena take the place of genuine theory building. Methodological variations are more frequent now, but there is little attempt to validate. There is little consideration of anomalies in experimental results. Others are like Chapanis and Seminara (it is interesting how similar their views are despite differences in their backgrounds, the first being primarily an academic and researcher and the second primarily a practitioner). Both do not see theory as having much value for applied work.

It would be fair to say that those who have actually concluded their formal training and begun their careers no longer worry about university training because they have formidable learning problems while working on the job. Academic training becomes merely a hurdle to be overcome in securing one's degree. Outside of a few pertinent topics, like some engineering courses, computers, and statistics, little of what one has been trained in is of direct value (possibly of indirect value, as providing a foundation) unless one is into research or teaching (and yet there are innovative training programs [e.g., Hahn, 1996] that do relate to actual work). One gets the feeling that the sample felt that those who create HFE training programs have an inadequate acquaintance with actual applied work because there should be (and may not be) a match between the two. The feeling is this: Train for what one will eventually be doing. Learning about the processes involved in color perception, for example, is probably of no value unless one is called on to work on a project involving vision/perception.

Except for one or two of the more reflective people in the sample, the old timers do not worry about theory because theory has little relevance to their work. For example, Chapanis sees system development as essentially idiosyncratic (each design having its own set of problems), therefore general guidelines (presumably derived from theory or the combination of different studies) do not apply.

Significant Gaps in Theory, Methodology, and Practice

The determination of what the professional sees as a gap may suggest what he feels is important in the discipline. It is assumed that the nature of the gap, if there is one, is determined by the professionals' special interests and experiences. The experience factor is particularly important in those who have spent their entire career in system development and who have perhaps not been encouraged to think in terms of a larger context (this may appear to be a slam at practitioners, but it is not; see previous chapters). Those who are more research-oriented may have a somewhat broader concept of the discipline; even so, that orientation is limited by personal experiences and interests. For example, Kurke feels that there should be more effort supplied in forensics because that has been a primary area of endeavor for him.

A number of respondents indicated that they were not interested in theory. If they had worked in system development, they were much more concerned about developing methods to quantify cost–benefit consequences of design alternatives. This has been a long-standing interest from the earliest days. Professionals working in system development have always wanted to see improvements that can be directly related to system development (Woodson, Askren, Vreuls, Roebuck). The gap between researchers and application professionals, as noted in Meister (1985), is still in evidence.

New areas, particularly cognititive engineering, demand research. In this connection, O'Hare suggests that there are significant gaps in theory as it relates to multidimensional performance. He feels that we need new methods in cognitive ergonomics because reduction into simple functions and verbal protocols do not do the job. In his opinion, the fractionation of the field into new specialities has negatively affected basic HFE training. For those like Swain, who feel there are gaps in methodology, quantification seems to be the answer.

Many of the respondents were unwilling to discuss this topic or confined themselves to generalities and superficialities. What this says about the capability of many HFE professionals to deal at a sophisticated level with HFE problems is somewhat depressing.

Research in HFE Journals

This question was asked because publication of research in journals is one of perhaps the most obvious, overt manifestations of the discipline certainly so far as research is concerned. Therefore, the adequacy of what we read in the journals suggests what the status of the discipline is. Of course, what we are talking about is only one refereed journal; other HFE journals such as *Ergonomics* may supply different answers, but the author doubts this.

Only one of the respondents (Vreuls, who is a researcher/practitioner) brought up the most significant question: What is the function of the journal? Vreuls suggests that he is both satisfied and dissatisfied. Studies are more applied and useful than 30 years ago, he says, but they are not useful when faced with design problems. He says much research is politically motivated (by which one supposes he means governmental interests) and budgets are tighter. Journals are driven by input more than user need. Perhaps, he says, the problem is that of the journals themselves: what they are supposed to do.

Debons suggests that the journals are supposed to assign meaning as well as transmit information, although how one can do this is difficult to say. Howell (who is the editor of *Human Factors* and primarily a psychologist by his own designation) tries to establish a centrist position between academics and practitioners, each of whom, he suggests, demand more of their own type of work.

Overall, there is a feeling among the respondents that the research in the journals is somewhat inadequate, although there are a few who appear satisfied. It may be illustrative to summarize some of the responses received:

Wherry—The research lacks generalizability. Although HFE has broadened its interests, there are too few unifying concepts and gaps in what we know. Experimental techniques permit us to determine which design is better, but not why. Acceptance of statistical techniques like ANOVA and lack of interest in individual differences have hurt research. Real-world studies would be too long to publish, so articles are about simple problems that do not answer important questions.

Seminara—As a practitioner, most of the research is of little interest.

Obermayer—What is being published depends on what is funded and who is able to do research. Practitioners are unable to do research and thus assemble publishable data.

S. Parsons—The journal (*Human Factors*) covers limited areas. Like most practitioners, he looks for material he can use and often does not find it.

Strother—Applications are overlooked.

Neal—Research is becoming less theoretical and more applied.

Harris—The problem with the journals is the quality of the input, rather than editorial policy. Too many articles are narrow in scope and do not contribute to theory or useful information.

Erickson—There is not much in the journals that is useful to design engineers.

Swain—*Human Factors* is too much like the *Journal of Experimental Psychology.* He is not satisfied with the research published today.

Kurke—Most of the journal contents are boring and irrelevant to his interests.

Askren—Too many picky, little studies; we need more theory and relevant research related to major issues.

Heglin—There is a marked decline in research quality, particularly as shown in the *Proceedings.* There is a lack of breadth in theory, clear-cut experimental design, hypotheses. However, the earlier problems may have been simpler or else, with more specialties, people are publishing in journals other than *Human Factors.*

Wolbers—Research quality has improved in terms of methodological rigor. There is a much better job of interpreting findings in terms of applications.

Van Cott—We need more serious, naturalistic studies.

In general, if one can summarize the previous summaries and extrapolate to the HFE population at large, practitioners in particular seem to be largely uninterested in the content of the research because it does not speak to their interests. Academics are somewhat less dissatisfied because the research, while becoming more applied in orientation, is still heavily loaded on the academic side. There is a feeling that it lacks sophistication, breadth, and relevance. Obviously what is published depends on what is submitted, and this is in part a function of nontechnical variables such as funding and political (sponsor?) interests.

It may have been difficult for the respondents to attempt to assess research changes over time because the criteria (e.g., quality) are too vague to define concretely. There appears to be a feeling that the published research should have a broader base, talk to more important problems, and so on, but these may merely represent poorly articulated wishes.

What we take as defects in the published papers may really reflect deficiencies in HFE as a whole, the fractionation into specialties, the lack of unifying concepts, and the concentration on what can be researched (simpler problems) rather than on what should be researched. It may be that we need some overall vision about what we are trying to accomplish, and this seems to be lacking. However, even if professionals are dissatisfied with the journals and what these publish, they seem to accept what they have because there is no acceptable alternative.

None of the practitioners among the sample suggested that research was unnecessary, merely that it lacked relevance to their special interests.

Research is what is published; this is an output-oriented process. If so much research is rejected for publication (about 80% for *Human Factors*), what is wrong? Hypothetically, every piece of research is potentially valuable. If research is important to the discipline, and no one has suggested otherwise, and so much research is rejected, and what is published, after all this, is not particularly satisfying, something is obviously wrong, but the only changes the respondents can suggest are general. Many of our sample felt unable to respond to the question, which strongly suggests that they had not thought or thought deeply enough about the problem, which may be, one fears, characteristic of HFE professionals as a whole.

What Is Distinctive About HFE?

This question was an attempt to discover more directly how HFE defines itself. Previous attempts were indirect (e.g., examining responses to published research). The answers to this question may be those that were heard almost 50 years ago, before the Society was first established in 1957. There is a great deal of heterogeneity in the views expressed. First there are those who refused to answer possibly because they had never thought of the question and it did not interest them (but how could so fundamental a question not interest professionals?). These are the professionals who deal not with the discipline, as we have defined it, but who are merely members of a profession. For others who failed to respond, the question (like a question asked by one's therapist) may have been too frightening to deal with. These two points of view may be two faces of the same attitude.

Then there are those who see HFE as being solely or largely psychology. For example, Askren says there is an 80% to 90% overlap with applied psychology. HFE emphasizes technology, whereas applied psychology places more emphasis on the human response.

In addition, there are those who see HFE as a branch of engineering. O'Hare's point of view is that engineering is becoming the parent discipline. His reasoning is that, in the early days, it was the HFE professional who was used to supply HFE to the system, so he had to learn the military system (e.g., radar or sonar) before applying behavioral principles. Now in fields like aviation or computers, training in the system for its engineers includes behavioral principles and methods. Hence, the HFE professional is not needed.

There are others, like Wolbers, who call HFE *multidisciplinary* and attach the pejorative term *philosophy* to it. The same point of view was expressed by a letter from the Dutch Ergonomics Society to the Human Factors Bulletin (April 1963). One can ask why the variety of responses—the lack of certainty in the definition of what one is and does. Some part of it may be an unwillingness to think deeply about what one is doing (intellectual laziness) or, because HFE activities are so varied, it is difficult to pinpoint them.

Undoubtedly there are professionals who shrug the question off as if it were irrelevant (and it may be for them). However, it is not irrelevant. An identification of HFE with other disciplines or a failure to identify the parameters of the discipline may lead one to ignore HFE problems that are crucial to HFE. This may in part be responsible for our relative lack of progress in certain areas. For example, if one thinks of HFE as part of psychology and psychology does not have a quantitative database, then it is inferred that there is no need for an HFE database.

However, there are more directly linked implications of a discipline's self-identification. Indeed, the real importance of the question can only be understood in terms of the consequences of thinking one way or another about the question. For example, for what and how does one train HFE professionals? Are they to be taught applied psychology, industrial psychology, engineering, or what? Often people mention the need for HFE professionals to be able to communicate with engineers in their own language. Do we teach them the engineering language?

The definition of HFE also determines what an HFE professional needs to know to be qualified. Is he to be taught only what is necessary to perform the requirements of a specific job (as some practitioners suggest)? If HFE is a science, does one teach the professional to be a scientist first (thus emphasizing fundamental information first) and then a practitioner, or vice versa? The answers to these questions require a prior self-definition.

If one thinks of HFE as something other than a distinctive discipline, why have the trappings of a distinctive discipline—meaning a special society like HFES—which implies distinctiveness? Why do we not incorporate under the banner of a Society of Applied Psychologists (if such a society exists)? Many HFE professionals consider themselves as different from those in other fields, but they are unable to pinpoint the elements that make them different. This is primitive thinking, much like that of a child, who knows it is a boy or girl, but is not clear about the characteristics that differentiate him from her.

Rudov makes the point that no matter how we conceptualize ourselves, others (meaning the general public or professionals of another discipline) may think of us as something entirely different from our own concept. However, one must know who and what we are before worrying about how others see us.

Howell points out the tremendous tension in the discipline between academics and practitioners. As a psychologist, he sees practitioner thinking winning, but practitioners might have just the reverse point of view. He also emphasizes that there is associated with the tension much unfinished business, and he is perfectly correct. The Society's efforts over the recent years to develop strategic planning strategies are in part a reflection of that tension—of the inability to decide whether one is human, mechanical, fish, or fowl. Like Howell, the author believes these internal disciplinary tensions will not be resolved until there is a consensus about what we are inherently.

The Relationship Between Engineering and HFE

Engineering is the context in which HFE practitioners perform, and engineers are the people with whom they have to work. No two more different types could exist—one representing the behavioral domain and the other the physical one. How then do they relate to each other?

The initial interactions with engineering in the early days were somewhat antagonistic. It appeared to HFE professionals that engineers had no knowledge of or interest in human behavior. HFE professionals working in system development had difficulty communicating with engineers because few of the HFE professionals had any engineering background. Apparently the situation has changed somewhat. As Obermayer indicates, it is no longer a matter of *us* versus *them.* A number of the respondents even see HFE as a form of engineering, albeit a special branch, like civil or mechanical engineering.

Everyone in the sample assumes that a mutually supportive relationship with engineering is highly desirable—an attitude that could also be found in the views of the first HFE professionals. Indeed, engineering has the advantage because HFE must co-exist with engineering or stay out of system development. Some respondents interpreted this question as one involving the identity of the discipline. However, even those who feel that HFE is a form of psychology see a mutually supportive relationship with engineering as desirable. Behind this attitude, however, is also a feeling of some mistrust—a sort of technological ethnophobia.

It would be too simplistic to say that practitioners are engineering-oriented and researchers are psychology-oriented. However, practice pushes HFE professionals into design, which is the essence of engineering, and many researchers, from lack of opportunity to interact with engineers, do not possess the design/engineering orientation.

One point that was made by Nicklas deserves to be emphasized: that HFE and engineering are bound together through the system development process. Vreuls suggests that when it comes to design, HFE is a part of engineering. This is an important insight. Apart from system development and design, HFE is largely, if not totally, psychology. When HFE becomes part of the system development process, it becomes more like engineering. Without that process, the HFE research performed (if it were performed at all) would be indistinguishable from that of experimental psychology. Consequently, when research ignores system development, practitioners see no value in it.

One respondent (O'Hare) saw the engineering specialties, particularly computers, as potentially co-opting the HFE database and presumably performing their own HFE design. (Indeed, some HFE professionals working in the computer industry identify themselves as computer engineers perhaps because this makes them more acceptable to engineers.) HFE professionals have always denigrated the hypothesized engineering ability to perform HFE development work, based on their notion that, if one is human, one can solve human problems in design. This may have been an assumption of engineers in the early days, but one would not expect engineers today to express such a crude sentiment.

If there is a significant change in the attitude of engineers toward HFE, it is in part because of the ceaseless propagandizing of the early cohort. It is possible that the most important activity performed by the initial generation of HFE professionals was to raise the engineer's HFE consciousness level by informal and formal educational efforts.

What Will Happen in the Future?

This question asked respondents to predict the future of HFE in terms of the changes they saw coming. Prediction is often an extrapolation of present feelings and attitudes, of hopes and fears. In that respect, it may cast some light on those feelings and attitudes. However, whether a prediction has any validity can only be determined by experience, and those who make the predictions are unlikely to be alive in 50 years to determine how valid those predictions were.

One trend in the predictions that the sample made is for greater acceptance of HFE by other disciplines. Obviously our respondents feel that acceptance at present is not complete and they are heartily sick of having to explain and justify HFE. For example, Woodson hopes that HFE will become mainstream. Harris has a dual prediction: one optimistic and the other pessimistic. The optimistic prediction sees greater acceptance by other disciplines. The pessimistic one sees other, more aggressive disciplines, like industrial engineering and safety, taking over HFE processes and HFE disappearing. Askren is pessimistic in the sense that, with tighter budgets, HFE will have less impact, although the possibility of user litigation may make managers make greater use of HFE. O'Hare sees a continued need for HFE in high-tech environments. HFE training would be more closely associated with specific engineering specialties. Wolbers sees the expansion of HFE into additional areas of work, with the demand for HFE increasing. Consequently, there will be continued fractionation of the discipline into specialties.

As can be seen, one prediction is for an expansion of the profession into new venues (e.g., architecture, horticulture, medicine, etc.) and, as a corresponding phenomenon, a trend we see already—the increasing fractionation of the profession into specialties. Fractionation can be of two types. The first is the voluntary breakout of specialties on the basis of the individual interest of HFE professionals: That is what is occurring now. The second form of fractionation is when HFE methods, data, and so on are absorbed by other disciplines so that the need for a distinctive HFE profession no longer exists. There are both positive and negative aspects in this absorption.

It has always been an implicit and explicit goal of HFE that other disciplines would recognize the behavioral elements in their disciplines, thus enabling the HFE specialist to work more effectively. However, there is the fear of some HFE specialists that, if this occurs, these disciplines will assign HFE work to their own unskilled, unqualified engineers—something that happened occasionally in the past. In this line of reasoning, which is somewhat neurotic in its anxiety, assimilation would represent the complete success of the HFE philosophy, which sees behavioral elements in all technology. However, it might be too great a success because it could lead to elimination of the need for an independent discipline. The reasoning is neurotic because such a degree of assimilation will never occur, considering the interests and backgrounds of other disciplines. There are those who would like to eliminate HFE practice in system development, with the HFE professional being solely a researcher and translator of research results into guidelines for engineers. This would indeed fossilize the discipline because it would eliminate the rationale for HFE research. Even this, although perhaps theoretically appealing, is not likely to happen in the short span because present HFE research is, as is shown in chapter 6, not geared or directed toward system development questions.

Respondents also showed a preoccupation with computer technology not only as an environment in which HFE must function, but as an integral part of HFE methodology. Again, this is an expectation or hope that computer technology will in some mysterious way solve major HFE methodological problems. Reliance on computers (like reliance on models) reflects faith in a *deus ex machina*, which would make it unnecessary for HFE to work through its problems on its own.

An individual prediction is of course only a reflection of the professional's interest and past experience. For example, someone completely absorbed in aviation like Strother would see the future of HFE in terms of changes in cockpit design. An increased role for HFE in litigation was mentioned by respondents who probably had some experience with this type of activity. This is part of the feeling that an increased emphasis on the user in product design will result in the expansion of HFE.

There is no mention in these predictions of what HFE professionals can do to improve the effectiveness of their profession. Most HFE professionals may be other-directed, in the sense of seeing the future determined by external events and processes, such as more high-tech, more computers and automation, and so on. Possibly this is because of the way in which the question was phrased. It would be interesting to examine the responses if the author had asked this: What can and should HFE do internally, in its own operations, to become more effective? Of course, one might get more of the same responses. To change oneself by self-scrutiny is always more difficult and demanding than to look to external agencies. The HFE experience may have formed professionals in this mode. It must be remembered that HFE did not originate because of its own efforts but because of the demands of World War II and later governmental requirements. In any event, suggestions by the author and others that HFE can develop itself as a result of its own efforts have generally been ignored.

Most predictions were optimistic. Only one or two were blatantly pessimistic, although even the optimistic scenarios suggested an awareness that potential dangers exist. There is a strong element of uncertainty in the HFE mentality, which may be a consequence of the uncertainty arising from early vicissitudes.

CONCLUSIONS

It is difficult to develop general conclusions because of the many individual points of view and the fact that respondents were asked extremely difficult questions (because the questions were abstract and covered experiences of many years). Reading the recollections (not the summaries, which are much more stark) gives one the impression that the early years of HFE were an exciting time simply because there were so many unknowns to be resolved. It was by necessity a time of exploration, but the first generation may not have known they were explorers because their time was taken up by the mundane tasks they had to perform to do their jobs. The excitement may exist only in retrospect; it may require the perspective of time for participants to realize what they had done and undergone.

The transition between psychology graduate school and the new discipline was greatly aided by the fact that industry as well as government had little knowledge of or experience with human factors. If one had a graduate degree in psychology (or, to a lesser extent, engineering or physiology) and some prior employment in a human factors-related capacity, industry in particular welcomed one. No one questioned then—or for that matter now—whether the paradigmatic structure of psychology—principally S–O–R and the experiment—was completely appropriate to new human factors variables. Because the paradigm was successful (in the sense that no one questioned it), it was incorporated into the concept structure of the new discipline. This process was aided by the fact that few recognized that a new discipline was developing. The major intellectual preoccupation was showing the relevancy of HFE to other disciplines and contexts (Meister, 1995b). To a certain extent, this is also true of the present.

Behavioral scientists had blundered into a new discipline. However, because they could not think of an alternative and were comfortable with the psychological paradigm, they maintained it (in silence, of course, because no one raised the question). In consequence, much of what HFE professionals do today is psychology under the rubric of HFE.

There are those who will say this: If the S–O–R paradigm and the experiment have served so well in the past, why try to discover some other tenuous paradigm that may not be as useful as the old familiar one. There is a point to this objection. However, this is to say that the conceptual structure that represented the discipline in 1950 is just as useful in 2000. This despite that (a) HFE professionals deal with much larger entities (systems and equipment rather than individual controls and displays), and (b) there has been a giant swing away from simpler human functions used with proceduralized equipments to much more complex cognitive enterprises in a computer-driven society.

Although the old familiar paradigm may be found to be just as effective in 2000 as in 1950, HFE will never know what it can achieve in the 21st century if it does not examine other alternatives. Almost certainly it will be impossible to dump S–O–R and the experiment, but it is possible that these need to be supplemented by other concepts and methodologies. The author is just as confused as anyone else as to what these might be; he can only suggest that we must discuss the possibility to see what emerges.

There are still many unknowns and many problems to be solved. Even the definition of the discipline is problematic. HFE professionals are still divided on whether what they are doing is an aspect of psychology, a branch of engineering, or merely a multidisciplinary approach. It is fascinating that the Revised Strategic Plan issued by the HFES in September 1996 lists as its first goal, "Articulate the definition and boundaries of the discipline of human factors/ergonomics and its unique technology. . . . Formulate first draft of basic 'boilerplate statement.'" The Society had previously adopted a definition (Christensen, Topmiller, & Gill, 1988, p. 7), but obviously this was not completely satisfactory.

There is then a continuing need to define the discipline, which the Society represents, but the Strategic Plan recognizes this only in the context of taking the message to other societies/disciplines. For example, target groups include IIE, ASSE, American Board of Industrial Hygiene, Risk and Insurance Management Society, and so on. This is other-directedness with a vengeance.

The problems the Society sees, which is a reflection of what most of its professionals perceive, are phrased primarily in terms of communication with and persuasion of others. This is undoubtedly related to the early problems that HFE faced in terms of acceptance, by engineering in particular. There appears not to be a recognition that the discipline faces systemic problems that are inherent in the discipline and cannot be solved solely by communication with and persuasion of others. This is not likely to be a popular viewpoint because it puts the responsibility for HFE progress on its professionals and not on others with whom they interact.

To return to the memoirs, it is apparent that there is an insufficient recognition of the distinction between a discipline (a conceptual structure consisting of theories, methodology, and data) and a profession (employment opportunities, job activities, a Society, journal publication, etc.). Disciplinary problems are generally perceived in professional, rather than disciplinary, terms.

If the sample is representative of the discipline as as whole, HFE recognizes the need for both research and practice, but does not fully understand how they are related and the relative emphasis to be placed on each. There is tension between researchers and practitioners who see little value in the products of each other's activities.

The early HFE professionals saw many of their problems in terms of their acceptance by representatives of other professions, managers, and so on. The Society still maintains that attitude. However, the more perceptive respondents also saw (and see) that much of HFE theory, methodology, data, and principles did not and still cannot be well utilized in practice.

HFE is not a monolith if only because HFE professionals divide into two major groups on the basis of their work, experience, and interests: researchers and practitioners, with an amorphous and intermediate group of researchers whose research is of an applied nature. There is a consensus on certain issues and great divergence on others. There is general agreement that progress has been made (in such things as greater acceptance by engineering, earlier inclusion of HFE in the system development process, computer technology that has fostered much more simulation than was available earlier), but that many problems still persist. There is general recognition that nontechnical factors (e.g., availability of funding, governmental sponsorship, management attitudes, etc.) play too great a role in what happens to HFE.

APPENDIX 5.1

First-Generation Participants in Survey

Askren, William B., Ph.D., 1951
Brown, John Lott, Ph.D., 1952
Champney, Paul C., B.A., 1952
Chapanis, Alphonse, Ph.D., 1943
Debons, Anthony, Ph.D., 1954
Erickson, Ronald A., M.S., 1963
Guttmann, Henry E., M.S., 1962
Harris, Douglas H., Ph.D., 1959
Heglin, Howard James, Ph.D., 1955
Howell, William C., Ph.D., 1958
Irwin, Charles Holmes, M.A., 1957
Jones, Daniel B., Ph.D., 1965
Klemmer, Edmund T., Ph.D., 1952
Krendel, Ezra S., M.A., 1949
Kroemer, Karl H.E., Dr. Ing., 1965
Kurke, Martin I., Ph.D., 1963
McGrath, James J., Ph.D., 1960
Mead, Leonard C., Ph.D., 1939
Meister, David, Ph.D., 1951
Neal, Alan S., M.S., 1964
Nicklas, Douglass R., M.A., 1950
Obermayer, Richard W., M.S., 1956
O'Hare, John J., Ph.D., 1958
Parsons, H. McIlvaine, Ph.D., 1963
Parsons, Stuart O., Ph.D., 1958
Pearson, Richard G., Ph.D., 1961
Revesman, Stanley, L. Ph.D., 1953
Ritchie, Malcolm L., Ph.D., 1953
Roebuck, John A. Jr., M.S., 1965
Rudov, Melvin H., Ph.D., 1964
Seminara, Joseph L., M.A., 1952
Semple, Clarlence A., M.A., 1964
Sheridan, Thomas B., Sc. D., 1959
Snyder, Harry L., Ph.D., 1961
Strother, Dora Dougherty, Ph.D., 1955
Swain, Alan D., Ph.D., 1953
Teel, Kenneth S., Ph.D., 1950
Topmiller, Donald A., Ph.D., 1964
Uhlaner, J.E., Ph.D., 1947
Van Cott, Harold P., Ph.D., 1954
Vreuls, Donald, M.S., 1965
Warrick, Melvin J., Ph.D., 1961
Wherry, Robert J., Jr., Ph.D., 1964
Wolbers, Harry L., Ph.D., 1955
Woodson, Wesley E., B.A., 1941
Zeidner, Joseph, Ph.D., 1954

APPENDIX 5.2

The Human Factors and Ergonomics Society (HFES)

The reason for including the Society as part of the informal history of HFE is that the Society represents the professional face of the discipline, and in fact is the formal expression of that discipline. The profession (as distinct from the discipline) cannot be understood without reflecting on the Society, which was formed in 1957 in Tulsa, Oklahoma. It was the product largely of the informal deliberations of the mostly West Coast aerospace HFE professionals, among whom the names of Arnold Small and Max Lund are representative. Tulsa became the birth place of the Society because the Douglas Aircraft division in Tulsa provided facilities for the first meeting.

The Society is administered by an elected president and Executive Council, members of which have staggered 2-year terms. The Executive Council is primarily responsible for Society policies because the president has no special powers of his or her own. There are various committees, of course, and the editors of the documents the Society publishes (a position of considerable prestige and influence).

The goals of the Society are expressed in its Strategic Plan of September 1996 (in a very abbreviated form):

Goal A: Articulate the definition and boundaries of the discipline.
Goal B: Promote the dissemination of HFE information.
Goal C: Support the advancement of the discipline (primarily through the development of standards).
Goal D: Promote the discipline to the outside world.
Goal E: Enhance HFES decision making.

A review of the details of the Strategic Plan makes it quite clear that the Society emphasizes communication, much of which is propagandizing. Little of the plan deals with analysis and review of the status of the discipline with suggestions for modifications. Almost everything in the Plan relates to the profession. This has been characteristic of the Society since its inception. Rather than think of this as a deficiency, its other-directedness is perhaps inherent in the nature of professional rather than scholarly societies.

Two of the major functions of the Society is to put on the annual meeting of the Society and to publish journals, *Human Factors* and *Ergonomics in Design,* and *monographs*. It should be pointed out that, although the Society reflects the member's preoccupations and the attitudes of its members (around 5,000), it also affects these simply by the topics of the material it publishes.

The Society has a number of technical interest groups like Aerospace, Test and Evaluation, and Cognitive Engineering. Members who are interested in one or more of these groups may join them by paying a nominal additional fee. The primary activity of the interest groups is to publish two Newsletters for its members in each year and select papers that are presented at the annual meetings. The interest groups that reflect the specialization of the discipline are discussed in chapter 7.

The annual meeting activates many of the professionals who in their everyday work may be largely unaware of the larger concerns of the profession. The meeting is of course an opportunity to meet colleagues whom they have not seen since the previous meeting. At the annual meeting, they may be exposed to something other than individual specialties, but this is modified somewhat by the fact that papers at the annual meeting are organized in specialty sessions and members listen primarily to the papers describing their specialties. This was not originally the case. This procedure, in which the technical interest groups are allocated a number of sessions to use as they will and are permitted to select the papers to be read, symposia to be conducted, and so on, came about as a result, in part, of the increasing specialization of the profession. It may be that *corridor communication* (informal conversations among attendees) is useful, but anyone eavesdropping on these conversations would consider most of them trivial.

The Society and the profession as a whole rely largely on activists (i.e., professionals who have a marked interest in some aspect of the discipline and who are aggressive in nature). There are relatively few activists; this is, one supposes, true also of other disciplines. The Society functions primarily as a communications medium through its annual meeting *Proceedings*, the journal *Human Factors*, and the magazine, *Ergonomics in Design*. The first two are discussed in chapter 6.

The Society, as indicated by the comments made in its presidential addresses, is fairly self-conscious, and that self-consciousness has been manifested in the last few years by semi-annual meetings of the Executive Council, at which attempts are made through discussion to discover what it should do and to develop plans to conduct these activities. The Strategic Plan referred to previously is the product of such discussions.

It has been pointed out that one function the Society could perform is that of monitoring the progress of the discipline and making its evaluations known. It does not do this primarily because it does not conceptualize its role in this way. It has functions other than communication, including development of standards, liaison with other societies, accrediting new HFE university programs, running an employment service, and providing recognition by awards to outstanding professionals.

The Society does monitor and evaluate its communications efficiency. It notes the back log in papers published by the journal. It periodically surveys the demographic characteristics of its membership (salaries, degrees attained, etc.) and the attitudes of its members toward Society procedures. For some time, there has been a feeling on the part of many members that *Human Factors* was too academic for them, too removed from the concerns of practitioners. Because of this, the Society in 1993 established a new journal (*Ergonomics in Design*) supposed to cater to the interests of system development and applications professionals. Whether this has completely satisfied the need is not known, but, in view of the dichotomy between researchers and practitioners, probably not.

The writing style of the new journal *Ergonomics in Design* is more sprightly than *Human Factors* and the *Proceedings* (the latter has no consistent style at all because, once a paper is accepted on the basis of its abstract, the published paper is written entirely in the manner preferred by its author).

Through the years, the Executive Council has maintained a rather conservative position, and this has set the tone for the Society as a whole. Many things the Society might do it rejects because it considers these as too daring. For example, it has always refused the push on the part of a substantial number of members to certify professionals (i.e., ensure that they can perform as required by their jobs). Although the Society provided seed money to a private organization that eventually initiated this effort (the *Board of Certification in Professional Ergonomics*), it was left to that private organization to perform the function. The reasons for the Society's refusal may be the fear of lawsuits if someone is rejected (there were threats in the past from individuals who were refused membership in the Society) and the feeling that certification makes the HFE professional something less than an academic (who is certified only by his or her advanced degree).

In the early days, the Society was much concerned about finances, naturally so. Finances appear to be no longer a problem, but they remain a significant consideration for the Executive Council, particularly the disbursement end of it. There is reluctance to spend large sums of money, although the Council is quite liberal in making available small sums for a variety of purposes.

The Society does think of its having a tutorial function, as represented by the development of a series of monographs on various HFE topics. It also sponsors workships at the annual meeting on a variety of technical topics, primarily the methodologies involved.

In summary, the Society is important because it is the professional face of HFE. It is a reflection of its membership's attitudes, and it also influences these attitudes by what it publishes in its documents. The Society recognizes and responds to a primary need: to represent the profession to itself and, more important, to represent HFE to other disciplines and the general public. Like its members, it usually does not engage in soul-searching about the effectiveness of the discipline. What it does (putting on the annual meeting and publishing journals), it does fairly well. In comparison with what it could do (principally examination of the status of the discipline), it is deficient, but it must be remembered that neither its members as a whole nor the Society that represents them is very scholarly (although of course a few of its members are).

APPENDIX 5.3

The following letter illustrates far better than the author can describe the factors leading to a career in HFE.

October 13, 1996

Dear David,

This letter is a belated response to your request for reminiscences about the early days of human factors profession for your history project. I have been storing them up for you, so with your list of questions in hand let me fire away:

As a high schooler in the mid-1940s, I was confused. WWII had concluded and we ended up on the right side, but what to do in life, what did the world need? Somehow I got interested in Bauhaus architecture and industrial design, but when I traveled from my home in Cincinnati up to the terrifying city of Chicago to interview some of my industrial design heroes, I found myself disappointed. They turned out to be slick marketeers, seemingly without the idealism of "technology to improve society" that I had somehow conjured up. A deviation from engineering into architecture as an undergraduate proved similarly disappointing.

Then after finally graduating as an engineer, I was suddenly a brand new ROTC officer (the Vietnam war was on, and if you didn't go ROTC you got drafted!). I specifically asked for appointment to the Aeromedical Lab at Wright Field, not far from home. Here, though I wasn't a very good soldier (I fought the war from a chair) I was exposed to what I was really looking for—experimental science with human subjects to make systems work better. As I was young and eager and unmarried, I was volunteered by my bosses to test out equipment first hand: floating around in ice water in a rubber suit with a thermocouple up my you-know-what, riding a vertical test track for ejections (which may be blamed for current occasional back pains), jumping out of airplanes to test parachute opening devices, steering an old B-24 while strapped in the prone position in the nose gunner turret to evaluate controls for a proposed new super-skinny fighter airplane (a real pilot was in the real cockpit of the B-24 to keep us from crashing), and repeated rides on the Aeromed Lab centrifuge (I was a point on their lateral G tolerance curve: 10 G for 2 minutes—when squashed flat you have to force yourself to breathe—it was much easier not to!)

Julian Christensen, Walt Grether, Jim Baker, and some other Founding Fathers worked nearby, but since I had no psychology training I couldn't quite figure out what they did. But I resolved then to learn more about psychology—it looked interesting.

After service I did my master's with John Lyman at UCLA who taught me to love being a free thinker and never to be threatened by narrow specialists (at the time UCLA had done away with specialized engineering departments). John remains a good friend and mentor. Unfortunately UCLA couldn't stand the interdisciplinary challenge and they retreated back into the shells of traditional disciplines.

Somehow the East beckoned me back, and I had a chance to be a research assistant at MIT and do a combined doctoral program in engineering and psychology. Curiously, MIT had no psych department at the time, though some wonderful people were there then: J.C.R. Licklider of auditory fame, Bill McGill (who taught me that information theory and analysis of variance were the same thing and himself went on to be the president of Columbia), George Miller of Magical Number 2 fame, and John Swets and Dave Green, the original signal detectors. Sadly at the time MIT decided against forming a psychology department (psychology didn't have partial differential equations with funny boundary conditions to snow people with), and all these towering figures split within a couple of years for greener pastures. I was advised to go down the street to Haaavaaad if I wanted psychology, which I did.

Here the story was different (and so was the culture—I had to learn to read books, not just solve equations!). Miller had by now moved to Harvard, and also here were S.S. Stevens, George VonBekesy, E.G. Boring and of course B. Fred Skinner. I got exposed to them all, and it was exhilarating. About then Skinner was having his debates with Noam Chomsky about whether language is learned by conditioning or in the genes (Skinner lost the debates), but I gained great respect for Skinner as a scientist and as a person, and I retain that respect today (though operant conditioning is way out of fashion today—psychology is given to fashions to a silly extreme).

At about that time (mid-1950s) Duane McRuer and Ezra Krendel had published their first work modeling the human operator, and this looked like a chance to combine the psychology with the control engineering I had been struggling to make sense of. You couldn't design a stable airplane unless you knew how to make the COMBINATION of pilot and machine stable in a closed loop. Various people were struggling with the problem, including Jerry Elkind, Henry Jex, and others of our early human factors colleagues. I chose to study time-variability of the transfer function—could it be measured over a short period? (Answer: barely.) We started getting together once a year for what came to be called the "Annual Manual" meeting, a kind of floating crap game which was informally sponsored by different universities or government labs once a year. It went on for about 20 years until the participants finally decided the problems of pilots as controllers in simple loops had been pretty much solved (now there were computer-based autopilots). The annual meeting died a peaceful death. But while it was going it was a great ad-hoc community.

Given that simple human-in-the-loop control was mostly settled (and has been one of the few areas of human factors where a concerted effort to develop a predictive model and then apply it to hardware systems has really worked well) it seemed time to look more broadly to human engineering needs. In the late 50s Sputnik went up, the Americans got into the act, research dollars became available. My first Ph.D. student Russ Ferrell and I found ourselves struggling with earth-to-moon round-trip time delays of three seconds and how to cope with them to control lunar robots and roving vehicles from earth. Russ' thesis laid the foundation for understanding "move-and-wait" and got us both thinking about "supervisory control", which of course is the flag I have been waving now for over 30 years.

Has MIT embraced human factors? I wish I could say it has, but in truth it has been an uphill battle. Certainly a few individuals like Vannevar Bush and Norbert Wiener made huge impacts on the MIT landscape and fertilized the intellectual soil here to motivate me and others in related pursuits which combine human and physical sciences. But at the same time MIT is a curiously conservative place, not given easily to "soft science" or to even engineering design when not tied to well-codified quantitative analysis. There has been no human factors program as such and never even a department of industrial engineering here. The feeling seems to have been that empirical methods are OK but insufficient for engineering at MIT. Newton's laws are old stuff, and any science or engineering worth its salt should have at least the equivalent of Newton's laws well established. Psychology has been in the same boat ("cognitive science" based on computational models and "brain science" based on neurophysiology were finally accepted).

Apologies to your readers for dwelling on MIT, but to some extent and for better or worse we do set an example for others. Clearly the engineering profession as a whole still regards human factors in much the same way as does MIT. I suppose we should compare ourselves more to medical doctors than to physical scientists and hardware/software engineers. Medicine is the engineering of the health of individual humans, and the methods have been largely empirical. Human factors is the engineering of the health of systems of people and hardware/software, and the methods have been similarly empirical. To develop theory is the great challenge, just as in medicine. People have come to accept medicine because they know their individual health is at stake. Eventually, for the same reason, they will come to accept human factors, because their social health is at stake.

All the best,
Tom
(Thomas B. Sheridan)

APPENDIX 5.4

It may be difficult for someone who has not lived through the very early days of HFE (the 1950s in particular) to experience what it was like then to work as a HFE specialist. The following letter written by one of the first generation and reprinted without change presents a picture that may help. The author is indebted to Dr. Steven M. Casey for a copy of the letter written to him.

5 May, 1997

Dear Steve:

I have just had the opportunity to read your very interesting article in the April issue of the HFES Bulletin on "The Business of Human Factors." It brought back a lot of memories of the early days, some fifty years ago, when we and the world were young. In those days the idea of designing jobs for people rather than selecting people for jobs was still relatively new. Accordingly, the point of departure for addressing "human factors" design issues many times evolved from the more traditional selection and training approaches of Industrial Psychology. Industrial psychologists (there weren't any formally trained engineering psychologists then) began to include in their repertoire, the consideration of the human as a key system element having unique "impedance matching" requirements for the optimal operation of man/machine systems. I think that is why many of the human factors organizations that have spun-off over the years had their roots in more traditional Industrial Psychology consulting organizations.

In addition to Dunlap and Associates and Harry Older's Psychological Research Associates whom you mentioned in your excellent article, two other organizations that were prominent in the late 1940s and early 1950s were John Flanagan's American Institute for Research in Pittsburgh, and Floyd Ruch's Psychological Research Center in Los Angeles.

The Psychological Research Center was a partnership formed by Floyd Ruch and Clark Wilson in 1948 to offer services to business, industry, and government in areas of applied psychology. Clark Wilson had been a Lt. Cdr. in the Submarine Force during World War II. During his Naval Service he was exposed to a government published version of Floyd Ruch's best selling textbook "Psychology and Life". He corresponded with Floyd Ruch and when the War was over he enrolled at the Univ. of Southern California receiving his Ph.D. in Industrial Psychology in 1948. At that point Ruch and Wilson formed the Psychological Research Center (PRC). As a side note, apart from the clinicians, most of the faculty members of the Psychology Department at USC, J.P. Guilford, Milton Metfessel, William Grings, Neil Warren, et al had been involved in the War effort in the Navy or in the Army Air Forces Aviation Psychology Program. In addition most of the graduate students at that time were returning veterans who had had first hand experience with the problems of human interactions with complex systems during the War years. Some, such as Wayne Zimmerman, had even worked for J.P. Guilford in the Army Air Force Aviation Psychology Program. The University of Southern California represented a great resource at that time in terms of faculty and graduate students with wartime field experience in operational environments. In addition to the traditional programs in industrial and experimental psychology and in psychometrics offered in the Psychology Department, I organized and taught a graduate course called "Human Factors in Engineering Design" in the Department of Industrial Engineering.

Anyway back to my story. Floyd Ruch had many contacts in the local business community and Clark Wilson had many contacts in the Navy Department and opportunities for business rapidly materialized. Ruch and Wilson initially hired four graduate students, Bob Mackie, Jim Parkinson, Jim Myers, and myself to work on the projects as they came up. Since Bob Mackie and I had been in the Navy during the War we worked primarily on research and development programs under ONR contracts. As the business grew more people were hired and it provided a great opportunity for graduate students at USC (like Stu Parsons for example) to gain experience and earn money at the same time. Research projects ran the gamut from driver safety, maintenance design for Navy Submarine Personnel, troubleshooting programs and aids for Air Force Bombardiers and Navigators, Army Combat Leadership Research during the Korean War, to personnel selection and training programs for business and industrial clients. At its peak there were probably 50 to 60 people employed at PRC. Unfortunately in 1951 philosophical differences arose between the principal partners regarding the direction the future of the organization should take. Clark Wilson left PRC and formed the Management and Marketing Corporation, taking several key personnel including Bob Mackie with him. The Psychological Research Center was renamed Psychological Services Inc. and remains in business today under the direction of Bill Ruch, Floyd Ruch's son. Today it is an internationally known consulting company specializing primarily in selection, test development, training, and organizational issues.

I remained at Psychological Services Inc. as Vice President and Director of Engineering Psychology until 1954 when I joined Douglas Aircraft Company to work on aircraft cockpit design and other human factors issues. Clark Wilson was President of Management and Marketing Research from 1952 to 1956 and concurrently formed, and from 1953 to 1958 was President of, Human Factors Research. In 1959 Clark Wilson relocated to the East Coast becoming a Vice President of Batten, Barton, Durstein, and Osborne and Bob Mackie became President of Human Factors Research. The rest you know.

The 1950s were an exciting growth time in the human factors field. In 1950 there were a relatively few people (less than 100), across the country, with similar interests. For the most part everyone knew one another. By January of 1960 the Air Force had compiled a directory of over 900 names and addresses of people engaged in human factors work.

In the early 1950s, those of us in the Los Angeles area formed a small informal group comprised of people working on human related problems at the various aircraft plants, companies such as Psychological Services and Protection Inc. (a manufacturer of crash helmets and safety devices) and at USC and UCLA. We called it the Aeromedical Engineering Association and generally met monthly at one of the local restaurants to have dinner and compare notes.

By 1955 there was interest in forming a more permanent organization so a committee was appointed from the Aeromedical Engineering Association consisting of: Stan Lippert, John Poppen, and Don Hanifan from Douglas Aircraft; and Larry Morehouse from UCLA to get together with Arnold Small, Don Conover, and Wes Woodson from the Human Engineering Society in San Diego, to develop plans for a National Society. The result was the HFES of today. The Constitutional Convention and First National Meeting was hosted by the Douglas Aircraft Company's Tulsa Division in Tulsa Oklahoma because it was a central location between the East and West Coasts. Max Lund from ONR agreed to hold ONR's Fifth Annual Human Engineering Conference in Tulsa in conjunction with the Human Factors Society of America's Organizational Meeting as an additional incentive or inducement for people to attend.

It is interesting to note how the Leadership of the Society has evolved over the years from West Coast dominance (Morehouse, Small, Roscoe, Lippert, Lyman, Mac Parsons, Woodson, Hopkins, Meister, Grace, Conover, Hornick, Burrows, Stu Parsons, Holly, Rabideau, et al.) to a broader more heterogeneous National involvement of people with interests far beyond the aerospace/military focus of the early years.

I hope these comments may be of interest to you Steve in your capacity as HFES Historian. The last fifty years have been an exciting time for those of us who have been privileged to work on a broad spectrum of man/machine systems from submarines, to high performance aircraft, to space stations, to advanced space systems. It is important, I think, to preserve some sense of history so that those who come after have a sense of where we have been. Keep up the good work!!!

Sincerely yours,
Harry L. Wolbers

6

Characteristics of HFE Research

One way to analyze the characteristics of a discipline is to examine its technical papers for what these tell us about the conceptual patterns of its professionals. The analyst works backward from the papers to infer the concepts that professionals applied to their work. This is a specifically historical analysis rather than an empirical one because it is not feasible to ask the authors of these studies what they were thinking about when they performed a study. As a reasoning process, this may seem somewhat tenuous, but the conclusions are anchored in evidence derived from the actual statements made by the authors of these papers.

This chapter is based on a content analysis of more than 621 papers published by the Human Factors Society during the years 1965 to 1995. The papers were divided into two classes: empirical (those in which at least one individual's performance was tested) and nonempirical studies (those in which performance was not measured, although reference might be made to previous performances). Nonempirical papers consisted of such material as presidential addresses, reviews of previous research, and descriptions of equipment and methodology.

There were two types of analysis. The first was longitudinal—tracking changes in research characteristics over the years. The second considered research characteristics implicit in the papers themselves, disregarding chronological changes. The content analysis of the empirical papers determined the primary and secondary themes of the papers; their source; the measurement venue; type of subjects used; the methodology employed; the motive for initiating the research; whether theory or modeling was involved in the research; the unit of analysis; and whether any hypothesis or applications were presented. These content analysis categories describe the context in which professionals worked (e.g., industry, university) and the assumptions with which they approached their work (e.g., the motive initiating the study, the type of study performed, any hypotheses or applications). The content analysis of the nonempirical papers only dealt with the primary and secondary themes of these studies because it was felt that the context information provided by the empirical papers also applied to the nonempirical material.

The conclusions derived from the analysis of the empirical papers suggest that HFE research stems mostly from the university (although this was not true of the early days of the discipline); it uses the experimental method mostly, but with increasing use of subjective techniques to support experimental findings; it involves more students as subjects than others; its research is more problem-oriented than technology-centered; and it is largely unrelated to theory or modeling, with few applications. Using a river as an analogy, HFE empirical research is broad in the range of topics addressed, but extremely shallow in terms of the frequency with which each topic is studied. Nonempirical research is much more concerned with design, measurement, and methodology issues. HFE research demonstrates a common tendency in primitive disciplines—an increasing *academicazation* and professionalism, which has tended to alienate the discipline from the system development and applications questions that were originally the source of the discipline.

EMPIRICAL HFE RESEARCH

Introduction

This section describes empirical HFE research as published in two documents: *Human Factors* and the *Proceedings* of the annual meetings of the Human Factors Society (now called the Human Factors and Ergonomics Society, but henceforth referred to as the Society). Empirical studies are defined as those in which human subjects are tested; nonempirical studies are those in which no human subjects are tested specifically for the study being reported. This is admittedly a crude distinction because nonempirical studies may include those in which previous research (in which personnel performance was measured) is reviewed. Nevertheless, although crude, the distinction is serviceable and easily made, which is why it was employed. A later section of the chapter presents the analysis of nonempirical research.

The papers in *Human Factors* and *Proceedings* were analyzed separately because there are distinctive differences between the two sources. Papers in *Proceedings* provide a more comprehensive picture of HFE research because a wider variety of studies is included than in *Human Factors* because of the latter's more professional editorial policy.

A major difference between *Human Factors* and *Proceedings* is that, after the first few years of publication, the latter imposed a limit on length of text (five pages), whereas this restriction does not apply to the former. Therefore, the author published in *Human Factors* has a greater freedom to expatiate. However, selection criteria for *Human Factors* are so rigorous that approximately 80% of papers submitted are rejected. Although selection criteria for *Proceedings* are not so rigorous, there are some. These criteria are understandably not well defined. One would hope that all published studies would represent significant contributions to the scientific knowledge base with accompanying technological applications, but very few studies satisfy these criteria. Criteria for less extraordinary, more routine studies are difficult to specify verbally.

Published HFE papers represent what can be called *intellectual history,* which is a history that can be examined to determine what HFE researchers were thinking about in the past. The methods of historical research can be applied to scientific research because published reports can be treated as historical artifacts like archeological remains.

The author's general hypothesis underlying this study is that there are certain relatively invariant (over the years) themes and trends that underlie HFE research and that discovery of these can provide a useful view of the discipline.

There are two aspects of the present analysis: historical and epistemological—historical because it may be possible to trace changes in these characteristics from 1965 to the present, epistemological because the study examines such items as the rationale for the research performed, the effects (if any) of theory, and the applications of the research. When we combine history with epistemology, we can see whether time has affected that epistemology. To the extent that what HFE personnel as scientists are doing in their research involves the production of knowledge, this study can also be considered a contribution to the sociology of knowledge. The sociology leads us to consider such factors as who did the research, from what institutions did they come, what kinds of subjects did they use, and so on.

The context of this study can be described in a series of questions that the study attempted to answer:

1. What are the characteristics of past and present HFE research?
2. Have there been any significant changes in the nature of HFE research over the years?
3. What factors influenced the nature of that research?
4. What motives induce or initiate HFE research?
5. What institutions have affected the nature of that research?
6. What topics have dominated HFE research? Has technology had a significant effect on the selection of topics?

7. What is the role of theory and modeling in affecting HFE research?
8. Is this research useful and to whom?

There must be some qualifications and caveats to begin with. Only research published by the Society is considered and that limits our purview, but not too substantially. Because of the immensity of the task of analyzing every paper published in the last 30 years, the author sampled by selecting certain years for examination. Of course, sampling is inherent in behavioral science; although it always leads to some error, the error can be quite small—the author hopes so. Within each year sampled, all papers were reviewed.

METHODOLOGY

Introduction

The methodology employed was that of content analysis, which is simply the analysis of a text by asking questions about that document. Content analysis is frequently employed by psychologists, HFE analysts, and historians. For example, task analysis, which is at the heart of HFE analytic procedures, is merely a special form of content analysis. The specific content analysis categories were derived from the questions asked of the study as presented in the previous Introduction.

The analytic categories used in this study do not exhaust all possible descriptive categories for a study, but a totally comprehensive taxonomy would be unwieldy. For example, one can analyze research in terms of experimental variables, hypotheses, dependent variables and measures, and statistical test design. Although these were considered as categories, they did not seem to be productive.

From the standpoint of the researcher, content analysis has certain advantages: It requires only a single individual and needs no equipment. However, unless the results of the content analysis are exposed to criticism by subject matter experts, it is an entirely idiosyncratic effort, subject to the vagaries of the analyst. Because the author was aware that the possibility of bias with subjective criteria and a single analyst always exists, he attempted to compensate when faced with an ambiguous situation. The analysis included two types of categories: purely factual items (e.g., the number of authors of each study, the source institution from which they came) and more judgmental items (such as motives for initiating the research and the overall type of study described in the paper).

Because judgmental criteria may change over the course of study, the original analysis was repeated after it was completed. To restate for purposes of emphasis, after all the papers had been analyzed, the process was repeated;

where criteria had perhaps unconsciously changed, the original analysis was modified to satisfy the final criteria. The author feels that this process undoubtedly improved both validity and reliability (although it would be hard to determine that reliability unless the study was repeated by another researcher). This last is of course not possible. From the standpoint of historical research, this poses no problem; although historians consult each other about the analyses they perform, they are not held to the same quantitative standards of proof as are HFE researchers.

One other point should be made in connection with methodology. This study did not attempt to evaluate the adequacy or significance of any individual study analyzed. This is a task that is so idiosyncratic that it cannot be addressed in a supposedly objective study. Obviously some papers appeared to be more important than others, but the presumed importance of a paper was not allowed to influence judgmental categories.

Data Sources

There were two data sources: (a) the published *Proceedings* of the annual meetings of the Society in the years 1973, 1984, and 1994; and (b) the papers published in the journal *Human Factors* in the years 1965, 1968, 1970, 1973, 1984, and 1994. Six hundred and twenty-one papers were analyzed. The total number of papers in each source and the number and percentage of those selected for analysis by year are presented in Table 6.1.

Certain conclusions can be drawn from Table 6.1. The most important is that the percentage of empirical papers increases markedly as the years progress. The increase is most obvious in *Human Factors*, where, as of 1994, almost all papers published are empirical. The same progressive increase is noted in *Proceedings* as well, but the percentage increase is not as great as in the journal. The interpretation of the differential between the two documents can be made that the journal was intended to adopt a more academic format, which largely excluded nonempirical papers, whereas the *Proceedings* was supposed to accommodate a greater variety of papers. Therefore, in *Human Factors* at present, there are few conceptual *review* papers. (Although, as of the book publication date, there has been an increasing effort to include these.)

Although nonempirical papers are found in *Human Factors*, these were almost invariably published as part of Special Issues, which were not under the control of the formally designated journal editors. Nonempirical papers published in *Proceedings* obviously do not fall into the Special Issue category, but are often associated with symposia.

If the author is allowed a personal impression, the early *Human Factors* papers were more innovative than they are now, in the sense that they asked fundamental questions like, what is human factors as a discipline, what

TABLE 6.1
Number (#) and Percentage (%) of Empirical (E)
Papers in *Human Factors* and *Proceedings*

Human Factors								
1965			*1968*			*1970*		
TOTAL	*# E*	*% E*	*TOTAL*	*# E*	*% E*	*TOTAL*	*# E*	*% E*
60	39	65	70	40	57	60	35	59
1973			*1984*			*1994*		
TOTAL	*# E*	*% E*	*TOTAL*	*# E*	*% E*	*TOTAL*	*# E*	*% E*
54	32	59	64	56	87	49	42	86
Proceedings								
1973			*1984*			*1994*		
TOTAL	*# E*	*% E*	*TOTAL*	*# E*	*% E*	*TOTAL*	*# E*	*% E*
84	38	45	218	146	67	254	189	74

should it be doing, and so on. (These questions, which may be considered now as somewhat naive, have not yet been answered but they are being asked only infrequently, except possibly with regard to individual specialty areas.) It is possible that those who engineered the shift of HFE literature to a more academic model assumed that speculative papers would now be published in *Proceedings*. The latter does publish such papers, but again infrequently. Again, there appear to be recent efforts in the *Proceedings* to redress the balance. The academic model appears to dominate *Proceedings* as it does the journal, but less so.

Content Analysis Categories

The content categories listed in Table 6.2 are defined in the following discussion.

Primary Topic. This is the overall topic that the study addressed. Almost always this topic was system-oriented, like aerospace or automotive, or descriptive of a higher order human function, like visual performance. Almost every study has more than one topic and it is a matter of judgment which one is primary. There appears to be a hierarchy of study themes—lower level ones being subsumed by higher ones. Of course, the most common theme would be human performance, but this is so nondescriptive because

TABLE 6.2
Content Analysis Categories

(1) General theme
(2) Specific theme
(3) Source: (a) university, (b) government, (c) industry, (d) contractor
(4) Venue: (a) laboratory, (b) simulator or field test, (c) operational environment, (d) classroom, (e) irrelevant, (f) unspecified
(5) Subjects: (a) students, (b) industrial workers, (c) general public, (d) operational personnel, (e) unspecified
(6) Methodology: (a) experiment, (b) performance measurements only, (c) subjective (1. interview; 2. questionnaire; 3. ratings; 4. observation), (d) operational data, (e) physiological measurements, (f) accident data, (g) case history, (h) environmental measurement
(7) Research initiation: (a) theoretical test, (b1) empirical problem, (b2) conceptual problem, (b3) design problem, (c) missing data, (d) new technology, (e) product development/evaluation, (f) methodological investigation/development, (g) model development/testing, (h) test development/evaluation, (i) training evaluation, (j) procedure development/evaluation
(8) Theory involvement: (a) none, (b) reference only, (c) theory testing, (d) model development/testing, (e) model reference
(9) Research type: (a) theoretical study, (b) methodological study, (c) equipment/system application, (d) study of variables affecting human performance, (e) product development/testing, (f) model testing, (g) determination of performance characteristics, (h) analysis of operational data, (i) problem solution
(10) Unit of analysis: (a) individual, (b) workstation, (c) system, (d) organization, (e) sensor measurement, (f) grouped operational data
(11) Hypothesis: yes/no; confirmed/disconfirmed
(12) Design application: yes/no

of its generality that it was discarded. Color coding is a subset of visual performance, so that the latter would be considered primary, the former secondary. The more inclusive category was always considered primary. The most inclusive topic in general was the system to which the study applied. Thus, any study that dealt with aviation or space automatically was assigned to aerospace as a superordinate category, although the study might deal with such subtopics as navigation or pilot decision making, which then became a secondary topic. If a superordinate system was not discernable, as was the case with more psychologically oriented studies, then the superordinate human function was assigned (e.g., visual performance, decision making, or tracking). No attempt was made to break the study topic into more than two categories, primary and secondary, although it would have been possible to do so. Obviously most studies have multiple descriptors.

There were two purposes in describing studies in terms of primary and secondary topics. One wishes to know which research themes were most common and repeated most frequently over the years. The frequency with which certain topics were addressed would indicate how intensive research was in any given area. The greater the number of research topics addressed, the broader the range of research interests. The greater the frequency of research on any particular topic, the more intense the research. The number of topics and their frequency could also be analyzed chronologically.

Secondary Topic. A primary topic presupposes a secondary one, with the superordinate topic encompassing the subordinate one.

Number of Authors. A study was performed by one, two, or more authors, and this was noted. This category was adopted because of the hypothesis that, with the increasing sophistication of the discipline, research would be undertaken by teams rather than by individuals and this might throw some light on the sociology of HFE research. One might also consider the gender of the researcher, but this analysis was not undertaken. Obviously because there were few women in HFE in the early days, there would have been a significant rise in their numbers over time.

Source. Research is performed in the context of some institution. It is extremely rare for an individual to perform HFE research out of his or her home, although there are one or two examples in the data sample (Kelvin, 1973). In general, research emanates from a university, government agency (laboratory, test station), industry, or a contractor. It was hypothesized that in the early days of the discipline more research would emanate from industry and government facilities, but that as time went on the influence of the university would be increasingly manifested and become dominant.

Venue. Where the research was performed was also considered important. The number of alternative locations is limited: laboratory, simulator, field test site, operational environment, or classroom. In some cases, locations would be irrelevant (e.g., if the study involved a mail survey). The research might also fail to provide information about study location, in which case it was noted as *unspecified.*

It was hypothesized that, because of the need for control over experimental conditions, the laboratory would at all times be the choice of the researcher. As is seen, this is only partially true. Use of two such different test environments as the laboratory and field test site in the same study makes it likely that discrepancies will arise between their data. This raises the question of how to compromise divergent results (e.g., see Kline & Beitel, 1994).

Although the terms of the subcategories are in common usage, they must be defined. As a minimum and fundamentally, a laboratory provides controlled conditions for presenting stimuli and recording responses. It does not pretend to reproduce an operational system or situation, although the stimuli and the situation may reproduce certain aspects of the operational system or environment. When an attempt is made to reproduce the operational system and/or situation in its entirety, one is dealing with a simulator. Sometimes it is a judgment call as to whether one is dealing with one or the other (e.g., in partial simulations). Laboratories may vary from the very complex to the very simple; many laboratory situations are so simple as not to deserve the term.

The field test site moves the test situation to a controlled environment outside the laboratory (e.g., a test driving range). This is a situation intermediate between the highly controlled laboratory and the less well-controlled operational environment. If one were considering a driving study, the operational environment would be city streets or a highway. If the research situation were aerospace, the operational environment might be an instrumented aircraft in flight or even a routine commercial flight.

In certain simple situations, a classroom could be used for testing. The test location would be irrelevant in the case of a mail or interview survey, unless the survey included questions about the operational environment in which the survey was taken. Of course, in a number of studies, the researcher might fail to indicate the study location.

Subjects. Obviously the number and characteristics of the subjects used in a study are critical factors in determining the results secured. Some of the available categories appear obvious, but others must be defined. Students are those whose primary descriptive characteristic is that they are affiliated with a university and study as a major vocation. Industrial workers are essentially factory employees, assemblers, and inspectors—those who engage in manufacturing activities. There may not be too many of these because they overlap with the category *operational personnel,* meaning personnel who are engaged in skilled work other than manufacturing and who are not students. Examples are typists, office workers, process control operators, and pilots. There are occasional obscurities. An individual with a private pilot's license but who is not a commercial pilot would still be considered operational personnel in an aerospace study because of the special relevant skills he or she had. The general public are those subjects who have no special qualifications relative to the task required by the study. They may be picked at random in a shopping mall or be on the researcher's list of paid volunteers. Students are also without special skills, but their relationship to the university requires their special categorization. At the start of this study, it was considered possible that the dominance of one subject population over others might change over the years and that the nature of the subject population might reflect the nature of the research being performed.

The adequacy of the number of subjects used in these researches is not addressed in this study. The author assumed that the number of subjects was appropriate for the statistical design employed and, in most cases, this is probably correct. In psychophysical and biomechanical studies, the number of subjects is quite small, but that is consistent with the requirements of such experiments. Authors almost never discuss whether they had sufficient subjects for the statistical treatment they used, and it is better that no rigorous examination of this point be made.

Methodology. To analyze research tendencies, it is necessary to consider the methods used. These break out into two general types: objective and subjective, each with several subcategories. The primary objective method employed is the experiment, although it is possible (but relatively infrequent) to perform an experiment with subjective methods only. Without going into a detailed definition of the experiment, one can accept that any situation in which two or more treatment conditions are compared represents an experiment. The experiment may utilize a within-subjects design, in which all subjects experience the same treatment and the conditions being compared represent differences in stimuli, task characteristics, or the subjects. The term *experiment* is broad, subsuming everything from the classic one-treatment condition (experimental and control group) to multiple treatment conditions with multiple groups and elaborate statistical designs. The category entitled *performance measurements only* represents the situation in which nothing is compared. Obviously such a situation occurs only infrequently, as the reader sees in the results section.

Within the subjective category are to be found the interview, questionnaire, ratings/rankings/paired comparisons, direct observation, audio/video recording (considered as delayed observation), and verbal protocol analysis (Ericsson & Simon, 1980). None of these requires more detailed explanation. The analysis of operational data (i.e., historical performance data almost always in grouped form, like demographics, age, gender, or location) is another type of methodology. Operational data are distinguished from accident data. The latter are indeed a form of operational data, but accident data are sufficiently distinctive to require a separate category. Physiological measurements (including anthropometric and audiovisual testing) are another category. The case history represents a single test situation, a single subject, or the report of a single experience. In all case histories, some performance data were gathered. A somewhat infrequently utilized category is that of environmental measurements (e.g., measures of temperature, lighting, noise, etc.).

Any single study may involve more than one of these methods. This means that, in most cases, an experiment is performed in which subject performance is measured more or less objectively (e.g., tracking a visual target, navigating). This is combined with, among other techniques, a posttest interview, the subject's rating of some quality or phenomenon, and so on.

Research Initiation. A fundamental question is: What causes the research to be performed? This goes to the heart of the discipline. For example, if a study is initiated by a design problem, the discipline is fulfilling its responsibilities as a design-oriented discipline. If a study is initiated to resolve inconsistencies in the results of preceding studies or to fill a gap in knowledge, the discipline is responding to its needs as a branch of knowledge.

Manifestly, no study is performed as a result of a single motivation; everything in life is multidetermined. Because of this, it is almost impossible to determine which motive was most important. In consequence, all motives that can be discerned were noted. In addition, one has only what the researcher has written from which to infer his or her motivation. The researcher's own words were used as much as possible; it is sometimes necessary to infer, but one avoids this if one can. Sometimes the inference is easier to make than at other times. For example, if the author of a study says that the study has been performed to investigate carpal tunnel syndrome or solve a training problem, one can probably take his or her word. The motivation is less clear when the author indicates no purpose for the study but simply wades into an investigation of how Variable X affects Performance Y. The inference here is a little more tricky. In many cases, an explicit rationale for the study is not clearly expressed.

One cannot get involved with social and financial motives for research. Obviously if one's employment is as a researcher for, say, a government agency or as a contractor, one must perform the research dictated by government or employer policy. The relationship between employment and the selection of a research topic is more tenuous in research stemming from the university. Because of the *publish or perish* malaise, the academic researcher must research and publish, but what causes him or her to select a particular field (e.g., Human–Computer Interaction) and the particular variables involved? Obviously there is a cultural Zeitgeist surrounding research, and certain topics are undoubtedly viewed with greater approval by one's peers than are others. The purpose of the research may be specified in the written paper, but does this reflect the actual motivation? Is it because certain topics are hot at a particular time, as situation awareness is at present, and therefore more likely to be accepted for publication? Because the source of the data in this study is the printed word and the researcher cannot be interrogated, one must accept what he or she says.

Certain definitions are necessary.

1. *Theoretical test.* This is a study that explicitly seeks to test a theory, someone else's or one's own. The author of the study must explicitly state that this is a test of so and so's theory. The paper cannot merely refer to that theory.

2. *Empirical problem.* The impetus for the study is a problem arising in the real world that has certain HFE implications and must therefore be

investigated. Examples are: a high accident rate in driving automobiles or piloting aircraft, the occurrence of visual and fatigue complaints in office workers who use word processors, and the incidence of muscular trauma. The author must explicitly state in the paper that this was at least a background reason for studying this or that variable; it is not enough to infer this. Empirical problems may be of two types: general and specific. The latter represents actual safety problems actually encountered, whereas the former is more general and potential (e.g., the need to measure operator performance in nuclear power plant control rooms). It is sometimes difficult to tell whether an actual empirical problem exists because the existence of such a problem is so often assumed by authors as a rationale for a study. The empirical problem that initiates a series of studies is rarely the direct initiator of an individual study. The original empirical problem becomes part of the research context, but the specific problem that initiates the specific research is one or two steps removed from the overall problem. The empirical problem may be the difficulty of training pilots; the specific problem that initiates the individual study is the problem of how effective flight simulator cues are or which transfer of training principles ensures the most effective performance. If we think of factors initiating research in the same manner as physical forces, there are no easy, direct relationships among research forces.

3. *Conceptual problem.* A conceptual problem may or may not be derived from an empirical problem. It is much easier to infer a conceptual problem that has an empirical problem as its basis. A conceptual problem always involves the researcher's concept structure or interpretation of the potential causes of a difficulty. Is it too literary to say that a conceptual problem is inward looking, whereas an empirical problem is outward looking? The distinction between empirical and conceptual problems is also a matter of degree—of emphasis.

For example, fatigue in office workers is an empirical problem widely reported in the literature. Taking this as the distal cause of the study, a more proximate cause of a particular study may be the researcher's development of a hypothesis that the inability to adjust chairs may have at least contributed to the fatigue problem. The latter is a conceptual problem. Hence, the study might contrast fatigue claims by workers before and after the introduction of ergonomically designed chairs. Where there is no obvious underlying empirical problem, one looks next for a strictly conceptual problem. Logically, there should be either an empirical or conceptual problem, but sometimes neither is evident. The impetus for a study may be simply missing data or a new technology investigation. One tries one category and then another because there must be a rationale for performing a study.

4. *Design problem.* The category is assigned if the motivation for the publication of a paper was to demonstrate how a specific design problem was solved and to show how the solution could be applied to a variety of

design situations (e.g., see Kelso, 1965). The design problem is obviously a subcategory of the empirical problem, but it is called out specifically because of the special interest HFE has in design.

5. *Missing data.* The need to resolve inconsistencies in preceding data and the need to provide additional data because too little is known about a particular topic or variable are often cited as a rationale for the study. Some of the protestations about inadequate data may be *pro forma.* Because there is always insufficient data, the rationale is scientifically acceptable. Missing data are recognized by the following representative terms: "Questions are unanswered by the available literature . . ."; "few data have been presented . . ."; "there is a lack of studies . . ."; "advance our understanding . . . "; "received relatively little attention. . . ." The missing data category was used only when the researcher specifically called attention to a lack of data.

6. *New technology.* Occasionally the development of new technology requires behavioral study or the behavioral implications of already existent technology serve as a rationale. This category was applied whenever the study was obviously and closely tied to technology.

7. *Product development/evaluation.* If a new product is developed (with the new product being anything that related to a system, equipment, training course, set of software, etc.), it requires behavioral evaluation. In the early days of HFE, such evaluative studies were sometimes published in *Human Factors* as a tutorial contribution or the process of developing a product was described, also as a tutorial mechanism.

8. *Methodological investigation/development.* There are two ways in which one can conceptualize methodological development. Specifically, the researcher tests two or more alternative ways of accomplishing a function in terms of their effects on operator performance. An example is the comparison of inner and outer attitude displays in aircraft design (Roscoe, 1968). More generally, a study is performed that has direct or indirect implications for how one conducts research in a particular subdiscipline (e.g., Fowler, 1994). An example of this type of methodological study might be the comparison of the effectiveness of alternative ways of measuring work load. It must be noted that the comparison of two or more treatment conditions does not make a methodological study. If it did, all studies would be methodological in nature. For example, comparison of age or gender in driving involves a comparison of treatments, but it is not methodological because it does not supply a method of accomplishing a task or a new/better way of performing HFE research. Rather, such research merely studies the variables affecting human performance (in driving) or the determination of the performance characteristics of age- or gender-related personnel.

9. *Model development/test.* If a study is performed to support the development of a mathematical or other type of model, or if the model is already

developed and the study tests the adequacy of the model, this category was applied. Because the term *model* is often indiscriminately applied, the study must explicitly call the subject of the study a model and indicate its nature in terms of mathematical (or other) operations.

10. *Test development/evaluation.* The development of a specific test to measure performance (e.g., development of the Behaviorally Anchored Rating scale; Smith & Kendall, 1963) is a function performed by HFE research, hence the category.

11. *Training evaluation.* This category includes study of the training effectiveness of some equipment such as a flight simulator, a procedure (e.g., adaptive training), or a particular procedure for measuring training effectiveness, such as transfer of training.

12. *Procedure development/evaluation.* This category describes the development or testing of a specific procedure for performing a task or accomplishing some goal.

Theory Involvement. It is popularly believed, even by those who are not behavioral scientists (Bailyn, 1994), that theory is indispensable to research and research is performed to test theory. Hence, it is meaningful to inquire how much influence theory and modeling (the two are quite closely linked) have on HFE research. The question of how research is utilized in the development of theory was not addressed. There are several subcategories:

1. *No involvement.* The published paper contains no mention of theory, either in terms of testing or by reference. This category is not difficult to ascertain. If theory is not cited, there is no theory.

2. *Theory reference only.* The paper contains a reference to an already existent theory, but the study has not tested the theory in any way. The mention of the theory must be related to the purpose of the study or the way it was conducted. For example, it is not sufficient to use d' as a performance measure and, to cite Green and Swets (1966), if signal-detection theory is not directly related to the study.

3. *Theory testing.* The study specifically attempts to test the author's or someone else's theory; the researcher specifically cites testing of the theory as the purpose of the study.

4. *Model development/testing.* The same is true of a model.

5. *Model reference.* The same is true of a model.

In evaluating theory involvement, if it is said that there is none in a study or in HFE research in general, we cannot of course prove a negative. However, we can say that if there is no mention of theory in a study, researchers did not consider any theory important enough for it to be mentioned. Theory may have had a contextual effect (realistically impossible to demonstrate), but not one that was critical enough for the researchers to note.

A theory must be differentiated from a hypothesis. An hypothesis is specific to a question in the study; a theory is a conceptual structure that proposes to explain certain behavior (e.g., Gestalt theory, the theory of operant conditioning, signal-detection theory). Reference in the paper being analyzed to previous research in the area of investigation is not regarded as involving theory.

Research Type. This category overlaps, to some extent, that of Research Initiation because, in some cases, the purpose of the study also describes the type of study performed. The difference between the two is that Research Initiation explores the rationale for performing the study; the category Research Type describes what the study is after it has been performed. Although the two are closely linked, it appeared worthwhile to the author to discriminate between them.

The subcategories are:

1. *Theoretical study.* A study that tests a theory is considered a theoretical study (e.g., Wickens & Goettle, 1984).

2. *Methodological study.* A study that contributes to the way in which HFE research is performed or that tests which of several alternative equipments, displays, modes of operation, and so on is more effective is considered a methodological study (e.g., Hart, Hauser, & Lester, 1984).

3. *Equipment/system application.* This is a study that specifically relates some aspect of technology to human performance. For example, in one study (Konz & Streets, 1984), the difference between various hammer handle shapes was tested; in another (Kao, 1973), the effect of the shape of pen points on performance was investigated.

4. *Study of variables affecting human performance.* Such a study relates Variable X (age) to Variable Y (stress) in terms of their effect on human performance (e.g., fatigue, reaction time) or which describes how certain factors function in terms of that performance (see, e.g., Kumashiro, Hasegawa, Mikami, & Saito, 1984).

5. *Product development/testing.* See Beringer (1984).

6. *Model testing.* As the term suggests, the study tests a model (Karwowski & Mital, 1984).

7. *Determination of performance characteristics.* The study reports how people in general or in particular (as represented by the subjects of the study) perform in relation to the behavioral variables being studied (e.g., the muscular strength of young males as a function of practice time; Caldwell, 1970). The difference between this category and Subcategory 4, study of variables

affecting human performance, is that with the latter the researcher has determined that Variable X is indeed significantly related to operator performance, but the extent of the relationship is not fully explored. Let us hypothesize a biomechanics study that determines that male and female adult grip strength differs significantly, but presents the data only in terms of aggregated means and standard deviations for the two genders. This would represent a study of variables (gender) affecting human performance (grip strength), but not one of determining human performance characteristics because it presented only overall data. If the hypothetical paper reported grip strength in terms of age categories within each gender class and range or standard deviations for each age subgroup within the gender group, the study would be considered as having determined human performance characteristics because one could use the data for classification purposes. It is entirely possible for a study to be classified in both ways because, as indicated earlier, the two categories are closely linked.

8. *Analysis of operational data.* In such a study, operational data, such a demographics or accidents, have been analyzed to determine relationships (Laughery & Schmidt, 1984).

9. *Problem solution.* This study describes how a specific problem has been solved. The problem must be explicit and its solution must be adequately described.

Unit of Analysis. This describes the nature of the measurements reported in the study. The subcategories are:

1. *Individual.* The performance of the individual subject is what has been measured. The fact that the individual scores have been aggregated is of no significance. If the original measures were of individuals, this category applies.

2. *Team/workstation.* The unit of measurement is the workstation, which also implies team performance because if only one individual were involved it would be categorized as individual measurement.

3. *System.* Without going into detail as to how a system is defined (see Meister, 1991, for this), the unit of measurement may be a sonar or radar or comparable system unit. Obviously individuals or teams are involved in the human performance associated with the system, but the measures are reported only in terms of the system with which the individuals are associated.

4. *Organization.* Same as Subcategory 3 except that a total organization is involved (e.g., company, department, ship).

5. *Sensor measurement.* In addition to human performance measures, sensor measurements are reported (e.g., aviation parameters such as aircraft speed).

6. *Grouped operational data.* The study reports data such as demographics, accidents, or human-initiated failures in terms of classes (e.g., elderly

males between ages 70 and 85) and not in terms of individuals even if the data were originally collected on individuals.

Hypothesis. Scientific studies are supposed to be performed in terms of overtly expressed hypotheses (hypotheses that are presumably derived from theory). The subcategories here are: yes (one or more hypotheses were formed before the study was performed) or no (they were not).

Design Application. This describes the application of the study results to any real-world situation. The application must be explicitly specified by the researcher. Although design is emphasized, the application is to any potential intervention to or influence of the study findings on real-world phenomena or situations. This category has two subcategories: yes (such an application is suggested in the paper) or no (it is not suggested).

RESULTS

It is impossible because of space limitations to provide tabular data and other details. The results of the study are described in terms of the individual content categories discussed in the previous section and in that order. Data for the *Proceedings* and *Human Factors* were analyzed separately because it is entirely possible that the material in the two documents have dissimilar characteristics and, if combined into a single whole, might suppress meaningful differences.

It was hypothesized that the number of topics (the range of the discipline) would increase over the years. This is confirmed by the data, although the increase is by no means constant. It is apparent that only a few topics have been intensively researched. This suggests that HFE research in general is broad but shallow. Although a variety of topics was researched, only a few were studied with a frequency of more than 2 or 3. Considered as a whole, HFE researchers appear to want to know a little about a great many topics, but not very much about any one of them.

The major research topics are organized primarily around type of system (e.g., aerospace, computers) and to a lesser extent around human function (e.g., visual performance). In terms of system type over the years, the primary systems studied and reported in *Human Factors* are aerospace (primarily aviation), automotive, computers, and underwater (although this last appears only in 1970, which may make it a fluke). Among human functions, the ones most often studied are visual performance, biomechanics, information processing, vigilance (related to visual performance), and communications. There is little difference in major topics as between *Human Factors* and *Proceedings* papers. What we have then is a romance with aviation, the automobile, and computers as systems, and visual performance as a human function linked to displays. With the exception of computers, nothing much has changed in terms of research emphasis since 1965.

The emphasis in the HFE literature on aviation, the automobile, computers, and displays mirrors the intense interest generally felt by laypersons—the general public. The passion with which Human Factors people dove into computers (to the extent that a few ergonomists gave up their identity as HFE professionals to call themselves human–computer engineers) mirrors that felt by people in general. Scientific analysis thus follows popular perception. Of course, this does not invalidate the emphasis because these are important technological instruments, but it does show a historical connection. If ergonomics had developed in the latter part of the 19th century, when popular attention was focused on railroads and steamships, the same HFE research emphasis would have been shown in relation to railroads and steamships.

Visual factors play a significant role, as does biomechanics (strength, lifting). The reason for the frequency of biomechanics research in the literature is somewhat difficult to understand. As systems are automated, one would suppose that the importance of biomechanics would diminish. However, as more than one paper mentioned, there are still many jobs that remain fundamentally physical.

One topic category is significant by its absence—maintenance. There are two ways in which one can look at technology: in terms of the way in which it (and the human) functions routinely and the way in which it (and the human) performs or fails to perform when the technology breaks down (as it invariably does). Psychology as the study of the normal (abnormal psychology is at best a minor specialty, at least academically) has influenced our discipline to emphasize the normal functioning of technology. The author's viewpoint is that there should have been much greater emphasis on the human in technological malfunctions. Both approaches are important, and the second has sometimes been implicit in studies of the first, but maintenance and maintainability have not been given their due.

Number of Authors

A count was made of the frequency of multiple authorship of these papers. A few papers had as many as five or six authors, but the median number was two. The significance of this figure is unclear. There was always a possibility of a change over the years, but the frequencies do not change markedly and the significance of multiple authorship is obscure. One might conclude that American HFE research is often a team effort, but whether this is culturally significant is unclear.

Source

Certain changes occurred over time that have significant implications for the nature of HFE research. In the early years of the discipline, the university contributed only a small percentage of the papers published in *Human Factors*, with a proportionately greater influence of government agencies, industry, and contractors. The percentage of university-sponsored studies has risen significantly from 1965 (19%) to the present, with the percentage in recent years varying between 56% to 70%. In the same period, studies from other sources decreased progressively, with the greatest hits occurring in industry and contractor sources.

In the early years, industry contributed significantly to the numbers of papers published in *Human Factors*. A glance at the early issues of that journal reveals qualitative differences between early and later issues. Many of the papers (not counted in the present study) were frankly tutorial and demonstrational. No such papers are presently published in *Human Factors* and indeed not since the 1973 issue of the journal. The year 1972 was apparently a watershed for published *Human Factors* research. Prior to this time, the Society had not published a *Proceedings* document, and many of the studies published in the journal were reprints of papers presented at the annual meeting for that year.

With the start of *Proceedings* publication, the practice of publishing selected annual meeting papers was abandoned. *Human Factors* began to adopt a more professional image, as suggested by the reduction in percentage of nonempirical papers (tutorial, demonstrational, review). The format presented by *Human Factors* came to resemble that of American Psychological Association journals (other than the *American Psychologist*). The *Journal of Experimental Psychology* is an example of the format followed. Starting with the 1973 issue of *Human Factors*, the proportion of studies stemming from the university increased progressively, whereas that from industry and contractors was much reduced.

It can be hypothesized that as the image of *Human Factors* became more professional or academic, HFE people who were academicians or more academic-oriented were more willing to send their papers to *Human Factors* (or editors became more eager to accept them). If the editorial policy of *Human Factors* changed around 1973 with regard to criteria for acceptance of papers, it would have tended to discourage the submission of papers from industry unless these were strictly experimental. The criteria changes in journal publication were probably not immediate; they changed subtly, and it is not clear that they were ever enunciated openly.

At the same time, as the discipline was maturing, it may have been felt by the editorial staff of *Human Factors* that the early tutorial papers were no longer necessary and a more academically respectable image was necessary for the discipline. In any event, although papers from nonuniversity sources have never dried up completely, their relative contribution to the total number of papers published has diminished.

Whether this is good or bad for the discipline—whether this was the result of a conscious decision on the part of "movers and shakers" in the discipline—cannot be objectively determined, but may be related to the continuing complaints by practitioners (as differentiated from researchers) that HFE research had little to offer them and their concerns. That this continues to the present is graphically illustrated by Sind-Prunier (1996). A later section reviews the attempt to suggest research applications, but in general it must be admitted that HFE researchers do a much better job of human performance measurement than of translating the results of their measurements into useful applications.

The dominance of the university in a discipline that should have a much heavier loading on application is somewhat surprising, but may be partly explained by the fact that the pioneers of Human Factors in World War II were primarily academics who applied experimental psychological methods to applied tasks.

It should also be noted that a certain amount of HFE research that does not reach the open literature is performed by government agencies, contractors, and industry. Many of the abstracts in the four large volumes published by Tufts University for the Army's Human Engineering Laboratories during the period 1960 to 1967 never reached the general public. Unfortunately, what is not published in the open literature loses much of its value without an audience.

Hence, what we may be seeing in the data from this study is partially the bias of editors toward university and university-dominated (i.e., academic-type) research. Also, there are many multiple-source studies; it is unknown what the influence of the university is on these.

As a corollary to the preceding, an analysis was made of the 1997 *Proceedings* papers to determine (where it could be determined considering many researchers do not indicate the source of their funding) which source received the most funding. Forty percent of the 124 papers published by a university source were funded, almost all by government agencies; 27% of the 29 contractor papers were funded. Multisource papers (those in which university authors were paired with government or industry authors) increased the influence of the university even more.

Venue

Within the *Proceedings*, 1973 to 1994, between 40% and 50% of all studies were performed in a laboratory, again defining laboratory in loose terms. In the journal (*Human Factors*), the percentage has been somewhat higher—

between 50% and 70%. Simulators have had a much smaller *piece of the action*—between 12% and 22% in the *Proceedings*, and roughly the same percentage in *Human Factors*.

What is encouraging for those who believe in measurement in the operational environment (OE) is the small but important percentage of studies that made use of the OE during conduct of the study. This percentage is quite variable year to year, ranging from 35% in 1973 to 11% in 1994 for the *Proceedings* and from 7% in 1965 to 19% in 1994 for *Human Factors*. This suggests that the OE as a data venue is decreasing in popularity in *Proceedings*, but correspondingly has risen in popularity in *Human Factors*. What these two contradictory trends mean is difficult to tease out with any confidence. If the OE is representative of nonacademic-type studies (as the laboratory is of academic ones), then the *Proceedings* is becoming more academic, whereas *Human Factors* is becoming slightly less so. In any event, it apparent that all HFE research is not being performed with special laboratory equipment or simulators, as one might suppose from the apparent need to maintain control over one's variables. Then, too, the simulator must be considered a reproduction of the OE, although it is not the OE.

The contribution of the OE to research is in large part the consequence of convenience and work carried out with aviation and automobiles. In both cases, it is possible to record data from subjects flying routine air routes and driving on the highway. Nonsystem-oriented research (i.e., on human functions, general displays, etc.) can be studied more easily in a laboratory. Where the venue is irrelevant, it is because the methodology employed was that of surveys, interviews, questionnaires, and the like. It is unfortunate that a significant part of the research did not specify the venue used.

Subjects

There is a popular supposition that most HFE subjects are university students. They do form a large part of the subject population, although the percentage decreases in *Proceedings*, whereas it increases markedly in *Human Factors* from 1965 (27%) to 1994 (45%). The rise in students in *Human Factors* may be linked to the corresponding increase in studies from the university in the same journal.

However, it is interesting that the general public and operational personnel (nonstudents) form a larger proportion of research subjects than students in *Human Factors*. Operational personnel are of two types: people who are specialists in the operation of the systems being tested (e.g., airline pilots) or people who work at jobs other than those they perform as research subjects. Thus, operational personnel may be office workers, management staff, or engineers. General public are people who are paid for their participation as research subjects or who are picked up at shopping malls and the like. Except in studies involving strength (biomechanics) or visual functions or in which personnel must have specialized skills to perform (e.g., chess players, pilots, divers), most subjects are not selected on any criterion other than availability. In a number of studies, students or volunteers were used to perform highly skilled decision-making tasks like military analyses, which in real life require much training and experience. Again, except when special skills or capabilities are required, the amount of information provided by researchers about their subjects is slight. Presumably the assumption is made that human qualities that might affect the research results are randomly distributed across subjects and therefore do not significantly affect test results. As to type of subject varying over the years, there is considerable variation, but it is impossible to discern any consistent trend in this variation.

In the early years, 1965 to 1973, somewhat more operational personnel were utilized as subjects than in later years, but operational personnel have always been a major source of subjects. The failure to specify such concrete facts as subjects or venue suggests a certain conceptual sloppiness—not only on the part of the researchers, but also on the part of journal editors.

Methodology

General. The most common method is the experiment, whether it employs the classic between-groups design of experimental and control groups or the within-groups design in which all subjects receive the same treatments but in different orders. Incidentally, although a separate analysis has not been performed of this aspect, it appears to this author that, over the years, the within-groups design has become increasingly popular because it eliminates to a great extent the annoying problem of variance due to individual differences.

Beyond the primacy of the experiment in HFE research, it is interesting to note that subjective methods (interview, questionnaire, ratings/rankings) are increasingly used to supplement the experimental methodology. In relation to the other methods listed, experimentation is progressively a smaller percentage of all the methods used. This does not negate the overwhelming influence of the experiment. Rather, it suggests the increasing importance of nonexperimental methods as a supplement.

Complexity. The number of techniques used in any one study is perhaps one measure (certainly not the only one) of the study's complexity. (If one were to analyze the intricacies of the statistical design utilized, this could be yet another measure.) There is certainly more than a suggestion that researchers increasingly make use of a larger number of techniques. If this is equated with increasing complexity, one might say that HFE studies are becoming more complex. However, so many factors affect frequencies that one must accept such a conclusion with great circumspection. The research topics addressed will obviously affect which techniques are used and how many. At best we can say that there *seems* to be a trend toward increasing complexity.

Research Initiation. As noted previously, the popular stereotype is that behavioral research, like all science, is initiated in support of theory. Therefore, it is necessary to test this proposition. It is probable that the motivation for research is multidetermined, thus it is desirable (if it can be done) to find all the factors responsible for HFE research. Certain hypotheses can be entertained: (a) Research is initiated to test or at least illuminate theory; (b) in the case of a discipline so closely related to technology, one might suppose that much HFE research would be initiated by new technological developments or design problems; (c) new product development might also be a factor; (d) a problem arising in the real world, related to HFE or a conceptual one, could initiate research; and (e) a lack of data or the need to resolve inconsistencies could be a rationale for research.

Of the various motives, we take each in turn:

1. *Theoretical test.* Research to test a theory is almost completely absent. At best, only 1% to 2% of all papers published seek to test a specific theory. The role of theory is examined in greater detail later.
2. *Empirical problem.* An empirical problem is noted in almost half the studies published in the 1973 *Proceedings*, but subsequent *Proceedings* show a marked reduction. In *Human Factors*, almost a quarter of the studies published reference a problem occurring in real life, but there is no consistent trend in either of the two documents. Where an empirical problem is noted as the research rationale, it is often because the study was funded by agencies that are oriented toward the solution of real-world problems (e.g., OSHA).
3. *Conceptual problem.* Conceptual problems are infrequent in the 1973 *Proceedings*, but increase markedly in 1984 and 1994. Conceptual problems are frequent in the early years (1965–1970), become infrequent in 1973, but increase in the 1984 and 1994 *Human Factors*, but not to the earlier level. The trends are therefore somewhat confusing; on the one hand, conceptual problems in the *Proceedings* become increasingly important, but in *Human Factors* they are variable. The 1973 *Proceedings* would be expected to carry over the earlier empirical emphasis and only gradually assume the more academic, professional orientation of conceptual problems. Because *Human Factors* was expected to be a professional journal, it would be expected to emphasize conceptual problems.
4. *Design problems.* For a discipline so closely linked to design, it might be expected that design problems would have initiated a fair amount of

research, but the data refute this. Only in the 1994 *Proceedings* and *Human Factors* is there as much as 4% of papers in which design is considered as a research rationale.

5. *Missing data.* This category is a major rationale for research. Of course, it may have become *pro forma* for researchers to justify their work (at least in part) by claiming the need to add to the database, but even so that claim reflects a strong interest in that database.

6. *New technology.* Only in the 1973 *Human Factors* and the 1994 *Proceedings* does this category have any weight and that weight is not large (15%–17%). Just as with the design problem category, the presumed relationship of HFE to technology does not seem to stimulate research.

7. *Product development/evaluation.* This rationale, which is also related to technology and design, has only a minimal effect in initiating research.

8. *Methodological investigation.* The methodological rationale decreases somewhat in *Proceedings* over the years or remains modest (11%–15%) in *Human Factors.*

9. *Model development/test.* A few studies are performed in pursuit of the development of a model or the testing of an already existent model. This rationale is much like that of theory testing; one would expect a conceptually sophisticated discipline to do more in the way of model development and testing, but it simply does not occur. The reliance of the younger generation of HFE people on models and theories (the expression is sometimes heard, why do we have to build a database when we have theories and models?) represents a pious myth.

10. *Test development/evaluation, training evaluation, and procedure development/evaluation.* These motivations have only nominal weight except for training evaluations: The frequency of studies initiated by these motives is almost nonexistent. In some cases (only a few, it is surmised), the opportunity to conduct a study arises and the researcher "goes for it" (Hart et al., 1984).

Theory Involvement

In most cases, there is no theory involvement at all. There are minimal references to theory, as when results are explained in relation to a theory or when a method based on theory is utilized, but they are very few. The same can be said about model development and testing and model references. Moreover, what theory there is (and the research that accompanies it) has been directed at the individual and his or her responses to technological stimuli. There is a subspecies of HFE research devoted to teams, but it is comparatively small in Society publications. With the exception of a few researchers like Rouse and his colleagues (e.g., Rouse & Boff, 1987), HFE theory and research has not dealt with the system either as a physical or conceptual structure.

However, some studies dealing with this theme are to be found in the journal entitled *Transactions on Man, Systems, and Cybernetics*.

As indicated previously, the popular stereotype of what Appleby, Hunt, and Jacob (1994) have called *heroic* science (completely objective, correctly representing what is out there) and what Kuhn (1962) described as processes interrelating theory and research, this stereotype is that theory directs and induces research (as well as vice versa). The actual fact, as shown by the preceding analysis, is that this is only a myth, at least as far as HFE is concerned. Only a few heroic scientists like Wickens and his colleagues (e.g., Wickens, Sandry, & Vidulich, 1983) manage to overcome the barrier to what may be called *theory in research.*

The relationship of theory to HFE research is a difficult problem that needs to be explored in depth; all this study can point out is that there is a gap between the two, as reflected in what is published. The problem can be subdivided into three parts: (a) the relationship between theory and research, the assumption being that theory should direct research; (b) the relationship between research and practice, the assumption being that research should provide guidelines for design and operation; and (c) the interrelationship among all three. The link among them may be that, if research is weak because theory is lacking, research cannot supply the guidelines that should direct practice.

If we have that much difficulty with the relationship between theory and research, what can one say about the relationship of research and practice? The effect of research on HFE practice is also extremely tenuous. If one reads books describing engineering psychology (e.g., Proctor & Van Zandt, 1994) and tries to tie theory and research to HFE methodology used in practice, the lack of relationship is astonishing. Few of the techniques used by HFE practitioners in their work have any close connection with what some people have ungenerously termed *psychobabble.* Of course, the various task analytic techniques are useful in system development, but task analysis is only descriptive.

Theory and research provide only background context for actual operations. If a practitioner examines a design drawing for conformance to HFE principles, the theory and research behind these principles is usually not apparent to him or her. For example, the assumption is that ineffective arrangement of controls and displays will negatively influence operator performance. However, the research that buttresses this paradigm hardly exists; only a few studies (e.g., Chapanis & Lindenbaum, 1959) performed in the early days deal with such relationships. What is most astonishing is that there has been so little discussion of the question and, in fact, that HFE people are not aware that the problem even exists.

Theory in engineering psychology has the same effect on HFE practitioners as knowledge of hydraulics has on the actual operations of a plumber. Undoubtedly, knowledge of hydraulics makes a plumber a more cultured plumber, but it is unlikely to make him a more skilled one. Only if the theory has a direct relationship to the job is it useful. Knowledge is both explanatory and instrumental. Instrumental knowledge (e.g., how to type this manuscript) allows me to do my job; explanatory knowledge or theory about what one does may allow me to understand *why* and *how* I am doing my job (e.g., I understand that typing is a learned skill and I can refer to research that tells me about the effect of practice on skill development and lack of it on skill decay), but when I am typing I do not utilize typing theory.

(Of course, the Devil's Advocate could ask: Even if the individual is not aware of the theory and research performed to answer questions related to practice, does this make the theory and research less valuable? The answer is, of course not. However, if one examines the theory/research and compares them with the actual methods used and only a tenuous connection can be found, what does this say about the theory/research?)

Whether the prior question is true for other scientific disciplines, the author does not know. Nor does the fact that theory is relatively ineffective in HFE make HFE less a science because theory, whatever else it is, is not the defining attribute of science. (*The Random House College Dictionary* [Stein, 1980] defines *science* as "systematized knowledge of any kind.") One would wish that there was a closer link among theory, research, and practice because this would justify the stereotype of HFE as a heroic science. Unfortunately, this study does not support the stereotype. There may be those who equate the review of previous research in the introduction to the published study or the discussion of study results with theory, but by anyone's definition of theory this is invalid.

The results of this analysis seem to suggest the disconnection of HFE theory and research. It would be useful if researchers, academics, and theorists were to probe the reasons for the disconnection to search for an answer and possible remedial processes, but no such examination and analysis will ever take place. That is because the science stereotype is too strong, too overlearned, and has too much emotion associated with it. Indeed, one would predict that the results of the study described in this chapter will be ignored rather than refuted. By all means, theory is, as Kantowitz (1989) said, necessary and important, but it may be that HFE professionals would prefer to keep it remote from practice—where it can do no harm.

The relationships among theory and research and research and practice may be fairly complex. Researchers may or may not be concerned about theory, and practitioners may be more or less oblivious to theory. Practitioners may or may not be concerned about the utility of research outputs. Logically they should be if they believe that research should lead to practical guidelines, but they may not believe that research can be useful in practice. We do not know what practitioners feel about research or what researchers feel about practice, but there seems to be a disconnect because, if there were much feeling about the relationship between the two, research studies would have much more to say about application. As we see, they do not.

Research Type

Few of the studies reported in the HFE literature can be considered theoretical (2%–4%). Although the motivation for the studies was not usually methodological, 25% to 35% of the studies are of the methodological type. That is, they are either contributions to the general methodology of the discipline or they seek to determine which one of several alternative ways of performing a task or alternative equipments is superior based on human performance. Other studies have equipment/system implications, but there are only a few of these and they seem to be becoming fewer over time. In the development of ergonomics, one would imagine that technology would be the directing force, *the hand of God* so to speak. However, it is possible that in HFE research technology has been transformed into a conceptual structure that becomes superior to the actual technology.

In our research, the explanation of human reactions to technology—important as it is—may have become more important than the technology (e.g., design guidelines), which is why there is so little emphasis in that research on design processes.

Many studies throw light on the variables affecting human performance. This is despite that almost never was research initiated for this purpose, although the study motives implied this purpose. Associated with this characteristic of the studies is the determination of human performance characteristics. For example, such a study determines that the elderly perform a particular function less well (by a certain percentage) than younger subjects. The point about this category is that it enables the researcher to say that, as a general characteristic of a subject type, these subjects ordinarily perform in such and such a way. The exploration of variables affecting human performance merely tells the reader that such and such a variable is involved in performance. Determination of performance characteristics tells what the nature of that involvement is. Other types of studies involve product development and testing and model testing, but the frequency of such studies is not large, nor is that of analysis of operational data or the solution of specific problems. However, use of operational data may make it possible to use the system or organization as the unit of analysis (see e.g., Baker, Olson, & Morisseau, 1994).

Unit of Analysis

In almost all cases, the unit of analysis is the performance of the individual aggregated for a statistical analysis. The reader's natural reaction may be: So what else is new? It seems obvious that if one measures performance, one does so individual by individual. Yet is there some way in which one can concurrently measure system effects? The individual performance results are ordinarily aggregated in HFE studies to determine such measures as mean and standard deviation, but the aggregates are almost never linked to a particular equipment, system, or organization.

To develop system measures, the researcher must be interested in how well the system performs. Of course, there is no system measure when only human functions like visual performance (which is inherently individual) are measured. However, when the human performance occurs in the context of system operation, it should be possible to secure a total system measure (as well as individual ones; e.g., accuracy of gunnery performance for an artillery or tank unit).

To measure system performance, the researcher must reorient his or her research interest from the individual to the system or to the system as well as the individual. This would not be difficult to do except for the overwhelming influence of learned methods of conducting research. In this connection, to the extent that the academic model (referred to previously) starts with the human and his or her performance and not the system, it is inevitable that the resultant research will be individual rather than system-oriented.

Hypotheses

It is a presumption, on the part of scientists and nonscientists both, that research is performed to test hypotheses previously developed by the researcher. If this is true, one would expect that the researcher in his or her report would indicate the hypotheses being tested by the study. The data show that, in the great majority of cases, no hypothesis was reported by the researcher. However, starting in 1973, there is a continuing tendency for hypotheses to be reported.

The absence of a hypothesis in a study suggests that the researchers were simply going on a *fishing expedition* to see what they could find. The absence of a hypothesis may suggest a lack of a conceptual structure that initiated the study, but again this cannot be proved. A hypothesis may be implied in what the study is about (e.g., someone tests alternative warning labels because he or she intuitively hypothesizes that some labels will be better or preferred by subjects). However, such inferred hypotheses are weak. The unwillingness to specify hypotheses may be an oversight or reluctance to *put your money where your mouth is.*

Design Application

This category is especially interesting to us because of the claim, by practitioners primarily, that HFE research does not lead to outputs that can be used by them (Norman, 1995). In most cases, the data show that there are far many more reports without application than with application, even considering the author's extremely liberal criterion of what an application is.

However, in all fairness, it may appear to the researcher that reference to the results of the study inherently implies an application (e.g., if a particular method of performing a job is shown to be superior, it is obvious it should be used). This may account in part for the infrequency of specified applications. Moreover, when the study involves a purely conceptual problem or merely seeks additional data (missing data), no application can be expected. However, there are many studies in which the application of the results is not at all clear. It is unfair for the reader—the research consumer—to assume the burden of inferring what the researcher had in mind.

One may conjecture that the absence of application in some studies also comes about because researchers cannot think of any application, which raises this question: Why did they do the research? To many researchers, application is perhaps unimportant. From the standpoint of the discipline as a whole, this is impermissible.

RESEARCH ADEQUACY

Criteria of HFE Research Adequacy

To evaluate the adequacy of HFE research, it is necessary to apply adequacy criteria. Criteria of research adequacy are usually found in texts on statistical design and deal with such topics as an adequate subject sample size and the importance of selecting subjects who represent those to whom results will be applied. These are single-study or *technical* criteria. This section discusses *strategic* or discipline-wide criteria, which apply to the single-study as well as the entire body of research.

The author proposes three major adequacy criteria:

1. First and most important, the HFE research should be closely linked with real-world operations. Some readers may feel that this is subsumed under the generally accepted criterion of validity, but the criterion in this section goes beyond that. The real world is the world outside the measurement situation and includes the engineering facility as part of the real world. *Validity* as it is ordinarily defined need not have any connection with the real world. As long as the test results produced in a laboratory correctly mirror actual performance (whatever that is), a study is valid. This is performance in the laboratory in which the study was performed. Hardly any researcher wonders whether that study performance mirrors performance outside the laboratory; if he or she did, there would be many more validation studies than there are. The criterion in this section also suggests that the

origin of the research should stem from real-world (as well as purely conceptual) problems and that the study results should generalize to and affect real-world performance.

2. Another potential criterion is that the focus of that research should be not only human responses to technology (e.g., how well operators perform when faced with a particular configuration of stimuli) but also study of those system factors in which human performance is embedded and which determine, at least in part, that performance.

The study of human responses to technological variables has some value because it enables us to suggest that certain equipment/system configurations are more desirable for human performance than others. However, because that study did not begin in most cases with system characteristics (the focus is on the individual and the stimuli), its effects relative to systems and their design are quite limited.

3. A third criterion is that HFE should be useful, in the sense that its outputs can be applied to the solution of system development and other operational problems. The application should manifest itself, most prominently, in the form of design guidelines.

In the listing of criteria that are usually applied to research (validity, reliability, generalizability, etc.), none deals with the question of utility possibly because it might be difficult to achieve a consensus on what utility meant or how to apply it to research. It might also be dangerous to have preconceived notions of what is good or bad research.

The definition of *utility* in this section is highly pragmatic. Researchers assume that their research is useful by contributing to knowledge as a whole and to other researchers. The author's criterion assumes this also, but extends beyond knowledge in general to those who must utilize the outputs of knowledge in their work. If research is not useful in this sense or is merely assumed to be useful, it should be viewed as having minimal value.

Does HFE Research Satisfy These Criteria?

Is HFE research closely linked to the real world? Reading hundreds of papers published over 30 years, one would hardly think so. The link was certainly closer in the early days of the discipline but has become more tenuous over time. What does it mean to say the link is weakening? Simply that, even if one can trace the research back to some original real-world problem, the way in which the problem is conceptualized by the researcher is expressed in somewhat abstract terms that are remote from the original problem.

The researcher will undoubtedly say, how else can one deal with the problem? The argument is that, in its original form, the problem may not be amenable to study by experimental methods and must be transformed to make it susceptible to experimental rigor. This means that it must be decomposed into its constituent variables if it is to be studied correctly. This is admittedly a valid point, but it becomes necessary then for the researcher to translate the variables (or rather what is found about these) back into the original problem. The reader of these papers is sometimes unable to recognize what the original problem was.

It is evident from the results of the present study that the individual is the focus of HFE research. The reader may ask what is wrong with that considering that the discipline is concerned with the human in relation to technology. However, the system in which the human performance is embedded is hardly mentioned as a system, regardless of how many technical details are presented in the written paper. If it is conceded that the characteristics of the system have some effect (and not merely an insignificant one) on human performance, it would seem that some consideration in the research should be given to these characteristics.

If these are ignored, in effect we say that they have no effect, which is manifestly untrue, and the picture of human performance that our research presents becomes distorted. The mere description of an equipment to which the subject responds is not consideration of the system. The system *qua* system has its own variables (see chap. 3); if these are not manipulated experimentally, the system is not being studied. In fairness, some HFE research does include a few system variables, but these are almost always merely incidental to the true purpose of the study, which is to investigate human responses.

The reader may say, how else is one to investigate the effects of variables except in terms of human responses to these? True enough. However, to study the system, one must begin with system variables, after which human responses to their manipulations can be recorded.

It is equally obvious that HFE research is not useful to what should be two of its primary consumers: the practitioner and designer working in system development. This has been and still is a point of controversy between researchers and practitioners (see Meister, 1985). Does the researcher have any responsibility for the application of research outputs? Many researchers will say they do not. If researchers do not have that responsibility, who does? Is it the practioner who has had nothing to do with the research? A third party? It is probable that if the researcher does not develop his research (or at least think about it) at least partially in terms of application, no application will be advanced at the conclusion of his or her paper. If one reviews the assembled published research with a view to developing design guidelines, few if any can be developed (see Okawaga, 1993). Simply phrased, the discipline has done a rotten job of applying its research to real-world problems.

The Management of HFE Research

In the discussion of the results of this study, HFE research was described as being broad but shallow, meaning that a great number of topics were addressed, but only a small subset of these received more than cursory attention. This is unfortunate because it suggests that the discipline is not doing a good job of assigning priorities to its research efforts. HFE research should be more intensive even if this means reducing the number of primary and secondary topics addressed. To assign priorities (if this can be accomplished at all) means that one must manage research and that HFE research should be better managed than it is presently.

Because researchers prize their apparent freedom to select their own research topics, managed research may sound vaguely dictatorial, even un-American. Nevertheless, American HFE research, like research everywhere, *is* managed by those who provide the funding for it. The author's study of funding in the 1997 Proceedings reveals that the money goes where one might expect it to go—the universities. Most of that funding is derived from government agencies; a little comes from wealthy corporations or subsidized think tanks. Those who provide funding ultimately decide what is studied; if they are uninterested in a topic, the funding for that topic dries up and research is no longer performed. However, if a new concept suddenly attracts attention (i.e., becomes hot), new sources of funding appear. What this means is that the money managers, who are rarely if ever HFE specialists, need to be educated about what is important and unimportant in research. Based on their own individual interests, researchers seek to convince the money managers of the importance of those interests; this will probably continue, but researchers must make more intelligent choices—not so much for their own individual benefit, but for the benefit of the discipline as a whole. It may be too much to expect the individual to abandon self-interest, but professional societies like the Human Factors and Ergonomics Society might attempt to do what they can.

The Problem of Application

Another recommendation is that HFE research should be application-oriented to a much greater degree than it is presently. It cannot be emphasized sufficiently that the function of research is not only to discover and explain relationships among variables, but also to control, where the term *control* means to have an impact on the researcher and, to an even greater extent, on technology as a whole.

Over the past 30 years, HFE research, as reflected by both of the source documents analyzed in this study, has developed a professional image, and no one would wish it to regress to a less sophisticated state. However, the characteristics of that image are those of academic psychological studies. There is nothing wrong with that except that psychological research of a nonclinical nature has never been noted for its applications and utility. Because the majority of HFE researchers have been trained in departments of psychology, it is natural that they would carry on what they have learned. If utility is a criterion for HFE research, and it should be, then research must diverge from its psychological origins, at least by considering in advance of a study its possible usefulness in terms of application. This is hard to do, but the effort must be made.

Theory

It was noted previously that there is practically no theory involvement in HFE research, and this is a consistent tendency rather than one that has changed over the years. This despite the constant drumbeat to support the need for theory and models (Kantowitz, 1989). It would be incorrect to say that HFE has no theory. It has, as represented by the resource theory of Wickens (1992); the attention theory of Lindsey and Norman (1992); the skill, rule, knowledge theory of Rasmussen (1983); and, perhaps the most well known of them all, the signal-detection theory (SDT) of Green and Swets (1966). Unfortunately, these theories have had little discernable effect in terms of producing design guidelines. Is it possible that this is because they are psychological theories, relating to the individual only, when design guidelines must relate to the physical system as well as to the human? Those theories in which the human reacts to stimuli (some technological, most not) are theories derived from psychological sources that are not concerned with the fundamental basis of HFE—the system. As a consequence of research focus on the individual, the system becomes merely background for the study. Unless this point of view is changed, it is doubtful that there will ever be a behavioral system theory.

Perhaps the initial focus should be on why theory does not have a greater relationship to HFE research. Perhaps HFE people are more pragmatic than others and this leads them to denigrate theory (although this is speculation only). Whatever the cause, the matter must be addressed, although theory development and testing are daunting tasks. Whatever the specifics of the theory, it must be a system theory organized around system elements, of which the human is only one.

Historical Changes

The changes in the kind of material *Human Factors* published, after the society decided in 1972 to publish *Proceedings*, have been referred to previously. The earlier papers were much more applications-oriented, whereas the later ones have largely lost this orientation. To explain relationships between Variable X and Performance Y does not automatically lead to application: One must actively conceptualize research in terms of potential applications.

One of the discouraging aspects of HFE research is that its initiation is not closely tied to new technological developments. At best, 10% to 15% of the published papers had this as motivation for or an implication of the research. The reader may say, what about the effect of computerization on HFE? If much HFE research has been performed in support of computer design (software and hardware), it does not appear in HFE journals, although it may have been published in journals devoted more specifically to computers than are the documents published by the Society. Certainly the *International Journal of Man Machine Studies* and the *IEEE Transactions on Man, Systems, and Cybernetics* contain many more computer-oriented behavioral studies than does the Society's journal. To the extent that such papers were published by researchers who thought of themselves as computer engineers rather than as behavioral scientists, it would be entirely natural that they would submit their work to publications in their primary field of interest. Some evidence for this is indicated by the bibliography in a review chapter on interface design by Williges, Williges, and Elkerton (1987), which contains slightly over 150 references, of which only 11 were published in HFES documents.

Recommendations and Thoughts for the Future

The difficulty of attempting to make recommendations for an entire discipline must be recognized. How can one push a discipline in another direction even if the directional change is only slight (particularly when so many professionals see nothing wrong or inadequate in what they are doing)? It is like trying to push a 50,000-ton aircraft carrier by manual means alone.

There are several major players in the history of HFE research: the individual researcher, the community of researchers as a whole, the cultural context (Zeitgeist)) in which research is performed, and the funding agencies. The individual researcher is usually aware of the funding agencies and the general community of HFE scholars, but is unaware or at least only partially aware of the cultural context. There is some limited interaction among the players. The individual researcher can influence funding agencies and the HFE community slightly, but in most cases cannot influence the Zeitgeist at all because the latter is so amorphous. Speaking only somewhat fancifully, the Zeitgeist is the intellect and attitudes of the discipline, which is composed of the collective intellects and attitudes of all its professionals. It is a conceptualization by the research community of the way in which their people operate, although it is impossible for any individual researcher to know precisely how others of the ilk function, except by empathizing with them at occasional meetings and in correspondence. Despite this, the ultimate purpose of this chapter is to challenge the Zeitgeist to change itself.

The picture presented by this section is that HFE research does not live up to its potential. If that is true, the entire discipline is not living up to its potential. No one would suggest that the discipline is in danger of dying out, but it is in danger of becoming somewhat irrelevant to the technological challenges of the new century.

HFE is hampered by the intellectual heritage of its psychological prehistory. It is still a young discipline and old research habits learned in departments of psychology die very hard, if they die at all. It is not clear how many HFE professionals recognize (feel empathically, that is) that they are working in a new discipline and not merely in a special application of psychology. How many HFE people feel that there is anything inappropriate about the conceptual baggage their psychological predecessors left them? Without necessarily jettisoning all its history, can HFE be persuaded to look at its underlying concepts in a different way? Perhaps.

CONCLUSIONS

From the preceding analysis one can conclude the following:

1. Over the years, papers have tended to become more professional, in the sense that they have become more academic in orientation. The change in orientation appears to have begun in 1972, when the Society first published annual meeting *Proceedings*.

2. Initially more papers came out of industry and government facilities combined than from the university, but this has been reversed. Now more papers derive from the university, with far fewer industry papers.

3. The percentage of papers reporting empirical studies has increased progressively to the point that, in *Human Factors,* there are few nonempirical (e.g., reviews of research, comments, or tutorial) papers.

4. The laboratory is the preeminent location for data collection, but it is being challenged by the simulator and the operational environment.

5. The dominant themes for HFE research over the years have been the aerospace, automotive, and display systems and the human functions of visual performance and biomechanics. The range of topics researched is quite broad, but the frequency with which each topic is studied is low, so that HFE research can be characterized as *broad but shallow.*

6. Subjects are most often students, but operational personnel and the general public are, when combined, just as frequent.

7. The method most frequently employed by HFE researchers is the experiment—increasingly of the within-subjects design. However, over the

years, subjective methods have been increasingly used to support or verify the experiment. If the number of techniques employed is used as a criterion of study complexity, there is a slight tendency for studies to have become increasingly complex over the years.

8. Research is usually initiated by several motives. Most often these include an empirical (real-world) or conceptual problem and a lack of data. New technology runs a distant third. Almost at no time is a design problem the reason for performing a published study.

9. Contrary to popular stereotypes of the nature of scientific inquiry, theory and modeling have little impact on HFE research. There is almost no theory testing and only a little more theory or model referencing.

10. HFE research is most commonly a methodological study and/or a study of variables affecting human performance and/or the determination of human performance characteristics. Most studies are of more than one type. Little work is published dealing with product development or testing, modeling, or training evaluations.

11. The unit of analysis is almost always the individual. Few studies suggest any hypotheses on which the study was based. There is little attempt on the part of researchers to develop applications for the study results they have produced.

The effect of the changes noted previously is progressively to divorce HFE research from its real-world sources and make that research somewhat irrelevant to real-world problems.

NONEMPIRICAL STUDIES

Introduction

The previous section of this chapter only dealt with empirical studies—those in which the performance of at least one subject was measured specifically for the study. The other study category is nonempirical, in which no personnel are tested for the purpose of the study being reported. This last is an important qualification because nonempirical studies, as these have been defined in this chapter, may include references to previous studies in which subjects were tested or may include demographic data like accident frequencies. Nonempirical studies may also include: presidential addresses, review and critiques of previous research, tutorial papers, case studies (as long as no one was tested specifically for the study), principles and theories, models, descriptions of equipment and their design, and symposia. The only criterion for inclusion of these varied studies in the nonempirical category is that no one was specifically tested for the study that was published. This may be a crude distinction, but it is operationally discriminable.

Methodology

The analysis performed of these studies dealt with the primary theme (what the study, considered as a whole, was about) and a major secondary theme (a factor that intersected with the primary theme). For example, in Van Cott (1984), the major theme was technology and the secondary theme was the model he described. None of the analytic categories previously employed in the content analysis of the empirical studies (e.g., source, venue, methodology) was investigated for the nonempirical category because most of these items did not pertain to nonempirical studies.

There were, as with the empirical studies, two sources: the journal *Human Factors* and the HFES annual meeting *Proceedings*. All of the nonempirical papers in the *Human Factors* issues starting from and including 1965 to 1995 were analyzed. Nonempirical papers in the *Proceedings* of 1973, 1975, 1980, 1985, 1990, and 1995 were also analyzed. For each source, the primary and secondary themes of each nonempirical study were determined.

The data were analyzed to indicate the total number of papers (empirical and nonempirical) in each source document, the number of nonempirical papers, and the percentage of all papers that were nonempirical. In *Human Factors*, this percentage is quite variable—ranging from a high of 45% in 1977 to a low of 11% in 1984. The overall average for *Human Factors* is 26.3%.

In *Proceedings*, the nonempirical percentage ranges from a high of 52% in 1973 to a low of 22% in 1995. There is a consistent reduction in the percent of nonempirical papers in *Proceedings*, with the greatest change occurring between 1985 and 1990. In this respect, the content of *Proceedings* papers becomes progressively more like that of *Human Factors*. Overall, one can say that approximately a quarter of all papers are nonempirical.

Nonempirical papers can be directed to the solution of problems that are of a general disciplinary nature (e.g., what is the status at a particular time of HFE) or problem themes that stem from specialty issues. Because of the fragmentation of the discipline, most nonempirical themes were associated with individual specialties, although a small number (e.g., presidential addresses) related to the discipline as a whole. Although initially an attempt was made to separate disciplinary-related from specialty-related themes, this was dropped because most nonempirical papers were of the specialty type.

The initial impetus for the analysis of nonempirical papers was to answer this question: Are HFE professionals concerned at all about the major problems that beset the discipline? These problems are listed here:

1. The nature and status of HFE and the philosophical aspects related to HFE.
2. HFE as it is performed in other countries and contexts (abbreviated as History).
3. The system development process.
4. The role of theory and research in HFE.
5. The role of HFE practice and its relationship to research.
6. Design factors and development of design guidelines.
7. Measurement processes in general and how these affect HFE.
8. Experimental designs.
9. Subjectivity and its effect on HFE research data.
10. The effect of automation on HFE.
11. The operational environment and what it means for HFE.
12. The need for quantification in HFE and how it can be achieved.
13. The HFE database (what it is and the factors that affect its development).
14. HFE data and how they are handled.
15. The prediction of human performance.
16. HFE relationships with other disciplines.
17. The Human Factors and Ergonomics Society and its role in HFE.
18. Technology and its effects both generally and on HFE in particular.
19. The question of research utility.

Results

Six of the themes with the greatest frequencies—aviation, automotive, biomechanics, environmental factors, computers, and training—also had high frequencies in the empirical papers. Other themes are peculiar to the nonempirical papers: applications, nature/status of HFE history, design guidelines, measurements, and methodology. These were not found to any great extent in the empirical papers. This suggests that HFE professionals are concerned with such issues as measurement methodology and design, but only as conceptual abstractions.

There was some interest during the 1970s in the HFE of other countries (Seminara, 1975, 1976), but that diminished quickly. There is some interest in technology and automation (Williges, 1984), but no interest in the relation of theory to research or of research to practice. There is little or no discussion even on a conceptual level of the operational environment—of the need for quantification, prediction, the development of a database, or research utility—the necessity for which has been expressed in an earlier book (Meister,

1989). There is some concern for the Society and, a staple of earlier papers in particular, application of HFE principles and methods to other disciplines or fields of endeavor.

It was pointed out in connection with the empirical papers that there appeared to be little concern for theory or models, design, and applications. However, these show up in the nonempirical papers and redress the balance. These theories, models, and so on are not tested, which may make one think that professionals are not sufficiently serious about these areas to engage them in empirical work. The theories and models that are developed (even if not tested) relate almost exclusively to specialty areas like biomechanics (e.g., Genaidy & Asfour, 1987). There is little concern for discipline-wide problems like validity (or its absence), subjectivity, and so on.

The determination of the themes that are not addressed casts almost as much light on the research preoccupations of HFE professionals. These absent concerns included such matters as the relationship between theory and research, system development issues, the relationship between practice and research, subjectivity, the operational environment, quantification in other than a statistical sense, concern for data or an HFE database, any ability to predict, validity, criteria, ethics, forensics, human reliability, social factors and technology, teams, and decision making. However, there is slight concern for automation and technological change.

RESEARCH RELEVANCE AND REPRESENTATIVENESS

Relevance

To round out our analysis of HFE research, it is necessary to examine it from the standpoint of its relevance and representativeness. These are of course evaluative criteria, and there may be those who believe that the evaluation of research is unwarranted because the only research obligation is to add to the science knowledge base. However, this criterion does not discriminate between good and poor research unless one believes that all research, however trivial, is worthwhile.

The first question is this: relevance to what? Relevance implies some strong interest on the part of the research user: What does he or she want the research to tell? Two interests immediately suggest themselves: (a) relevance of the research to the design and development of equipment and systems, and (b) relevance of the research to the description and explanation of human–technological interactions in the operational environment (OE). If one could use the research as material for developing behavioral design guidelines, the research would be relevant in the sense of (a). If the research helped to explain human–technological performance in the real world, it would be relevant in the second sense.

The difficulty of determining relevance is that complicated judgments must be made about individual studies. Almost no researcher indicates in a published study what its intended relevance is, and certainly not in the terms specified earlier. The study may indicate the problem that initiated the study, but this is not the same as relevance, except possibly to other researchers. Relevance may be high in applied studies, which are funded to answer specific questions, but such research is usually not published in the general literature.

The author attempted to make relevance judgments for 170 papers in the 1996 HFES *Proceedings* and the 198 papers in the 1997 document. His estimates were that 33 (19%) of the 1996 papers were marginally relevant, as were 48 (24%) of the 1997 papers. Although these are interesting, although dispiriting, values, the author does not invest these judgments with any great significance considering the highly subjective and idiosyncratic nature of the judgments.

Representativeness

Relevance Option (b) endeavors to explain how and why real-world phenomena occur. Most HFE research attempts to be relevant in this sense, at least by implication, and it may be more successful with regard to (b) than for the design/development Option (a). Relevance in Option (b) depends, however, on a construct called *representativeness*, which can be defined as the degree of similarity between the measurement situation and that of the OE. In previous chapters, it was determined that the ultimate referent for HFE research was the OE—the *normal world* in which people ordinarily interact with technological things.

Another word for representativeness is *fidelity* to the real-world situation because the ultimate goal of HFE research is to explicate that situation—to understand it. From that standpoint, any deviation from the OE represents an element of artificiality, which degrades relevance.

Representativeness is linked to Heisenberg's principle (the act of measurement influences the output of the measurement). Because the measurement environment is part of the measurement act, where, what, and how one measures significantly affects the resultant data. In contrast to relevance, representativeness can be defined and scaled in largely objective terms. The construct is composed of four dimensions that have certain ordinal scale characteristics (see the numerical identifiers presented later), although the intervals between scalar units cannot be considered equal.

1. *Measurement environment* (where the research was performed). In terms of decreasing fidelity to the OE, the alternatives are: OE (6); OE

reproduced in film and/or sound (5.5); test site (5); high-fidelity, full-system simulator (4); medium/low-fidelity simulator or part-task simulator or mockup (3); laboratory (2); other (e.g., classroom, survey, computer game [1]); demographic data (0).

2. *Measurement unit size* (how much of an operation has been studied). Ideally this would be full-system operation (5); subsystem operation (4); task/workstation (3); part task (2); individual control or display (1); no operational equipment or use of a synthetic system (0).

3. *Performance measured* (what was studied). Ideally the mission (5); function (4); task (3); subtask (2); nonequipment-related task (e.g., peg test [1]); other (e.g., analysis of accident reports, body movements, environment measures [0]).

4. *Task characteristics* (fidelity to real-world activities). Real-world activity (4) is preferred followed by: simulation of real world activity (3); abstraction of real-world activity (e.g., computer graphics [2]); nonequipment-related task (e.g., board game [1]); nontask (e.g., anthropometric measures, body movements [0]).

The same 1996 and 1997 *Proceedings* papers were examined for each of the four representativeness dimensions.

Individual values were collapsed into the dimensions. For example, all studies in which the OE was utilized were combined (e.g., $f = 23$ in 1997) and the OE percentage of the 198 papers was determined (11%). This was done for each category in each year.

Assuming that the ideal situation would be for all studies to be performed in the OE with full-system and mission operation, it appears that only about 10% of all 368 studies utilized the OE as a measurement environment, 81 or about 22% involved full-system operations, 48 or about 13% dealt with full missions, and 54 or about 15% involved real-world activities.

The most common measurement environment was, as one would expect, the laboratory. The most common measurement unit size is one in which there is either no operational equipment or a synthetic system, such as a computer game, was used. The most common performance measured was either a nonequipment-related task or something other than a task. The most common task characteristics were those of a simulation of a real-world activity or a nonequipment-related task.

The overwhelming propensity to use the laboratory as the measurement environment makes it more attractive to use nonequipment or synthetic tasks because of the difficulty of moving OE characteristics into a laboratory. Because nonequipment tasks that have no mission or system relationships are utilized, performances are made more molecular so that few large-scale operations are measured. In addition, there are certain specialties such as industrial ergonomics (biomechanics) or visual performance that are not closely tied to the OE and that do not require the OE or OE characteristics to measure the performances they are interested in. These specialties add their weight to the tendency to nonrepresentativeness in HFE research. As was seen earlier in this chapter, on the whole, HFE research is highly artificial even when not trivial.

Recommendations

What can be done to make HFE research more useful? An increase in its relevance and representativeness can perhaps be secured by requiring any research report to include at least one paragraph specifically outlining the design or other application uses of the research reported. Should such a proviso be implemented, there would undoubtedly be considerable opposition on the basis that it would chill publication of studies that do not suggest a specific link to system development or the OE. Although it might be more difficult to get certain papers published, such a requirement might force the researcher to consider what the utility of his or her study actually is and this might actually improve the quality of HFE research. The application statement might also include an explanation of why a particular measurement environment other than the OE was selected.

It should not be inferred that the author considers HFE research to be useless. All HFE research is performed to satisfy some interest—either of the researcher or the researcher's funding agency. Although each study may be useful to that researcher or agency, from the standpoint of the discipline as a whole, the collected body of research may have little value for the discipline.

Studies have greater or lesser utility; what is needed is a deliberate effort to extract the utility of all these studies in individual specialties. This is the translation function referred to previously; an example as it applies to the quantitative prediction of human performance is given in chapter 7. The translation function requires a deliberate effort by professionals who consider this a necessary research task. It will not be performed by the individual specialist whiling away his or her time reading papers in the library. The translation effort must be promoted by the discipline's professional societies because only these are concerned with the discipline as a whole. However, the discipline, represented by its professionals, may be unwilling to apply the requisite effort. In that event, HFE will be unable to utilize its full potential. Unfortunately, it seems as if few HFE personnel are concerned about maximizing their utility.

7

Special Interests Within HFE

As we have seen, every discipline begins as a more or less integrated whole and then fractionates into individual specialties. This chapter describes the individual HFE specialties, with emphasis on two: macroergonomics or, as it is often called, organizational development and management (ODAM); and human reliability (HR) or, as the author prefers to call it, quantitative prediction.

The specialties described in this chapter roughly follow those that have been formally accepted by the HFES. A technical group within the Society is accepted when a certain number of Society members petition to be accepted as a technical group. As of 1997, the following are technical groups within the Society: aerospace systems, aging, cognitive ergonomics/decision making, communications, computer systems, consumer products, educators' professional, environmental design, forensics' professional, individual differences in performance, industrial ergonomics, medical systems and rehabilitation, ODAM, safety, surface transportation, system development, test and evaluation, training, and virtual environments. The author's breakout of specialties based on research papers closely follows the HFES classification, but is not identical with it. It should be noted that there are other special interests among HFE professionals that have not been formalized by acceptance by the Society (e.g., HR).

Fractionation into specialties occurs for two interactive reasons: the internal interests of individuals and the external interests of funding agencies, which draw those with compatible interests further into the specialty. The importance of funding cannot be exaggerated. We have seen (chaps. 4 and 5) that HFE would not have developed without the interest of the government, which supplied the funding to draw professionals into certain types of research and controlled and fostered their activities in system development by establishing standards that required industry to employ and make use of HFE professionals.

The content analysis process described in chapter 6 was again applied to the research papers published in *Human Factors* and *Proceedings*. The goal was to determine the characteristics of the research performed by the various specialties within the discipline. It was found that the papers fall into three categories—system-oriented (50%), process-oriented (27%), and behaviorally oriented (18%)—and that each of the specialties falls into one of the three categories. System-oriented papers are primarily concerned with the equipment aspects of the system, process-oriented papers are concerned with how problems are addressed, and behaviorally oriented papers are, as the name implies, concerned with the behavioral aspects of systems.

Not all the technical interest specialties accepted by HFES are well represented in the research papers that were content-analyzed. Although ODAM is a strong interest of many professionals, it tends to publish outside the publication channels of the Society. This is true of another specialty emphasized in this chapter—HR.

Another name that has been applied to ODAM is *macroergonomics*, suggesting that this specialty is concerned with larger system entities than traditional HFE, which it calls *microergonomics*. As the acronym ODAM suggests, macroergonomics (the specialty was renamed in 1997) is concerned with the organization and structure (e.g., the company) within which personnel work. In terms of content and methodology, it has a great deal in common with industrial psychology, out of which it appears to have metamorphosed. Indeed, there are those who question whether there are any differences between macroergonomics and industrial psychology (Smolensky, 1995). In any event, macroergonomics is one of the behaviorally oriented specialties. Its importance lies in the fact that it is the single HFE specialty that is most concerned with the total system. If, as we maintain, HFE is a system-oriented science, then macroergonomics deserves increasing attention because of its emphasis on the system.

Elaborating chapter 6's content analysis of HFE research literature, that analysis took the original 621 papers in the chapter 6 sample, subdivided them into specialty categories, and then compared the specialty categories with the original content analysis categories, plus a few others. The major conclusion derived from this additional analysis is that similarities among specialties far outweigh their differences. The reason is perhaps that all specialties in HFE utilize the same top-level conceptual structure (the S–O–R paradigm), although that structure shows differences as one descends to a more detailed level of analysis. Most of the specialties are system technology oriented, a lesser number are process-oriented, and a small number are behaviorally oriented.

The specialties differ in terms of at least two dimensions: orientation and molarity. Technology oriented specialties, those described in chapter 6 as system development ergonomics (SDE; e.g., aviation psychology), are primarily oriented toward equipment (actually human reactions to equipment). HCE specialties are more human-centered in the sense that they relate to many more facets of humans: health, work environment, anthropometry, safety, special characteristics such as age or disability, and training. There is no suggestions that one cluster is better than the other, although SDE contributes much more to application than does HCE.

The specialties also operate at various levels of functional molarity. For example, biomechanics functions at a molecular (physical) level; naturalistic decision making functions at a more molar (cognitive) level. This has implications for the type of measurement environment they tend to employ. For example, biomechanics and visual perception more often utilize the laboratory because their functions are not so influenced by the system (e.g., operational) context. Naturalistic decision making is strongly influenced by context, which means that this function is more likely to make use of the OE or simulations of the OE. The nature of the behavioral function involved determines the particular variables studied.

The interesting question for the historian of science is not why specialization and fragmentation of disciplines occurs: This is ultimately the effort to control one's area of knowledge, which can only be done by specialization. The more interesting question is why the breakout in HFE into system/technology, process, and behavioral types of specialties.

System/technology is easy: Technology is one half of the human–technology relationship. Besides, HFE specialists are like laymen in being fascinated by engines of great power, like aircraft, automobiles, and computers. Specialists who concentrate on the process type of specialties seek to solve a puzzle: How does one incorporate the human into technology? This is essentially a methodological problem. Those who concentrate on one of the behavioral specialties are fascinated not by technology but by the human response to technology.

This seems a neat explanation, but may only be a way of restating the obvious. There is, as in all disciplines, a chance factor (winning a grant for a particular research topic, getting a job in a particular industry) that may determine why one individual chooses to emphasize one specialty and another individual a different one.

Funding is of course all important. Absent a particular empirical problem, which the government (the ultimate source of all funding) wishes to solve, no speciality will exist.

If HR were one of the specialties well represented in HFE literature (and it is not, for which there are reasons that are discussed), it would be either system-oriented (because of its emphasis on process control industries) or process-oriented (in its attempts to develop a methodology for quantitatively predicting human performance).

Efforts at quantitative prediction began quite early in the history of HFE largely because of the military's concern for the negative effect of human errors on system performance. This concern was reinforced by government regulation of the nuclear power industry, especially after the Three Mile Island and Chernobyl disasters.

Of the HFE specialties, HR is the only one specifically directed at quantitative prediction. Although one can call it an HFE specialty because its concern is with human performance and many of the people engaged in its research and application are HFE professionals, it also has close ties with reliability engineering. Although its subject matter is behavioral (human errors), its application has been almost exclusively to human performance in an industrial setting.

Unfortunately, HR has had almost no effect on HFE as a whole. Although one of the major purposes and functions of HFE is quantitative prediction, none of the other HFE specialties endeavors to develop a methodology that will permit it to predict performance quantitatively. From a research standpoint, all the HFE specialties other than HR are engaged in the traditional experimental process, with no attempts to predict (qualitatively or quantitatively) other than by deriving conclusions from their research. It may be that the external force that energized HR (governmental funding) does not exist in the case of these other specialties.

The absence of quantitative prediction in HFE severely weakens the effectiveness of the discipline. Quantification exists in HFE, but almost exclusively in the form of experimental statistics, which has little relevance to anything outside of the research arena. In terms of its emphasis on quantitative prediction, HR could serve as a role model for the other HFE specialties, but it has not; it is almost segregated.

Governmental interest in the form of funding and regulation of system development has had a great effect on HFE, but primarily in practice rather than research. HFE is responsive to the special interests of its professionals, but not in a well-organized or effective manner.

A COMPARISON OF HFE SPECIALTIES

Although it has a core content consisting of assumptions, principles, and methods (these have been explored in chaps. 1 and 2), to most of its professionals HFE is the individual specialty they research or apply. Inevitably questions arise: What are the similarities and differences among these specialties? How do they reflect HFE as a whole? Obviously on a superficial level, specialties are differentiated by their major themes (e.g., aging, aerospace, visual performance). However, these are only their outward appearance; what we want to do is look behind the surface to see if there are clusters of specialties that are the same within the cluster and different from other clusters.

Once again we must go back to the research literature and examine that literature in terms of its special interests. Although the Society's technical groups suggest something of what we will find, the research should be a more valid guide to the characteristics of these specialties. The 621 papers in chapter 6, which were analyzed in terms of generalities and similarities, were now examined in terms of individual differences.

Methodology

The papers fall into three overall types: system/technology-oriented, process-oriented, and behaviorally oriented. System-oriented papers are primarily concerned with the physical, equipment aspects of the human–machine system they describe; process-oriented papers are concerned with how equipment/system problems are addressed and solved; behaviorally oriented papers are concerned primarily with the behavioral aspects of the system. It must be emphasized that these three types are not pure (exclusive of each other); to be assigned to one type does not mean that elements of other types are automatically excluded in the individual paper. In other words, a single paper may have elements of other types, but certain elements are dominant and therefore determine the classification it will receive.

The three types and 18 subspecialties are divided as follows.

System/Technology-Oriented Specialties.

1. Aerospace: civilian and military aviation and outerspace activities.
2. Automotive: automobiles, buses, railroads, transportation functions (e.g., highway design, traffic signs, etc.). Also includes ships.
3. Communication: telephone, telegraph, radio, but also including direct personnel communication in a technological context.
4. Computers: anything associated with the hardware and software of computers.
5. Consumer products: other than computers and automobiles, any commercial product sold to the general public (e.g., pens, watches, TV).
6. Displays: equipment used to present information to operators (e.g., HMD, HUD, meters, scales, etc.).
7. Environmental factors/design: the environment in which human-machine system functions are performed (e.g., offices, noise, lighting, etc.).
8. Special environment: this turns out to be underwater.

Process-Oriented Specialties. The emphasis in these papers is on how human functions are performed and methods of improving or analyzing that performance. The papers fall into the following classes:

1. Biomechanics: papers describing human physical strength as it is manifested in such activities as lifting, pulling, and so on.
2. Industrial ergonomics (IE): papers related primarily to manufacturing processes and resultant problems (e.g., carpal tunnel syndrome).
3. Methodology/measurement: papers that emphasize ways of answering HFE questions or solving HFE problems.
4. Safety: closely related to IE but with a major emphasis on analysis and prevention of accidents.
5. System design/development: papers related to the processes of analyzing, creating, and developing systems.
6. Training: papers describing how personnel are taught to perform functions/tasks in the human–machine system.

Behaviorally Oriented Specialties.

1. Aging: papers describing the effect of this process on technological performance.
2. Human functions: emphasizes perceptual-motor and cognitive functions. The latter differs from training in the sense that training also involves cognition but is the process of implementing cognitive capabilities. (Since this writing, an HFES specialty called *cognitive ergonomics/decision making* has been categorized.)
3. Visual performance: papers that describe how people see. They differ from displays in that the latter relate to equipment for seeing, whereas the former deals with the human capability and function of seeing.

Results

Number of Papers by Specialty. There is a wide disparity among the various research specialties. As was found in the earlier study, over the years aerospace has been the most researched topic (100 papers), with approximately twice as many published studies as any other category. Certain specialties are very recent entrants, such as aging. Others, like special environments (underwater), only appeared in a single year and were then dropped. For those specialties with publication frequencies less than 10 to 13, it is impossible to make any valid judgments. In terms of the three types of specialties, 50% of the papers represent system/technology-oriented specialties, 27% process-oriented, and 18% behaviorally oriented. The numbers do not add up to 100%, but the discrepancy is the result of dropping fractions and, in any event, is 5% meaningful?

The scientific playing field (as represented by the type and number of papers published) is not a level playing field. A research specialty waxes and wanes by the interest it holds for researchers, but nonscientific factors influence and distort that field. One example is biomechanics, which was pulled into HFE research, although superficially it appears to have little to do with technology except in such aspects as muscular trauma. Of course, anthropometry is an HFE concern as it applies to design, but much of the published biomechanics research is not design-related.

The ease with which a study can be performed contributes, in part, to its frequency in the literature. One example is that of warnings in the safety literature, whose frequency is much higher than one would otherwise expect, whose study requires little instrumentation or little reproduction of the operational environment, and which can often be performed in a classroom. It is obviously easier to run experiments with nonsense syllables on cards as stimuli than studies requiring a nuclear power plant control room simulator. This does not mean that in every case the more difficult study situation will be rejected. However, given two studies of equal worth or interest, if one is easier to perform than another the easier one will probably be preferred.

If a professional society announces, directly or covertly, that it has space in its publications for a particular research category, it will receive papers on that topic—perhaps not immediately, but eventually. To that extent, a professional society can encourage research on a particular topic perhaps by calling it by a new name. For example, suppose everyone called *situation awareness* (which is a new term for a supposedly new phenomenon) merely by the old-fashioned name of *attention*; would the latter have produced the excitement of a new research topic?

Frequency of Publication Over Time. If one plots frequency of publication of an individual specialty over time, there appears to be no discernable progression.

Source. The nature of the system or research specialty often determines the predominance of one source or another. The predominance of the university persists as it did in the previous study, but the individual variations are interesting. For exmple, the greatest number of consumer product papers is published out of industry (61%), as one might expect. One often finds automotive papers much from the university (65%), but aerospace papers come more from government laboratories, industry, and contractors, when combined, (66%), than from the university (34%). Aerospace (16% of all papers sampled) tends to shift the balance away from the university. If this specialty were removed, the balance would be more heavily weighted in favor of the university. Why system/technology-oriented papers—such as computers, industrial ergonomics, safety, or displays—should come mostly from the university is not clear.

Venue and Subjects. The laboratory as a research venue maintains its predominance except in two specialties: aerospace and automotive, where it is challenged by the simulator (47% in aerospace, 17% in automotive), field test (23% in automotive), and operational environment (16% in aerospace, 26% in automotive). The low percentage of the laboratory in safety research arises because this specialty makes extensive use of operational data (e.g., accident reports). Simulators are used much more than the laboratory in aerospace because of the cost of experimentation with actual aircraft. Similarly, in automotive research, the general availability of these vehicles makes it much easier to conduct research in the field and the operational environment. Behaviorally oriented specialties (human functions and visual performance) seem to be exclusively devoted to the laboratory, which may be a reflection of the difficulty of studying psychological variables in anything other than a highly controlled environment.

With so many papers being published out of the university, one would expect that students would be most frequently represented as subjects, and this is true for most specialties. The major exceptions are again aerospace, where operational personnel are most common (68%), and automotive, where the general public is used most frequently (51%). The reasons are quite logical. Because flying is, except for novices, a highly skilled activity, only those who have been highly trained for this work can be used as subjects. Indeed, even the use of as large a percentage as 21% of students as subjects in aerospace studies makes one wonder. However, because almost everyone except the very young, the very old, and the disabled drives an automobile, it is to be expected that the general public would predominate as subjects in automotive research. Other specialties in which subjects other than students are most frequently encountered are system design, methodology, industrial ergonomics, and computers. In most of these cases, the nature of the tasks to be performed may require a nonstudent subject population. Even in such cases, however, convenience is often a prime determinant of who is selected as a subject. The relative lack of workers as subjects except in biomechanics and safety results from the fact that the other specialties are not much concerned with heavy industry, in which workers predominate.

Methodology Employed

The experiment is, as one would expect, the overwhelmingly dominant methodology employed in all specialties. What is of interest is that several of the most technologically advanced systems—aerospace (25%), automotive

(30%), and computers (39%)—make the most extensive use of subjective methods of measurement. These techniques are often used to determine preferences, as in consumer products, or ascertain the extent of difficulty or stress (e.g., SWAT ratings in aerospace studies). Operational data (e.g., accident rates, demographics) are used most frequently in safety (23%), industrial ergonomics (15%), and anthropometric and physiological testing in biomechanics (21%). Again, the nature of the specialty calls for these types of data.

Research Initiation

As noted in the chapter 6 study, the professed motives for initiating a study are primarily empirical and conceptual problems as well as missing data. This is true of all the specialties, but the distribution of these motives varies. For example, conceptual problems are most frequent in training, methodology, visual performance, human functions, and biomechanics, all of which are oriented more to behavioral factors in technology than the other specialties. Empirical problems are to be found most often in system design, safety, industrial ergonomics, aerospace, automotive, and computers, all of which are oriented more to the physical factors in technology. The frequency of missing data rationales is roughly the same over all specialties, suggesting that this may be a *pro forma* motive. Note also that new techniques applications are only found in equipment-oriented specialties (aerospace, automotive, computers, displays, methodology, and system design research papers).

Theory Involvement

There is almost no theory or model testing in any specialty except biomechanics, but references to theory do show differences among specialties. The greatest frequency of these is found in training and human functions, both of which are, as noted previously, oriented toward behavioral factors in technology. Combined with data from research initiation, this suggests that behaviorally oriented specialties are more internal-looking, whereas system-oriented ones are more externally driven.

Research Type

Chapter 6 indicated that most of the studies reported were either studies of the variables affecting human performance, the determination of human performance characteristics, or methodological studies. The logic of these categories stems from the use of experimental methodology, which deconstructs the total measurement situation into more molecular variables. The question in the present study is whether any individual specialty evidences more of these characteristics than do other specialties.

Most of the methodological papers are methodological in type, entirely as one would expect, but training is not far behind. Biomechanics and visual performance have the lowest loading on this type of research. One category that should differentiate is application of new technology, which should be more characteristic of system/technology-oriented specialties. That is what one finds. Heavy hitters in this category are system design, displays, aerospace, automotive, and computers, although the frequency of this research category is not high. For the other specialties, the application factor does not exist. Except for methodological papers, all specialties study variables affecting human performance. This may be because, as mentioned previously, this type of research is most closely associated with experimental methodology. Because all specialties make use of this methodology, the category would not differentiate. Product development/test, which should be associated with application, is emphasized by computers, industrial ergonomics, methodology, and system design, all of which are either system/technology- or process-oriented specialties.

Unit of Analysis, Hypotheses, and Applications

The earlier study examined (a) the unit of analysis utilized in these studies (e.g., individual, system, organization, etc.), (b) whether hypotheses were described by authors in the published study, and (c) whether authors suggested any applications for their results. These categories did not differentiate among the research specialties because almost all studies were oriented around individual measurement (although data were of course aggregated) and there were relatively few hypotheses and applications.

Technological Involvement

Technological involvement is a new category (i.e., it was developed specifically to differentiate among the research specialties). The hypothesis to be tested was that behaviorally oriented specialties would make less use of technology than those specialties that were system- or process-oriented. The analysis of technological involvement (TI) was based on two criteria: (a) the extent to which measurement described performance in the operational environment, and/or (b) the sophistication of the measurement apparatus. For example, if in an aerospace study pilot performance was measured while the pilot was actually flying aircraft or flight performance was measured in a sophisticated flight simulator reproducing the operational environment, TI for that study was considered great or total (depending on how much of the operational environment was reproduced).

Although the two criteria are distinct, they could be combined; if so, the degree of TI (if either was less than total) was increased. The greatest degree of TI (total) occurred if performance was measured on the job, regardless of whether a sophisticated measurement apparatus was involved. If performance was measured in a nonoperational environment such as a laboratory, but highly sophisticated measurement equipment was used (e.g., a flight simulator reproducing the actual conditions of flight), TI was also considered to be total. If sophisticated measurement equipment was used, but without reproduction of the operational environment, this could result in a rating of some or great (but not total) TI. Lesser degrees of similarity to the operational environment and lesser amounts of sophisticated measurement equipment resulted in lesser degrees of TI.

System/technology-oriented specialties like aerospace, automotive, and computers have much higher TI than more human-oriented specialties (e.g., biomechanics, human functions, training, and visual performance). Essentially neutral specialties like methodology and system design are relatively evenly distributed across the scale. Again, the nature of the specialty seems to determine how much TI each has.

MACROERGONOMICS: A SYSTEM-ORIENTED SPECIALTY

Introduction and Summary

If we pay particular attention to macroergonomics as an HFE specialty, it is because it is a type of HFE specialty that is different from those discussed so far. The other HFE specialties are concerned with human performance in relation to relatively molecular units like the workstation, whereas macroergonomics (*M* henceforth in this section) or, as it is often referred to, organizational development and management (ODAM) is concerned with the structure (company or other organization) within which the more molecular work performance occurs. If one considers that organization as a particular kind of system, then *M* is a system discipline.

This cannot obviously be a comprehensive description of *M*; none of the following specialty discussions can be. To a large extent, this section is based on three papers (Hendrick, 1991, 1995, 1996b), although the conclusions and interpretations are the author's own. The most recent summary of the field is to be found in Hendrick (1997a). Hendrick (1991) defined *M* as a human–organization–environment–machine interface technology because it considers all four of these sociotechnical elements. However, its central focus is on the interfacing of the organizational design with the technology employed or designed to be employed, and all of this to optimize human–system functioning. Conceptually, *M* is a top–down sociotechnical system approach to organizational (first) and, ultimately, work system design and the design of related human–machine, user, and human–environment interfaces.

Dimensions of Organizational Design

Robbins (1983) defined an organization as "the planned coordination of two or more people who, functioning on a relatively continuous basis and through division of labor and a hierarchy of authority, seek to achieve a common goal or set of goals" (p. 5). This concept of organization implies a structure, which can be conceptualized as having three major dimensions: complexity, formalization, and centralization. *Complexity* refers to the degree of differentiation and integration that exist within an organization. Three major kinds of differentiation are found in the organizational structure: (a) vertical differentiation, (b) horizontal differentiation, and (c) spatial dispersion.

Vertical differentiation is operationally defined as the number of hierarchical levels separating the chief executive's position from jobs directly involved with system output. In general, as the size of an organization increases, the need for vertical differentiation also increases. The primary factor underlying this size–vertical differentiation correlation appears to be the span of control or the number of people that can be directly controlled effectively by any one supervisor, although optimal span of control is affected by other sociotechnical system characteristics.

Horizontal differentiation refers to the degree of departmentalization and job specialization found in the organization. Although it has the inherent disadvantage of increasing organizational complexity, the division of labor permitted by job specialization also has inherent efficiencies. Job specialization increases productivity—a characteristic noted by Adam Smith many years ago and by Henry Ford on his automobile assembly line. Division of labor creates groups of specialists. The way in which these are grouped in the organization is called *departmentalization*. The optimal degree of specialization from an ergonomics standpoint depends on other sociotechnical system factors.

Spatial dispersion is defined operationally as the degree to which an organization's facilities and personnel are dispersed geographically from the main headquarters. Three measures of dispersion are: (a) the number of geographical locations within the organization, (b) the average distance to the separate units from headquarters, and (c) the number of employees in the separate locations in relation to the number at headquarters (Hall, Haas, & Johnson, 1967). Increasing any one of these differentiation dimensions increases organizational complexity.

This definition of *organizational complexity* is peculiar to management and should be contrasted with system complexity because it impacts directly on the operator. System complexity to the author is defined by the number of informational states that the equipment displays to the operator as stimuli, whose function is to initiate operator actions and provide feedback. The stimuli are determined by the number of components in the equipment, the number and effect of their interactions, and the number of modes in which the equipment can be operated (Meister, 1996b). Organizational complexity functions as a context for personnel performance, whereas system complexity acts directly on the operator and determines his or her performance.

Integration refers to the extent to which structural mechanisms for communication, coordination, and control across the system elements have been designed into its structure. Some of these mechanisms are formal rules and procedures, liaison positions, committees, and information and decision support systems. Vertical differentiation also serves as an integration mechanism for horizontally and geographically differentiated units. There is a direct relationship between the complexity of an organization and the extent to which integrating mechanisms are required (Robbins, 1983). A critical *M* aspect of work system design is the inclusion of the appropriate kinds and number of integrating mechanisms.

From an ergonomics standpoint, *formalization* may be defined as the degree to which jobs within organizations can be standardized. In highly formalized organizations, jobs are designed to permit little employee discretion over what is being done, when, or in what sequence tasks will be accomplished. The management system includes explicit job descriptions, extensive rules, and clearly defined procedures. Of course, the design of hardware and software may restrict employee discretion. Organizations with low formalization allow their personnel more discretion; there is considerable autonomy and self-management. Behavior is relatively unprogrammed and personnel are allowed greater use of their cognitive facilities. It stands to reason that the simpler and/or more repetitive jobs are, the greater the utility of formalization, the higher the level of professionalism, and the less formalization.

Centralization refers to the degree that formal decision making is concentrated in an individual, unit, or level (usually located high up in the organization), thus permitting employees (usually low on the totem pole) only minimal input into job decisions (Robbins, 1983). Centralization is desirable (a) when a comprehensive perspective is required, such as in strategic (high-level), overall decision making; (b) when the organization is functioning in a highly stable and predictable environment; (c) for decisions where they can clearly be made more efficiently when centralized; and (d) when significant economies can be clearly realized by this type of organization.

Decentralization is preferable (a) when the organization is functioning in a highly unstable or unpredictable environment; (b) when a manager's job requirements exceed human information-processing and decision-making capacity; (c) when inputs from lower levels are desirable; (d) when attempting to stimulate greater intrinsic job motivation in personnel; (e) for gaining greater employee commitment to the organization; and (f) for providing more training opportunities for low-level managers.

The Sociotechnical Systems Model

The sociotechnical system concept views organizations as open systems engaged in transforming inputs into desired outputs. They are open because they have permeable boundaries exposed to the environments in which they function and on which they depend (Emery & Trist, 1960).

This transformation process is determined by technology in the form of subsystems, equipments, workstations, and so on and people in the form of a personnel subsystem. The design of the technical subsystem defines the tasks to be performed, whereas the design of the personnel subsystem prescribes the ways in which tasks are performed. (It might be said, of course, that the technological design of the equipment in many respects also determines the way in which tasks must be performed.) The two subsystems interact with each other, are interdependent, and operate under the concept of *joint causation*, meaning that both subsystems are affected by causal events in the environment. The technical subsystem, once designed, is stable and fixed; whatever adaptation the organization permits falls to the personnel subsystem to implement. Joint causation gives rise to an important related sociotechnical concept of great importance to ergonomics: *joint optimization*. This means that, because both technical and personnel subsystems respond jointly to causal events, optimizing one subsystem and then fitting the other to it will result in suboptimization of the joint system. Thus, joint optimization requires integrated design of the two subsystems.

Sociotechnical system factors are different from those of traditional HFE specialties. For example, the latter simply assume that technology exists and has an effect, whereas *M* attempts to define the essential characters of that technology. Perrow (1967), who incidentally is a sociologist by training, defined *technology* in terms of knowledge. He defined technology by the action one performs on an object to change that object (in sociotechnical systems terms, in the transformation process). Perrow noted that this action always requires some form of technological knowledge; hence, technology can be classified by the knowledge base it requires. In consequence, it is possible to identify two underlying dimensions of knowledge-based technology. The first is *task variability*, or the number of exceptions encountered in one's work. These can range from routine tasks with few exceptions to highly variable tasks with many exceptions. The second is *task analyzability*, or the type of search procedures one has available for responding to task exceptions. For any given technology, the search procedures can range from logical and analytical reasoning for well-defined tasks to experience, judgment, and intuition for poorly defined tasks. An example of the latter type of task is the one mentioned previously involving the monitoring of computerized equipment operations, with emphasis being placed on the recognition of problem symptomology and the solution of malfunction problems.

The combination of Perrow's dimensions, when dichotomized, yields a 2 × 2 matrix as shown in Table 7.1. Each of the four cells represents a different knowledge-based technology. Note that uncertainty (see chap. 3) intergrates all of Perrow's classes. Routine technologies, like mass production entities, have few exceptions and well-defined problems. Routine technologies are best accomplished by means of standardized procedures and are associated with high formalization and centralization. Nonroutine technologies like aerospace operations have many exceptions and difficult-to-analyze problems. Such technologies require great flexibility, and thus lend themselves best to decentralization and low formalization. Engineering technologies have many exceptions, but can be dealt with using well-defined rational-logical processes. Therefore, they lend themselves to centralization, but require flexibility that is achievable through low formalization. Craft technologies typically involve relatively routine tasks, but rely heavily on experience, judgment, and intuition for decision making. Consequently, decentralization and low formalization are required for effective functioning.

With regard to the personnel subsystem, two major aspects are important to organizational design. These are the *degree of professionalism* and *psychosocial characteristics* of the workforce. *Professionalism* refers to the education and training requirements of a particular job. Robbins (1983) suggested that formalization can take place either on the job or off. In the former case, formalization is external to the employee and consists of rules, procedures, and the characteristics of the human–machine interface that serve to limit employee discretion. These tend to characterize unskilled and semiskilled positions. The job formalization occurs through the internal professionalization of the employee through his or her education and training. From an ergonomics design perspective, there is a trade-off between formalizing the organizational structure and professionalizing the jobs and

TABLE 7.1
Perrow's Knowledge-Based Technology Classes

		Task Variability	
		Routine with few exceptions	*High variety with many exceptions*
Problem analyzability	Well defined and analyzable	Routine	Engineering
	Ill-defined and unanalyzable	Craft	Nonroutine

From Perrow (1967).

interfaces. Jobs requiring highly professional people should be designed to allow considerable employee discretion. Low education and training jobs require an evident supporting organizational structure and related human–machine interfaces.

Hendrick (1984, 1991) found that the most useful integrating model of psychosocial influences on organizational design to be that of *cognitive complexity*, as defined by the abstractness or concreteness of thinking. Concrete cognitive functioning is characterized by the need for structure, order, stability, and consistency in one's environment. Cognitively complex persons tend to have a low need for structure, order, stability, and consistency. Relatively concrete managers and workers function best under relatively high centralization, vertical differentiation, and formalization. Cognitively complex work groups and managers function best under relatively low centralization, vertical differentiation, and formalization.

M considers that critical to the success and survival of an organization is its ability to adapt to its external environment. Task environments vary along two dimensions: degree of environmental change and complexity. The *degree of change* refers to the extent to which a specific task environment is dynamic or remains stable over time; *degree of complexity* refers to whether the number of relevant specific task environments is few or many in number. As illustrated in Table 7.2, these two environmental dimensions in combination determine the environmental uncertainty of an organization.

Environmental uncertainty has repeatedly been shown to be the most important sociotechnical system factor that influences the effectiveness of organizational design. In conditions of high uncertainty (e.g., as exemplified by the advertising agency, with its constant demand for new ideas), the organization must be flexible and rapidly responsive to change. In conditions of low uncertainty (e.g., manufacturing common household appliances like toasters), maintaining stability and control for maximum efficiency becomes the criterion for survival. The greater the environmental uncertainty, the more important it is for the organizational structure to have low vertical differentiation, decentralized decision making, low formalization, and a high level of professionalization among its workers. Certain environments are ideal for high vertical differentiation, formalization, and centralized decision making—typical of classical bureaucratic structures.

TABLE 7.2
Environmental Uncertainty of Organizations

Degree of Complexity	*Degree of Change*	
	Stable	Dynamic
Simple	Low uncertainty	Mod. high uncertainty
Complex	Mod. low uncertainty	High uncertainty

From Hendrick (1991).

The Relation of Macro- to Microergonomic Design

Hendrick (1991) suggested that *M* design determines many of the microergonomic characteristics of human–machine interfaces. For example, horizontal differentiation prescribes how narrowly or broadly jobs must be designed. Decisions concerning the level of formalization and centralization dictate degree of routinization and employee discretion to be designed into jobs. Vertical differentiation decisions prescribe many of the design characteristics of managerial positions, including span of control, decision authority, and so on. In short, effective macroergonomic design drives much of the microergonomic design of the system components with the system's overall structure. (The other side of the picture, however, is that the nature of individual human–machine units may exert some, even if only a little, influence on organizational design.) Hendrick contrasted this with a purely microergonomic approach, which may create systems in which the personnel subsystem is forced to adapt to the system technology—and thus puts "square pegs into round holes."

M Methodologies and Applications

Hendrick (1991) cited a large number of specially developed *M* methodologies, of which one, participatory ergonomics, has received a great deal of attention recently. The reader who is interested in these methods should consult Hendrick (1991) for an extensive bibliography.

Hendrick also cited 89 studies in which *M* principles were applied, including 23 in production manufacturing firms, 3 of maintenance organizations, 5 of health care organizations such as hospitals, and 21 of administrative and managerial offices, including those related to the introduction of office automation. Four studies dealt with the structure or functions of R & D organizations, 9 with computer-based support systems, 13 with industrial accident analysis and prevention, and 4 involving building and office design. These studies were conducted in 12 countries.

M in Relation to Traditional HFE

M professionals, as differentiated from traditional HFE professionals, believe that traditional or what they call *microergonomics* has failed. To quote from Hendrick (1995),

> Traditional ergonomics has failed to significantly improve overall *system* productivity, worker health, and the intrinsic motivational aspects of work systems. As was noted several years later, progressively more examples were being seen where organizational systems with good traditional micro-ergonomic design

> were not achieving overall organizational goals because of a failure to address the macro-ergonomic design of the work system. (p. 1618)

Much of this kind of finger pointing can be ascribed to the strivings of a new specialty, which feels the necessity to bolster its self-confidence. There are indeed many failures of traditional HFE, as the preceding chapters have emphasized, but they are not the result of the failure to follow *M* doctrine and methods because, in real life, as opposed to theory, microergonomics and *M* do not interact.

It is important to note that *M* has a different set of predecessors than does traditional HFE. The latter is closely linked to experimental psychology, engineering, and physiology; *M* has close ties to and has in effect metamorphosed from the following: sociology, social and industrial psychology, industrial ergonomics, and management science. There are even traces of gestalt psychology in its background (the whole is greater than the sum of its parts). Much of this is foreign territory for traditional HFE specialties, and it is difficult to see the interrelationship between *M* and visual performance, for example, or aerospace. Nevertheless, a fundamental *M* premise is that traditional HFE cannot be successful unless the organizational context in which humans work with machines is taken into account.

In terms of historical perspective, *M* is a relative newcomer to HFE, having exploded into life in the 1980s, although it has strong ties to sociotechnical research going back to the early 1950s in England. A related permutation is something called *community ergonomics* (Smith & Smith, 1994), which has ties to urban anthropology. Its importance, at least as far as this author is concerned, is that it attempts to deal with units of analysis that are much larger than those ordinarily dealt with by the other HFE specialties. Traditional HFE research, as was seen in chapter 6, deals almost exclusively with the individual.

M assumes that without *M* there can be no significant improvement in human and system performance. Although this point of view is undoubtedly an exaggeration, it is obvious that there must not be a conflict between macro- and microergonomics design if the overall system is to perform satisfactorily. However, it is difficult to see how such a conflict could arise considering the difference in scale in which the two function.

The amount of effect on the individual worker and his workstation depends on what aspect or factor he concentrates on and the scale he is dealing with. In the design of an aircraft cockpit, for example, *M* may have little to say unless it attempts to analyze the organization and tasks of a three-man crew. However, crew system management is a relatively new aviation concept that does relate to *M*.

Despite the many studies Hendrick (1991) cited, it seems unlikely that most companies structure or restructure themselves readily, although there is certainly value in using *M* principles in the design of a new company. Older industries will not easily change because traditional managerial patterns are resistant to change. *M* operates at a level far more molar than that in which traditional HFE gets involved, which raises the question of how *M* exercises its effects on much more molecular system elements. If these effects occur, they must be explained in a conceptual structure that relates operator performance to company structure.

Conclusions and Speculations

Because of the relatively molecular level at which most ergonomists are involved and the molar level at which *M* functions, the descrepancy in scale between the two makes developing an integrated micro–macroergonomics theory or conceptual structure absolutely necessary. It is all very well to say that in a system everything in the system affects everything else, but the degree of influence from the top of the system structure down to the individual worker at his or her workstation may be so slight as to be imperceptible.

The question that a traditional HFE professional asks (and the author is certainly representative of the breed) is: How does *M* function in practice as well as research? Does it really affect hardware and software design and the design of equipment systems? How can one apply *M* principles to physical elements? Most design engineers, who have barely assimilated traditional HFE, may have difficulty coming to terms with *M* because they do not see management as their concern or that management is something that can be designed.

Implicit in *M* is an interest in the operator's motivational processes, which is far removed from traditional HFE research and practice. Presumably *M* wishes to make the worker's lot a better one by reorganizing the company in which he or she works. Traditional HFE would like to do this also by reorganizing his or her equipment or, in the case of industrial ergonomics, changing the work/rest schedule. A significant problem is that it is difficult to see how *M* principles of differentiation or cognitive complexity translate into specific design guidelines at a molecular workstation level. There is the implicit *M* assumption that *M* principles, when applied to company organization, will produce an effect that is not stimulated by the nature of the human–machine interface. Traditional HFE in system development seeks to make it easier for the worker to function with his or her console. However, ease of functioning does not translate into a positive motivation for work. It seems as if *M* actually attempts to influence motivation, and at least it speaks in Hendrick's (1991) terms of worker job satisfaction. If traditional HFE design guidelines for workstation design are vague (and they are), it appears to the author that any such *M* guidelines for this level of work activity are absolutely lacking. Of course, it could be said that motivational effects trickle down from company structure, and therefore it is unnecessary to produce these effects directly at the workstation. Regardless of whether this is true, the job satisfaction factor presents a serious problem to *M* (if *M* professionals really believe their theory) because of its elusiveness.

In contrast to traditional HFE, which began with and is still closely involved with military systems, *M* is almost exclusively concerned with civilian industry because the military has a traditional organization that cannot easily be changed to accommodate *M* principles. Nevertheless, there are newspaper stories (e.g., *Los Angeles Times* of December 26, 1996) that suggest that the Pentagon is actively considering new organizational structures to deal with a rapidly changing military technology.

Although *M* professionals may demur, it is the author's impression that *M* is much more concerned with people than with things (hardware, software), whereas traditional HFE is much more concerned with things than with people, although the term *personnel subsystem* was first coined by human engineers shortly after World War II in connection with military personnel. The system orientation in *M* is a highly desirable feature of the specialty, but its concentration on organization and management makes it just as narrow, perhaps, as the other, more traditional specialties. However, the traditions it draws from (sociology, social psychology, industrial psychology, quality control concepts, etc.) are much broader than those of microergonomics, which stem mostly from experimental psychology and engineering.

There are two criteria of *M* success: increased worker satisfaction with their jobs and increased productivity, which translates into profits. Because concern for worker attitudes may tend to improve these (the so-called *Hawthorne effect*), it may be easier to modify worker satisfaction, at least as expressed in attitude surveys, than it is to influence company profits, although the successes Hendricks (1991) reported presumably include the latter.

In the sense that technology, in particular the computer, will make significant changes in society (a common theme in magazines and books), traditional HFE will have to expand, possibly in the direction of *M*, if it wants to share in those technological effects. This is pure speculation, however. It remains to be seen how *M* deals with significant technological changes in computer networks, communications, and virtual reality.

The overriding question is whether the design of the company structure flows down to the design and efficiency of the individual system, equipment, and worker. Obviously if the sociotechnical system is an extreme one, structural changes will have profound effects at the worker level (imagine reorganizing the structure of a Roman warship if one is a galley slave), but at less extreme levels *M* effects may be much more muted. *M* professionals would say that *M* design of the company alone cannot be effective without consideration of traditional HFE factors (changing the length of the oar in the galley or increasing the number of rest pauses).

If *M* represents a somewhat alien strain that has entered the HFE structure, it is nevertheless a useful one. The author would like to see a closer melding of *M* with traditional HFE and this requires a movement of the two entities toward each other. That movement may have to begin with *M* thinking because microergonomics is highly traditional. The central question is how to translate macroergonomic concerns into effects felt at the microlevel. It has been repeatedly pointed out that HFE must cross domain (physical and behavioral) boundaries. Now it has another set of boundaries—macro and micro—that must be transcended.

It is inevitable that the HFE research specialist should become interested in *M* because, in studying the human–technology relationship, one inevitably bumps up against the higher order system even if one is only dealing with the single human–machine combination. That is because, even if for research purposes one arbitrarily eliminates all variables but those involved with the single combination, the latter is still part of the higher order system and the more molecular variables one wishes to study in the combination *may* be influenced by those higher order elements and dimensions studied in chapter 3. This question inevitably arises: Can one secure a true picture of the functioning of the single human–machine combination (or subsystems involving several such combinations) without considering the higher order system variables? The answer to this question can only be determined if one knows how strong the effect of these higher order variables and dimensions is on lower order elements. The effect of the higher order variables will of course be attenuated because these must permeate the boundaries of lower level entities, but to what extent? If there is any residual effect, is it important enough to require their consideration in studying the variables inherent in the single human–machine combination?

The picture that is suggested by the preceding is that of a giant eye that begins by contemplating the total system and progressively moves its gaze to lower and lower levels asking always, how are these entities at various hierarchical levels tied together? The rationale for this is that one always begins by deconstructing larger entities into smaller ones. One can do this by assuming that higher order systems have organizing dimensions that are peculiar to them (chapter 3 discussed dimensions like automation, complexity, autonomy, transparency, and indeterminism). One can then begin to interrogate the lower level subsystems and single human–machine combinations to see if these higher order variables and dimensions exert a measurable effect at lower levels.

One might also begin at the lowest system level and ask this question: How do these lower level entities combine to have an effect (if they do) on the higher order ones? For example, can the performance of a single workstation significantly influence the performance of the total system? It seems obvious that this is possible (e.g., if a radar-detection station malfunctions, the entire functioning of a ship can be affected, as apparently occurred in the Bay of Tonkin sea battle in 1964 that initiated extensive American involvement in the Vietnam War). Does this effect occur only with special equipments and under special circumstances? Manifestly, more research is needed to discover how the more molecular elements in the system combine to interact with the highest level (e.g., the commander's bridge or the CEO's office) and whether, in this process, new variables and dimensions emerge. Incidentally, to stimulate the meeting of macro- and microergonomics, there is a need to bring *M* to the attention of traditional HFE professionals. A review of the references in the three Hendricks papers previously referred to indicates that most *M* research appears in journals (and books) that the HFE traditionalist is not likely to read, although a few *M* studies are beginning to appear in the HFES annual meeting *Proceedings*. If it wishes to increase its influence, *M* should talk to the traditional HFE professional, not solely to its own professionals.

INDUSTRIAL ERGONOMICS

Again we must preface this description of the industrial ergonomics (IE) specialty by explaining why we emphasize it as distinct from the other HFE specialties. The reason for concentrating on IE is that it represents an important extreme of the HFE subject matter continuum. At one extreme we posit system-oriented specialties like aviation psychology and automotive systems (see chap. 6). At the other extreme there is IE, which is concerned largely, if not wholly, with the worker and his responses. There are other HFE specialties, like aging and individual differences, that are behaviorally centered, but none of these is as important (in terms of number of practitioners and involvement with factory processes) as IE.

Industrial ergonomics is a bit of an anomaly in HFE because, in contrast to the other specialties, its focus is not on system design and development, but on worker health and safety. Therefore, it has much more in common with European ergonomics than with the American variety. Actually IE straddles the fence between system development and health/safety because the concerns of the latter sometimes (often) require redesign of the work environment and manufacturing processes.

Perhaps a simple-minded way to describe what industrial ergonomists do is to list the major headings from the *Proceedings* of the 10th Annual International Industrial Ergonomics and Safety Conference held in 1995 (Bittner & Champney, 1995). These include: the effects of aging; cumulative trauma characteristics; methodology, models, and techniques; cybersickness; environmental hazard and stress effects; plant/organizational strategies; manual work; fitness for duty testing; hand tools; human–computer interaction;

macroergonomics; maintenance; manual material handling models; regulations, standards, and guidelines; rehabilitation programs and methods; safety; shift work and fatigue; and workplace, work, and equipment design. It is apparent from this list that the focal areas of concern are physical strength, manufacturing processes, the work (factory) environment, muscular trauma, hand tools, and manual operations. A few of the topics are relevant to system development, but most are not.

The goals of IE, as expressed in Mital (1995) are to: (a) eliminate or minimize injuries, strains, and sprains; (b) minimize fatigue and overexertion; (c) minimize absenteeism and labor turnover; (d) improve quality and quantity of output; (e) minimize lost time and cost associated with injuries and accidents; and (f) maximize safety, efficiency, comfort, and productivity. Mital went on to say that "the main components of ergonomics are anatomy, physiology, medicine and engineering" (p. 303).

It is obvious from the prior list that IE does not focus on system development or design of equipment to improve operator efficiency. In fact, the goals are essentially the same as those that would be claimed by industrial psychologists. By implication, system development can be inferred from the mention of engineering, but obviously this implication is weak.

IE is further distinguished by the experimental variables it measures. These are (again in the words of Mital, 1995): (a) work (task), workplace, and equipment factors; (b) worker (operator) factors; and (c) environmental variables. Typical of the independent variables involved in IE research are: task pacing, size of the object handled, weight/force, floor conditions, machine speed, tool characteristics, distances moved, posture, strength measures, age, gender, level of experience, and environment (lighting, temperature, etc). Almost none of these is found in the system-oriented literature (see chap. 6).

Inevitably the reader will ask this question (or we hope he or she will): Why in a work environment that is becoming increasingly automated and sophisticated does a specialty involved with almost *premodern* (one hopes the term is not considered pejorative) manufacturing processes play so significant a role in HFE research? As was seen in chapter 6, IE or biomechanics is a major theme in HFE research in terms of the frequency with which its papers are published in *Human Factors* and the number of models it appears to develop. The answer to the question why—any industrial ergonomist would respond—is that there are still and will in the future be many non-mechanical tasks to be performed no matter how much automation there is. Although this is quite true, this is not a satisfactory answer to the author.

HFE research can be approached in terms of two questions: (a) How does the human respond to technology and its characteristics, and (b) how can one change technology to maximize human performance, safety, and comfort? It is possible for one's research to seek to answer both questions, but most HFE research seeks to answer Question (a) because the answer to

Question (b) demands a somewhat nontraditional approach to that research. The majority of IE research and much of all other HFE research is focused on Question (a). It can be argued and with some justice that before one can attempt a change in technology, one must know how people react to that technology. However, the human-centered HFE specialties and unfortunately even the system-oriented specialties never get beyond the point of studying the first question.

It is possible to entertain the hypothesis that human-centered research like IE is firmly in the tradition of applied experimental psychology, and perhaps the specialists who practice it are more comfortable in that tradition than in system-oriented research, which has an aura of industrial engineering. To effect a change in technology (even if it is to improve human performance and comfort) is to assume in some small way the role of the engineer, whose job is specifically to change technology.

A question that arises is, how is one able to effect a change in technology if all one knows from one's research is how the human reacts to present technology? For example, if we know that the aged experience more difficulty perceiving labels and other visual stimuli, should this not suggest increasing the size of the type font? Is this not an attempt to influence technology? One can deduce from human reactions to technology certain changes that should be made in that technology. So where is the problem?

One can examine the relationship of the human to technology by starting from either end of the relationship: from the human or from the technology. If one starts from the human end, one examines one's responses to that technology. If one starts from the technology end, examines the characteristics of the technology end, one examines the characteristics of the technology, varies these, and sees how the human responds. The common element is, of course, the human response, which we cannot hope to avoid, but the difference is in the conceptualization of the variables with which one's research deals. If one accepts the technology for what it is and merely measures the human response to that technology, one is merely accepting that technology for what it is at the moment. Any program of redirecting HFE research toward Question (b) must involve some change in the variables it studies.

In IE and the other human-centered specialties, the focus is on the human, with the technology serving as a context for or source of stimulation. The determination of human performance characteristics—a major focus of HFE research (see chap. 6)—is an attempt to discover the capabilities of people, not the effect of variations in technology. In IE we see this most clearly in studies of human strength, lifting, manual operations, and so on (e.g., Harper, Knapik, & de Pontbriand, 1997). From knowing what the human can and cannot do, one can perhaps infer what the characteristics of the technology should be on the premise that any characteristics for which the human response is poor or stressful should be changed. However, it does not indicate how one should change that technology. This leads to the situation we have presently, in which behavioral design guidelines are largely lacking.

QUANTITATIVE PREDICTION

Introduction

This section describes the specialty commonly called *human reliability* (HR), but that the author prefers to call *quantitative prediction* because he sees it as much broader than the other specialties summarized in the preceding pages. The reason for giving greater attention in this chapter to quantitative prediction and HR than to the other specialties is that quantitative prediction is more than a special interest of certain segments of HFE professionals. It is a fundamental purpose and function of HFE as a whole, which means that it applies to all specialties, as well as being a speciality of its own, just as system development and test and evaluation also apply to all specialties but are specialties of their own. The same cannot be said of the other specialties, which are content-determined but not relative to the total discipline.

No scientific discipline can be truly scientific if it does not predict quantitatively the events and phenomena that it purports to describe. In that light, quantitative prediction is one of the purposes and functions of HFE.

It can be argued that the development of conclusions from empirical research is a form of prediction because it suggests that if the same conditions recur at a later time, the same performance on which the conclusion is based will recur. However, this type of prediction is weak because the conclusion relates to only one or two variables that have been studied and not to a total situation. More important, it is usually not quantitative. If it is objected that each study has at least one performance metric in the form of a probability significance level, the answer is that the metric describes only the probability that the effect found in the research occurred by chance. Such a probability does not predict performance.

Manifestly, therefore, the quantification inherent in experimental statistics does not satisfy the prediction requirement of a science. Statistical quantification is one part of scientific methodology and has been assumed by HFE as one aspect of experimentation. Because of that, the great majority of HFE professionals have rarely thought about quantification in other terms.

Starting shortly after World War II, individual HFE professionals began development of a predictive methodology. HR studies began about 1960 and must have had a prior history (in the 1950s), which is when the author first became aware of them. These studies and papers describing them continue to the present and have dealt with many topics. Various methods of HR prediction have been developed. Some of them, primarily that of Swain and

Guttmann (1983), have been used for many years to perform probabilistic risk analysis (PRA) as part of the planning for new nuclear power plants under the sponsorship of the Nuclear Regulatory Commission (NRC). A great deal of HR work has been done for the NRC at laboratories like Sandia and Brookhaven and under contract to consultant firms. Corollary efforts to understand human error and develop HR databanks have been pursued. Workshops and symposia have been held periodically, with papers published in prestigious journals like *Human Factors* and *IEEE Transactions in Reliability* (e.g., Regulinski, 1973). A number of books about HR have been published. Models and methods have been analyzed and compared and even articles published in nontechnical magazines (Williams, 1958). Attempts have been made to develop error probabilities for various systems like the X-15 experimental aircraft and Army machine gun systems (McCalpin, 1974).

The point of the preceding is that, although many, if not most, HFE professionals may be unaware of it, a continuing distinctive line of research and application—whose fundamental data are human errors and human behavior in a technological context—has proceeded for almost as long as the HFE discipline has existed. This is true despite that much of this has gone on outside of routine HFE channels.

This section is not a comprehensive review of HR issues and methods because the topic is complex. One short section cannot hope to describe the activities engaged in by many investigators over many years. This is especially so because the literature, extending as it does back to the middle 1950s, is quite extensive. It includes papers, reports, and books describing models of human machine functioning (e.g., Siegel, Wolf, & Lautman, 1974; Woods, Roth, & Hanes, 1986a, 1986b). At least 14 different methods of performing human reliability analysis (HRA) have been developed and compared (see Swain, 1989). Qualitative and mathematical analyses of human error (the primary HR metric) have been performed (e.g., Rouse, 1990; Rouse & Rouse, 1983; Regulinski & Askren, 1969). Particular attention has been given to the development of HR databases (e.g., Meister, 1993a; Payne & Altman, 1962; Topmiller, Eckel, & Kozinsky, 1982). Special journal issues have been devoted to the topic (Apostolakis, Mancini, van Otterloo, & Farmer, 1990; Regulinski, 1973) as well as articles in popular engineering magazines (e.g., Williams, 1958). A number of symposia dealing with the topic have been presented (e.g., Anonymous, 1967). Other topics have included methods of preventing or reducing error (e.g., Park, 1997), subjective methods of generating HR estimates (e.g., Seaver & Stillwell, 1983), as well as handbooks for performing HRA, the most important of which is Swain and Guttmann (1983). Human errors have been related to equipment malfunctions (Leuba, 1967; Lukas, Lettieri, & Hall, 1982), and various HRA models have been experimentally tested (e.g., Kirwan, Kennedy, & Taylor-Adams, 1996). Finally, a number of books have been published dealing with human error

(Reason, 1991) and HR as a whole (e.g., Dhillon, 1988; Dougherty & Fragola, 1988; Kirwan, 1994).

This description has been written to indicate how HR can be used as a sort of model for quantitative prediction efforts in HFE as a whole. It attempts to illustrate by reference to experiences with HR the kinds of problems that HFE will face if (or, more desirably, when) it attempts to quantify the discipline.

The author admits to a bias—that he is trying to make a case for HFE prediction based on HR experience. It would be desirable to develop a rationale for HFE quantitative prediction other than that it is a good thing to do. HR was and is driven by vital governmental concerns relating to nuclear power and fossil fuel plants. If the same concerns existed for HFE, the necessary quantitative prediction efforts would be performed.

To the extent that HR is a behavioral research specialty, it can be said that efforts to develop in HFE a quantitative prediction capability do have a history, although at the present time no formal predictive process is recognized and utilized by HFE scientists in general. Those who specialize in efforts to develop a usable, generally utilized, predictive methodology refer to their work as HR. This section continues to refer to it as HR, although the author prefers quantitative prediction as having a somewhat broader connotation and because some HFE professionals see the HR term as referring mostly to reliability engineering (which it does not). True, the term HR derives from the fact that within engineering there is a specialty known as *reliability engineering*, whose function is to measure the life cycle of equipment components and systems. It predicts, based on this measurement, the operational life of new components and equipments. However, HR is behaviorally based and depends on the same conceptual foundation as do the other HFE specialties.

Origins

When HFE specialists first entered industry, it appeared to engineering managers that reliability and logistics departments would be an excellent place to store their *human engineers* because reliability, logistics, and HFE are all functions that support design engineering. Some HFE specialists, such as the author, who worked in the reliability department, picked up a bit of the reliability methodology. Reliability picked up from us the concept that humans were responsible for errors that influenced the operating life of components and equipments. Therefore, it would be desirable to measure human propensity to error to make a more refined measurement and prediction than was possible if the prediction were based only on equipment variables. The connection between error and reliability prediction was therefore synergistic and gave rise to continuing efforts to develop behavioral predictive methods.

This tendency was reinforced when those who supervised the building of nuclear power plants insisted on prior mathematical analyses of the probability that such a plant might malfunction and cause an environmental catastrophe like Three Mile Island or Chernobyl. It was only logical (not all nuclear engineers realized this) that PRA should include an input from behavioral scientists, which was termed *human reliability analysis*. HR work (both research and practice) has therefore been driven by governmental concerns for reliability and safety.

Consequently, papers dealing with HR began to be solicited by those organizing reliability and maintainability conferences. Because many of these papers have been published in engineering journals, much of the HR effort has been disseminated outside of conventional HFE channels and it is doubtful that most HFE professionals have been aware of HR. However, because it deals with human performance, HR is definitely an HFE specialty and is accepted as such by the editors of *Human Factors* and other HFE journals. Although many of the people working in HR are engineers by training, the outstanding HR researcher in the United States was, until his retirement, Alan Swain—an HFE specialist. Considerable HR work has also been performed in England, particularly by David Embrey and Barry Kirwan, who are both psychologists by training.

The Rationale for Quantitative Prediction

If HR has been ignored by HFE, it is because the great majority of HFE professionals are psychologically oriented in their research. Experimental statistics, not quantitative prediction, has always been the hallmark of their methodology.

As long as HFE research is almost entirely experimental, statistical quantification is necessary. However, some (among them the author) would argue that experimental statistics is not sufficient because they do not predict performance; they merely assure the researcher that a given experimental effect has probably occurred not from chance.

It is at least conceptually possible that measurement can be performed solely for the purpose of prediction. Moreover, prediction permits many more things than can experimentation. For one thing, the prediction, if verified by collecting operational data, can validate (or fail to validate) the prediction and the variable interrelationships involved in the prediction. With quantitative prediction, it is at least theoretically possible to design more effective systems and anticipate (and thus counter) breakdowns of systems in the operational environment. (Hence, the concern for preventing or reducing error.)

Of course, it is important to test whether an individual variable has a significant effect on performance, thus determining which variables one should concentrate on in one's research. Because many variables (perhaps even most of them) have effects that are so slight as to be unmeasurable, it is conceivable that the purpose of experimental research is to determine which these inconsiderable variables are so that one can concentrate on the few remaining variables that are important.

However, the prevailing scientific methodology, as it is in other disciplines, is to stop at the point of determining which variables are or are not important. One could argue that, having determined which factors appear to be important, one should attempt to predict quantitatively how much importance the factor has—this is essentially what HR attempts to do. What is important in HR is the measurement, not the experimental design.

Conceivably, one can decide on the basis of logic and/or simple observation of behavior that some variables (call this a *cluster*) are more important than others and then perform a study that will provide data for quantitative prediction of performance involving that cluster. Such a study would not require the extraction of one or two variables in the experimental design, but rather the selection of a complex of variables that would appear to be logically important to the prediction. For example, in inspection tasks, skill level plus fatigue plus visual acuity might be one cluster of variables to be used in making a prediction. For a motor task, another cluster of relevant variables might be strength, age, and gender. The researcher would select subjects who vary on a scale developed to describe the cluster and then measure their performance on some relevant task. In contrasting people who are high and low on a particular cluster, one would of course be using a simple experimental design. However, the intent of the test would not be to determine if individual variables affect performance (this would already have been assumed in the variable selection process), but to determine exactly how important in terms of their producing error (or whatever other measure is selected).

Elements of a Predictive Methodology

Quantitative prediction has two major faces. The first face is the prediction of behavioral phenomena in the operational environment (i.e., the performance of personnel in operating and maintaining equipments and systems or doing any work related to technology). The second face is the prediction of the behavioral consequences of adopting one type of design configuration rather than another. To use an absurdly simplified example, what is the effect on the operator of installing two displays rather than one or one type of display rather than another?

The common element in these two aspects is the task. Task analysis is the foundation of the prediction as it is of HFE generally. Analysis may be performed in various ways: Conceptually by means of a verbal or physical simulation (walkthrough) of a procedure, by interviews with personnel, by observation of performance, by software performance modeling, and so on. The end result of the prediction can be used in various ways: (a) to predict, as does HR, the probability of error in performing a particular mission or job operation; (b) to discover potential difficulties in the scenario ("the greatest error probability is in Stage 3 because it is time stressed"); and (c) to decide among alternative design configurations ("which configuration has the least predicted error probability?").

The development of any prediction actually depends on and is a historical process because it assumes that performance that has occurred in the past in a particular context will probably recur in the future if the same conditions are repeated. (The same can be said of the experiment; its conclusions are correct to the extent that this assumption is justified.) Obviously if one is going to predict, one must record and analyze performance over multiple occurrences (i.e., over time).

A predictive methodology has certain elements:

1. the *purpose* of the prediction,
2. one or more data *sources*,
3. a method of *collecting and analyzing data* from those sources,
4. a quantitative *database*,
5. a common *metric*,
6. a *conceptual structure* (e.g., a way of viewing human performance; an error taxonomy for sorting data into certain classes that relate to performance; a taxonomy of performance shaping factors or PSFs, e.g., Fleishman & Quaintance, 1984),
7. a means of *combining* probabilities attached to molecular performance elements (subtasks, tasks) to determine a probability for more molar performance elements (mission, job operation),
8. a *format* with which one describes one's database, and
9. the *measurement* of the methodology's adequacy.

The remainder of this section deals with each of these prediction elements.

The Purpose of Quantitative Prediction. To predict, one must want to predict—an obvious statement that is usually overlooked. In this sense, purpose is will. The lack of a predictive methodology for HFE arises because, until now (and perhaps even now), HFE has not been interested in prediction, only in experimental testing of variables.

Both HFE and HR have certain common elements: the same conceptual structure and the task. What one wishes to do with these is what differentiates

HFE from HR. HFE is concerned with the variables that influence task performance, whereas HR is concerned with how well the task, mission, or job is performed. HR makes use of data gathered by HFE researchers and is certainly interested in which variables affect performance, but goes beyond HFE. Therefore, HR depends on HFE, although HFE, as it perhaps defines itself, apparently does not require HR.

The purpose of quantitative prediction is also determined by the uses one makes of one's research outputs. HFE uses its research primarily to develop conclusions related to explanatory knowledge. Except for a few feeble attempts to develop design guidelines, HFE finds no use for its data beyond explanation. (However, the author has heard that the government is massively supporting an effort to develop HFE design guidelines, but for only one specific technology—automated transport [Campbell, Carney, & Kantowitz, 1997]. It is not clear whether design guidelines for technologies in general can be developed—a question of specificity—nor whether these can be developed by the discipline as a whole without extensive governmental funding.) As was mentioned, HR has concrete uses for its prediction data, the determination of the source of error, the prevention or reduction of that error, and the evaluation of design or any other configurations.

There is no reason that HR and quantitative HFE prediction cannot function concurrently. The author sees HR as merely another of the HFE specialties, but there is no insuperable problem in general HFE prediction, provided HFE wishes to predict.

HR Data Sources. Because any prediction is based on historical data, it is necessary to examine available data sources. Obviously one cannot develop prediction without adequate data. For HR, there are five possible data sources, all of which, it must be said in advance, are somewhat defective, although it is possible to make some use of them individually and in combination. The five sources are:

1. *Published research.* These are the papers published in *Human Factors* and similar HFE journals.
2. *Studies* performed deliberately and solely to gather data for prediction purposes.
3. *Demographic data* (e.g., accident reports, compilations of malfunction reports, etc.).
4. *System test data.* These are data derived from system tests performed by companies developing systems for the government and from military acceptance tests.
5. Estimates in quantitative form made by individuals who are experts in individual systems.

Data published in research studies are almost always experimental in nature, which limits them, for reasons mentioned previously. Studies performed simply to collect predictive data may or may not be experimental. However, because HFE studies for this purpose are almost never performed, the question of the type of data they provide does not arise. Some demographic data (e.g., malfunction reports) are of some utility (see Leuba, 1967), but are limited because the categories of information these reports provide are few (i.e., the end result of an error, such as a workmanship defect). System test data describe human performance, but the performance data are often incidental to the test purpose and special arrangements must be made to record them.

Subjective estimates as a data source have been tried and found to be useful (Comer et al., 1984a, 1984b; Seaver & Stillwell, 1983). Because HFE scientists reared in the classic experimental methodology are usually doubtful about subjective data, which involve self-observation and observation of others, there has been little general use of this source.

As indicated previously, all sources have significant defects. Research published in the journals rarely contain the raw data needed to develop the HR metric because, in most cases, the data have been transformed into statistics of significance (e.g., ANOVA tables). Because of this, they are largely useless. Even when raw data become available by soliciting the researcher for his or her files, the conditions under which the data were collected may be unclear and irrelevant to the complex of variables the HR researcher is interested in. In such studies, there is often no relation to the operational situation, no data about equipment aspects, and the functions measured may be overly molecular and related only to psychological variables. It is possible to make use of this source, but a great deal of sifting and filtering is required, as can be seen by reference to Payne and Altman (1962).

Studies deliberately performed to provide HR data would undoubtedly be superior to any other source, but the practical difficulties are enormous. No single individual can mount a sufficient effort, and the effort must extend over years, not months. Funding must be made available from the only source that has real money—the government. These difficulties can be overcome if, for example, graduate students performing theses or dissertations were put to the task of collecting HR data. Considering the idiosyncratic interests of academics, this is unlikely.

Demographic data such as accident and malfunction reports could be useful, but the categories of information needed for a predictive database are unlikely to be found in governmental or institutional files because these have their own reasons for collecting data, which are not likely to parallel those of behavioral scientists.

Data collected on personnel performing in operational system tests conducted by engineering companies under contract to the government and the few specialized laboratories devoted to testing new military systems (e.g., Aberdeen Proving Ground) would be useful, but again access to these data would probably be quite limited. For example, many years ago, the author sought access to test data on astronauts and was told that the data were so highly confidential that even general officers in the military were denied access to them. There is also, as in the case of published data, no assurance that the complex of behavioral variables involved in the test performance will be recorded by those conducting the test because companies and test laboratories have their own agenda, which usually does not correlate with that of the HR researcher. In any event, the number of such tests is not large.

Quantitative estimates of personnel error and/or success made by experts—estimates that describe the customary performance of personnel in operational situations—have been investigated and show promise, but it is not completely certain that such estimates, which are highly subjective, are completely valid. To secure these estimates, special data-collection methods must be developed. However, in the event that one cannot make use of the other more objective methods, quantitative estimates can be used as supplements. The one great advantage that estimates have is that, compared with the other techniques, masses of data can be relatively easily secured. For confidence in the data, one would have to provide validation data, whose collection would present another problem. Needless to say, this source of HR data has been utilized only on an experimental basis in this country, although in England the SLIM-MAUD predictive technique (Embrey, 1983; Embrey, Humphreys, Rosa, Kirwan, & Rea, 1984) has made it the foundation of the method.

This review of HR data sources suggests that the researcher must attempt to scrounge data from wherever he or she can find these. The processes of collecting, analyzing, and refining data are described later. Data collection for HR purposes is a messy business and not as easy as performing the traditional experiment. The most available data source is published research, and experience in using this research (e.g., the Data Store; Munger, Smith, & Payne, 1962) suggests that it is a feasible source. However, the problem of making use of such data is so difficult in the minds of most HFE professionals that they reject the entire HR concept. It may appear to the reader as if the difficulty in securing appropriate data makes the whole HR question dubious, in which case, why do we address it? Although the difficulties are real and not to be ignored, they are not insuperable and the quantitative prediction of human performance is so important to HFE that these difficulties should not be allowed to derail the effort.

Data Collection and Analysis. Data collection depends on which data source the researcher intends to utilize. The two major sources are published research reports and expert opinion. In the history of HR, the most commonly employed source has been the published study, supplemented by observation and expert opinion data.

With published studies, the first task is to determine which studies provide relevant data. This presumes that the categories of data that one intends to use in the prediction have already been established in the form of an error taxonomy. If experimental studies are the data source, the independent variables must correspond to the error taxonomy. If, as in the case of the developers of the Data Store (Munger et al., 1962), errors are to be associated with control/display parameters, then obviously only studies dealing with controls and displays can be utilized. The study must also describe the parameters of the equipment (e.g., size of a display). If Study X satisfies this first criterion, a further set of criteria in the form of questions must be addressed:

1. Does the study contain useful data: raw data (unlikely), averages (much more likely), ranges of performance, and so on? Experimental statistics like ANOVA tables or *t* tests are unlikely to be useful.
2. Does the metric in which study results are expressed the same as the researcher's own metric or can it be transformed into a common metric like a *Z*-score?
3. Are all the conditions under which the study was performed adequate (e.g., more than two or three subjects, no apparent contaminating conditions)? Are any influencing PSFs described in the report?
4. Can the study data be classified into the error taxonomy categories?

Controversies over the answers to these questions may have to be settled by majority vote, which implies that more than one researcher is required to develop a database, which is what we are talking about to this point and which is the first step in the analytic process. Essentially the same process would be applied to any data derived from indirect sources like accident or malfunction reports or system tests.

In direct numerical estimation and structured opinions, the investigator has a direct role and certainly more freedom to establish his or her own description of the performance conditions. For example, in describing an operational situation to which an expert will supply a direct error rate estimate, the investigator can specify the conditions that will affect the estimate. For example, one might say that the estimate is for a highly skilled operator or novice, for a time-stressed or nonstress situation.

In paired comparisons, which is another method of collecting expert data, the investigator sets the situations for which a comparison is made, including PSFs, and asks the expert to indicate which situation produces a higher or lower performance error. There are statistical techniques for transforming the paired comparison results into error rates, but these are much too detailed to describe here.

The combinatorial process is a fairly complex process, as described in Swain and Guttmann (1983), because it means applying an error probability extracted from the database to each subtask and task and then combining in specified ways the individual molecular probabilities to derive a total error probability for the mission or job. When many subtasks and tasks are involved in the mission or job operation, the whole process can be time- and effort-intensive. The process can be performed for all the tasks that must be performed in a particular mission or for those few considered most critical to system performance.

Databases. A database is the source from which one selects the data from which a quantitative prediction is made. Before the selection can be made, however, the database must be developed. Many, if not most, HFE professionals talk about the HFE database as if it consisted merely of the volumes of published research studies on library shelves. With such a database (if one can call it that), one need do nothing more than read a paper and extract what one can from it. The reader is often disappointed that the papers in such a database cannot provide specific answers.

These published papers contain five kinds of material: (a) raw data (only rarely published), (b) refined data (e.g., means and standard deviations), (c) the results of the statistical analysis of the study data (e.g., tests of significance, ANOVA tables), (d) description of how the study was conducted, and (e) conclusions derived by the author(s).

The reader is asked to consider what one can predict quantitatively from each of these types of material. Of the five categories, only one—the raw data, which are almost never included with the published study—offers the possibility of making a prediction, and this is only after the data have been extensively refined. Therefore, the popular conception of the HFE database is completely incorrect.

A database requires development; it does not simply exist by the accretion of published papers. Once a database is created, its development is not over. Its effectiveness must be tested in usage. The purpose of the database determines how it will be tested. If it is to be used for prediction, for example, a prediction must be made and the results of the prediction must be compared with actual performance. The reader decides whether this is validation or utility. Development of a database is not a simple, easily accomplished process, and possibly this is what dissuades most HFE specialists from considering its development. It is an iterative process; additional data must be entered into the database periodically; the database never becomes quiescent.

Of course, there are compendia of facts, like Boff and Lincoln (1988), but compendia are not a database; they are simply summaries of materials published in individual papers and they cannot be used for quantitative prediction.

Database Format. Because a quantitative database, as HR specialists conceptualize it, is something specifically developed, rather than a compilation of qualitative conclusions, it must be organized into a format. It is possible to distinguish five types of database format. The first can be termed a *probability statement of task performance*. A hypothetical example might be: the probability of throwing a switch correctly and observing CRT signal quality is .9972.

A second type of database would consist of probability statements associated with specific equipment characteristics. Note that there is no single probability associated with the task of throwing the switch, apart from its parameters, although there is no reason that the first and second types of database should not be combined in the document.

A third type of database might consist of the raw performance values associated with particular parameters. The data are not presented in probabilistic form, although presumably if there were enough of them a probabiliity could be derived. However, for this type of database, for whatever reasons, no probabilities are shown. Presumably data would be selected to suggest the desirability of following one or other design guideline.

A fourth type of database could consist of quantitative, nonprobabilistic statements related to specific equipment characteristics. A sample item might be: Display Format X will produce 1.658 times more effective performance than Display Format Y (the two formats differing in specific ways as described in the database). The statement might be quantitative or qualitative, could use an arbitrary scale value to represent relative performance, or could use a rating scale or ranking.

A fifth type of database could be a melange of all the previous ones, and the particular kind of statement presented would depend on what type of data one had available to present. The author prefers this format because it takes advantage of whatever positive characteristics each type of database has to offer. It also makes it unnecessary to exclude any quantitative value because there are not enough data points to develop a probability or the data are more adequately represented as a graph depicting a relationship between variables. The database format illustrates the necessity of abstracting certain elements of the data to organize a database taxonomy. In qualitative compilations of HFE material, it is difficult to do this, which is why it is rarely attempted.

The HR Metric and the Role of Error. To predict quantitatively, one must have a metric—a way to express that prediction in quantitative form. For the prediction of human performance in operating an equipment, the most reasonable metric would seem to be probability of successful task performance because that is what one is interested in. The metric must be probabilistic because the individual variations in operator performance that one inevitably finds mean that a deterministic value cannot be utilized. Philosophically, we live in a probabilistic world; practically, because of individual variability, one can only estimate (i.e., make a probabilistic guess of the true performance value, which will of course vary somewhat from one operating cycle to the next). At best, one can supply only a range of values within which the actual value on any single cycle is likely to be found.

Success in task performance, as it relates to the operation of any single (or even several) control or display, translates into error because success occurs in operating an entire equipment, not the individual part of the hardware. Hence, if one wishes to determine which aspect of design is best, one must translate success probability into error probability. The same applies if one wishes to predict the performance of individual subtasks and tasks; success in performing these translates into success in avoiding error.

There is some reason then for substituting error for success: If one has success in avoiding errors in equipment operation, one has success period. Besides, it is possible to distinguish various kinds of errors, which makes these distinctions diagnostic. There are no different kinds of success; success is monolithic, and failure to achieve success is not diagnostic in terms of finding why success has not been achieved.

Moreover, in the earlier days, procedures for operating equipment were likely to be procedural (i.e., the designer laid down a step-by-step procedure to be followed in precise sequence). From this, it was conceptually simple to use error as a performance metric because any deviation from the procedure was automatically an error. Hence, the earliest error classifications were failure to perform a required procedural action, performance of an action not required by the procedure, performance of a required action out of sequence, and so on. With the introduction of more automated systems in which the computer operates the equipment and the operator monitors the equipment to ensure that it does not exceed parametric limits, the monitoring and diagnostic efforts of the operator could not be proceduralized or only partially so, thus a different set of error categories had to be devised. (See, for example, Meister, 1995a.)

HFE professionals working in prediction also took a reductionistic position in their concept formulations. Overall, task performance obviously consisted of activating controls, responding to displays, and performing individual actions, all of which combined over time to form the cycle of operation. Prediction was also associated with generic components of the human–machine interface (e.g., types of controls, types of displays) and generic human functions (e.g., perceptual, motoric, etc.). If one wished to make a prediction for a special component like the X control, one could do this only by assembling a history of performance with the specific X control. This would take much time and effort if one had to do this with each distinguishable control. Because there are many types of controls, displays, individual tasks, and so on, the only practical way to build up a database was to collect data on types and attempt to generalize to all the individual variations of those types.

The reductionist philosophy makes it necessary to combine predictive values, starting with more molecular items and progressively combining these to achieve a single predictive value for an entire equipment or process. Therefore, the predictive process has to have a method of combining predictive values. This topic is discussed later.

To return to the metric. One of the practical reasons for concentrating on error as a metric is that, given an error-inducing propensity (wherever it was, in the human, in the machine, in their interface), one would wish to reduce this error probability and thereby increase the probability of success. Error rates, if they could be associated with some sort of a causal factor, could be diagnostic of a condition that could be remedied. As Chapanis and his colleagues pointed out (Chapanis, Garner, & Morgan, 1949), error was either random (because of innate variability in the human) or systematic (meaning that some factor was causing the error; and if that factor could be found, it could be eliminated or changed). For example, in a production facility, if errors were classified by the particular stage of production or its venue, the stage and venue with the highest error frequency could be examined and the responsible factor found. Thus, the error measurement could be predictive as well as diagnostic.

Most HR specialists have used error probabilty as a metric, but time can also be used. In one outstanding example, the Data Store (Munger et al., 1962), base time was used and time was added to this base as a function of the type of control/display being activated or perceived. The highly molecular motion–time–method employed by industrial engineers has also used time as a metric, and the Russians make heavy use of time (Bedny & Meister, 1997). Most Western HR specialists have not found response time to be particularly useful in HR because response time has not been shown to be predictive of success.

In its simplest form, an *error rate* is defined by counting the total number of operations (whether these are tasks, repetitive actions, etc.) and using this number as a denominator for the number of operations in which one or more errors have occurred. The resultant equation is:

$$\text{Human error probability (HEP)} = \frac{\text{number of operations in which one or more errors occurred}}{\text{total number of operations}}$$

For example, if the total number of operations in a job is 100, and there have been 5 operations in which an error occured, the HEP is .05. If each of the five errors caused an operation to fail, the job success probability would be .95. Mathematicians can (and do) perform elaborate variations on this, but essentially that is what an error probability is. Askren and Regulinski (1971) developed a number of error metrics closely paralleling those used by equipment reliability engineers, but these never caught on. It is obvious that whatever predictive metric is developed, it must be measurable (i.e., it must be capable of being observed and/or counted). Where the metric describes overt motor functions, this is relatively easy to do. With perceptual and cognitive functions, there will be difficulties in measurement because the actions involved in these functions are covert.

It was early observed that all errors are not the same and could be classified in various ways. Errors differ on various dimensions:

1. The molecularity of the behavior in which the error occurs (e.g., errors of molecular motor actions as distinguished from complex cognitive errors). Obviously this dimension is linked to the human function involved.
2. The type of error (i.e., error associated with a work function: a design error, production error, error of navigation).
3. The human function associated with the error as a perceptual, motor, or cognitive error.
4. The operator's ability to recognize that an error has occurred and his or her ability to modify an erroneous action.
5. The correctability of the error once made and once observed; some errors, although recognized, may not be correctible.
6. The consequences of the error, which may be none, may be some delay in equipment operation or have more significant consequences, like a boiler explosion.

The HR researcher might not be interested in trivial errors or those that have no significant consequences, or he or she might be concerned only with cognitive errors. An error taxonomy that concentrated on simple motor actions would be different than one describing complex cognitive activity. Therefore, the type of taxonomy one develops depends on the type of performance one is interested in.

It is obvious that the definition of human error is becoming increasingly difficult because tasks are changing from routine (procedural) activities to decision making during abnormal situations. Rasmussen (1985) suggested that:

> the ultimate error frequency largely depends on the observability and reversibility of the error. . . . Error observability depends on the perception of a mismatch between the expected and the actual response of the work environ-

> ment to the human act. The information needed for control of actions and for observation of errors may be very different and be related to different time spans and levels of abstraction. (p. 1188)

Reversibility depends largely on the dynamics and linearity of the system, whereas observability depends on the nature of the task, which is obviously much influenced by modern information technology.

The HR Conceptual Structure. The conceptual structure underlying HR is the same as that of HFE as a whole—the S–O–R paradigm (stimulus–organism–response). In addition, special attention is paid to performance shaping factors (PSF). The S–O–R paradigm is common to psychology and HFE and reflects the rather obvious fact that the human responds to stimuli conceptually to acquire their meaning and responds (motorically, perceptually, and/or conceptually) to these stimuli. PSFs are aspects of the total situation in which the human and his S–O–R mechanisms are embedded; they modify the performance produced by the mechanisms involved in S–O–R. They are not independent or dependent variables.

The conceptual structure (S–O–R) determines the taxonomy or classification scheme into which the selected data are decomposed. The fact that the overall conceptual structure of psychology, HFE, and HR is the same suggests that the stumbling block in quantifying HFE data, which are largely experimental, is not their conceptualization of human performance. The problem is to decompose the data situation into the elements of the conceptual structure—in other words, to recognize independent and dependent variables and PSFs in the verbal description of the measurement situation. Descriptions of experimental situations in published reports are often not clear especially as they relate to the cognitive and PSF elements. The latter are not ordinarily described in the experimental report. There is often no information available about such PSFs as skill level, stress, how the subject viewed the experiment, and so on. PSFs are often as important to the performance as the independent variables the researcher measures, and quite often there is nothing known about them. These PSFs can be factored into structured expert opinions by describing the operational situations for which an expert judgment is requested. For example, one could ask an expert to modify a predicted error probability if the operator is a novice or performing under stress, but one cannot ask this of an invisible author.

Combinatorial Methods. The first major task in HFE prediction is to develop a database and metric and apply these to the unit of performance analysis. The second major task is to combine the data for the individual units into a probability for a total mission or job. The question may arise why a conmbinatorial method is necessary. If one collects historical data on a particular job and if it is assumed that job events repeat themselves essentially unchanged, then a prediction made for a given mission or job should be valid for all subsequent performances of that mission or job, taking into account, of course, individual differences among personnel and variability because of chance factors. This implies that the mission or job always remains the same and no PSFs are changed. For example, assuming that a bomber mission in a war is always performed in the same way and that none of the conditions affecting that mission (e.g., pilot skill level, disposition of enemy fighters, no increased flak) are changed, one could predict the aircraft loss rate with considerable accuracy.

However, in real life, conditions—particularly PSFs—rarely remain the same. The air raid on Ploesti in World War II produced a much higher loss rate than had been anticipated because the Germans became aware of the raid in advance and because serious errors in pilot performance occurred. When tasks and PSFs change, predictive data based on past history become much less predictive than they once were. Moreover, if one wishes to predict new missions or jobs, the previous mission/job probabilities will not apply unless there are commonalities between the new and old mission/job. Those commonalities are not at the total mission/job level, but at the level of the task or subtask. In other words, say that the old mission/job was composed of Tasks A, B, C, D, E, F, G, . . . ; the new or modified mission/job is composed of Tasks A, B, D, G, . . . ; and so on. Moreover, the interactions or dependencies among tasks are such that the new task combination in the new mission/job will interact in different ways than did the previous tasks and will therefore result in a different total predictive value.

If the commonalities between missions/jobs are at the task or subtask level, one can gather historical data for individual tasks/subtasks (i.e., nominal HEPs) and then, to predict for the new mission/job, combine them as the new mission/job scenario requires. This is what most HR combinatorial methods do.

The combination of predictive data for each task/subtask (these are, as has been indicated, commonly called called *human error probabilities* [HEP]) cannot be as easy as combination is performed in reliability engineering, which provided the first combinatorial model in Payne and Altman (1962). Swain (1968) showed that the simple multiplicative method (HEPs for Tasks $A \times B \times C \times D$, etc.) as used in reliability engineering could not be generalized to human performance. If one uses such a combinatorial model with human tasks, the probabilities one derives for a total procedure or job become ridiculously low, as the author found when he attempted to use the multiplicative model in actual practice. This is particularly the case as the number of tasks increases. Observation of the job or procedure when actually performed reveals that the true total error probability is much less than that derived with the simple multiplicative model. Some other method of combining task/subtask probabilities must be developed.

Another reason for developing probabilities at the task/subtask level is that overall mission/job success probabilities are not very diagnostic. If one wishes to find sources of potential operator difficulty, these will be found only at the individual task/subtask level. Thus, overall mission/job success rates are not very revealing. One needs these overall probabilities, of course, but they are not sufficient.

Because of the interactions and dependencies among tasks/subtasks, the development of a combinatorial method in HR has proved difficult. The probability estimates of the various techniques that result from the combination of task/subtask probabilities often have unacceptably large errors. However, comparison of the various predictive methods, as reviewed by Kirwan (in press), offer some hope.

The author does not intend to go into the subtleties of the various combinatorial techniques except to suggest that, if quantitative prediction is applied to HFE in general, HFE will have to face this problem also. Kirwan (in press) categorized the various methods into four classes, depending on data sources and models of operation:

1. *Unstructured expert opinion techniques,* including Absolute Probability Judgment or Direct Estimation (Seaver & Stillwell, 1983) and Paired Comparisons (Comer, Kozinsky, Eckel, & Miller, 1983);
2. *Data-driven techniques,* including the Human Error Assessment and Reduction Technique (see Kirwan, Kennedy, & Taylor-Adams, 1996) and Technique for Human Error Rate Prediction (THERP; Swain & Guttmann, 1983), which is probably the best known and longest utilized technique;
3. *Structured expert opinion techniques,* including Success Likelihood Index Method using Multi Attribute Utility Decomposition (SLIM–MAUD; Embrey, Humphreys, Rosa, Kirwan, & Rea, 1984) and the Socio-Technical Approach to Assessing Human Reliability (see Kirwan et al., 1996);
4. *Accident sequence data-driven techniques,* including Human Cognitive Reliability (HCR; Hannamann, Spurgin & Lukic, 1984) and Swain's (1987) Accident Sequence Evaluation Program.

There are other techniques as well, but these are sufficient to indicate the variety. All of them generate human error probabilities. They all have advantages and disadvantages and have been compared by Meister (1984, 1993b) and Swain (1989).

THERP uses a database of nominal human error probabilities, such as failure to respond to a single annunciator alarm. PSFs such as *quality of the interface design* (e.g., whether alarms are well human-engineered) are then considered with regard to this error rate. If a PSF is involved in the scenario being quantified, the nominal HEP is modified by the assessor by using an error factor, for example, of 10. Thus, if an initial HEP is 0.001, an error factor of 10 would increase the actual estimated HEP to 0.01.

Expert judgment techniques use personnel with relevant operational experience to estimate HEPs. It is assumed that these experts will have made errors themselves and seen others make errors, so that their information can be used to generate reasonably valid HEPs. Expert opinion methods may either ask their expert directly for an estimate or may use indirect methods (e.g., paired comparisons) to avoid biases associated with memory. Although the end result is a quantitative probabilistic prediction, the process used to achieve the prediction includes subjective as well as objective data.

Apostolakis, Mancini, van Otterloo, and Farmer (1990) and Reason (1991) criticized the various combinatorial methods; if efforts are made to incorporate quantitative prediction into HFE, it is likely that the same kinds of problems will arise. The experience of working with these HR techniques offers long-term promise that combinatorial problems can be overcome.

Measurement. Because over the years alternative taxonomies, databases, and predictive methods have been proposed, there have been many attempts to analyze and compare database taxonomies, test the adequacy of combinatorial methods (e.g., Meister, 1971 and Swain, 1989), and compare quantitative predictive methods (Kirwan et al., 1996).

There were and are also efforts to collect operational data, such as accident statistics and malfunction reports (e.g., Leuba, 1967; Lukas, Lettieri, & Hall, 1982) and to use these to derive models of malfunction performance and methods of prediction. There are two possible ways to approach empirical research in HR: (a) develop a methodology first and then try it out by applying it to empirically gathered data, and (b) work backward from empirical data as found (e.g., in nuclear power plants) and attempt to derive a model of human performance from the data.

Measurement in HR has the purpose of determining which of the several prediction methods is most effective—which is better than the other prediction methods tested. In the course of the testing, one may discover deficiencies that can be remedied.

The analysis and testing process has been the product of a small number of HFE people working in conjunction with engineers (e.g., Regulinski & Askren, 1969) operating under the stimulus of governmental agencies that felt they needed a prediction database and methodology. For example, starting in the 1950s, the Air Force was motivated to support studies like that of Leuba (1967), which dealt with failure reports and the information they contained. Still later, after the Three Mile Island and Chernobyl disasters, agencies supervising the nuclear power industry, like the NRC in the

United States, developed extensive behavioral programs that involved, among other aspects, attempts to develop a valid and reliable prediction methodology. The Electric Power Research Institute (EPRI) was established in Palo Alto, California, and supported much research. The NRC funded Sandia National Laboratory in Albuquerque, New Mexico, and Brookhaven National Laboratory in Upton, New York, to perform predictive studies. NRC made Human Reliability Analysis a required part of the Probabilistic Risk Assessment analysis performed prior to the development of any new nuclear plant.

Comparison studies involve a procedure in which a number of HR specialists (assessors) are given relevant information and asked to develop HEPs using various prediction methods. The author hesitates to call these validations except where external objective error rates serve as the validation reference. In the past, such comparison studies have not been very successful and a generally gloomy view is taken of the field by Apostolakis et al. (1990) and Reason (1991). However, Kirwan (in press), from which the following description is taken, reported a successful validation.

Recently, a large-scale validation took place in England to test three quantification techniques: THERP, HEART, and JHEDI. Thirty British practitioners, 10 for each technique, took part in this exercise, with each specialist independently employing only one of three techniques (because few HR specialists are skilled in more than one method). Each assessor quantified HEPs for 30 tasks. For each task, the following information was provided: (a) general description of the scenario (e.g., tasks required to be performed, conditions under which performed, etc.); (b) inclusion of relevant PSF information in the description; (c) availability of a simple linear task analysis of the scenario; (d) diagrams where necessary and relevant; and (e) a statement of the exact human error requiring quantification. Note that this validation process focused on the development of HEPs for individual tasks and not on the ultimate combinatorial process.

Each specialist had 2 days to carry out the assessments. Experimental controls were exercised so that they worked essentially under proctorship conditions. For each of the 30 tasks, the actual HEP was known to the experimenter but not to any of the specialists. Tasks chosen for this exercise were taken from the nuclear power and processing industries because all of the assessors were presently working in these two areas. A large proportion of the data was from real recorded incidents. (For a more complete analysis of the results, see Kirwan et al., 1996.)

The data analysis for the 895 HEP estimates generated by the specialist subjects (there were five missing values) showed a significant correlation between their estimates and the corresponding true values (Kendall's Coefficient of Concordance: $Z = 11.807$, $p < .01$). This supports the validity of the HRA quantification approach as a whole, especially because no specialist results or outliers were excluded from the data analysis. The analysis for individual techniques showed a significant correlation in each case, again using Kendall's Coefficient. All Z values were significant at $p < .01$.

Among the individual correlations for each of the methods and subjects there were 23 significant correlations, some significant at the .01 level, out of the possible 30 correlations. This is a positive result, again supporting the validity of the quantification approach. There was an overall average of 72% precision (estimates within a factor of 10) for all assessors, regardless of whether it was significantly correlated. This is a *reasonably good* result (in Kirwan's words). No single specialist dropped below 60% precision in the study. Kirwan (in press) stated that the overall results were positive and therefore lend support for the HEP quantification part of the predictive quantification process.

Other comparisons of the strong and weak features of the various methods were made using a special checklist containing criteria developed by Swain and his colleagues (Swain, 1989). In general, this critical comparison of 14 techniques, although entirely subjective, indicated that all the techniques had significant deficiencies, but that THERP, which was developed by Swain principally and is most frequently used in PRAs, was best. Therefore, the overall picture is somewhat ambiguous. The study by Kirwan et al. (1996) is most optimistic; other HR specialists have been somewhat less optimistic. The data do suggest, however, that there is something in HR that warrants further work.

Conclusions

This section began with the bald statement that HR should function as a model for HFE quantitative prediction. That statement did not intend to suggest that quantitative prediction should follow HR practices slavishly (particularly because there are serious weaknesses in the HR methodology), but rather to suggest that it is possible to mount a comprehensive effort to insert a bit more quantification into HFE than it has presently. One major deficiency in HR is the lack of appropriate data despite 50 years of research. A second deficiency is the lack of an effective method of combining component probabilities for subtasks or tasks into a total probability for a job operation or mission.

It is not necessary for HFE to mimic HR processes, although they offer much food for thought. It is possible to approach the problem of quantification in steps rather than as a totality, as HR did. On that basis, the first step in HFE quantification would be to attempt to integrate available data and secure additional data for development of a quantitative database. The goal would be to organize available data quantitatively to the extent that the characteristics of the data permit. An organized effort along this line would have a secondary effect, to suggest areas for which further research is needed, without leaving it to the idiosyncracies of funding agencies and individual researchers to determine the nature of HFE data.

Available data are presently unorganized and somewhat chaotic. If only because of aesthetics, it would seem desirable to organize these data so that HFE professionals can see what they actually have. It is possible that a great deal of the "reinventing-of-the-wheel" problem arises because personnel do not have a clear notion of what the past research has to tell them. Qualitative compendia like Boff and Lincoln (1988) are useful, but not sufficient.

The author believes that the effort to organize HFE data quantitatively would have, even if it were only partially successful, a highly stimulating effect on the discipline as a whole. True, there are practical difficulties of developing a common effort and securing funding for it, but the potential value of the effort merits the attempt. As a footnote, as of summer 1997, an effort is underway to establish a NATO research group that would attempt to address many of the issues addressed in this section (personal communication, J. Davies to author). However, like other NATO research groups, its life cycle is 2 or, at most, 3 years, which is too short a time to do anything but develop plans for a more ambitious effort.

8

HFE Practice

This chapter discusses practice. Previous chapters have distinguished between research and practice, and much of the book is about research. It is time to say something about practice because, as is seen, demographic data suggest that many professionals have been at one time or another both researchers and practitioners.

Practice takes place in a number of different venues. Practice is commonly associated with the performance of HFE specialists in an engineering facility devoted to development of new systems—military and civilian. Another venue is the contractor consultant facility; still another is the government research and development (R&D) laboratory. A fourth venue is the university. This chapter also questions whether there is a core set of knowledges that every specialist should possess; this is because the discipline is subdivided into so many specialties that one may lose track of its common knowledge base.

This chapter attempts to apply a dose of reality to the ordinarily somewhat artificial description of HFE. Any abstraction is, in terms of its details, bound to be false. In most HFE textbooks, which are oriented around research results, it is easy to forget that there is much more to HFE than research. The true HFE picture does not accord completely with the facade presented in textbooks and research papers.

To illustrate what we are talking about, we use a number of case histories derived from applications to the Board of Certification in Professional Ergonomics (BCPE). As a founding member of the board, for several years the author had the opportunity of reviewing 101 applications. These case histories illustrate the more general material. There is no pretense that they represent everyone's experience. The written applications on which these case histories are based are also somewhat idealized by their authors because the applicants were trying to persuade the reviewer of his or her worth to achieve the desired certification. Nevertheless, they do present a somewhat more concrete picture of what HFE professionals do in their everyday life.

A statistical analysis was also made of the 101 subjects in the applicant sample, covering such items as final degrees, areas of employment, and the kind of work performed. This analysis provides a context within which the HFE specialist can be viewed.

The various sections of the chapter describe the demography of the specialist: his or her activities in system development, work in the government laboratory, and HFE in contract research. These sections are followed by an analysis of what the specialist should be trained to do and should know when on the job. Four case histories, from which all identifications have been removed, illustrate what the chapter has endeavored to convey.

This chapter is about what HFE professionals do. Unfortunately, it cannot tell us what they think about what they do, whether they are satisfied with their work, or whether they are satisfied or dissatisfied, and to what extent, with the discipline. The concept structures referred to in chapter 2 contain affective elements because no discipline can be completely objective (and one would not wish it to be). However, the personal dimension of HFE presented must be somewhat skewed.

THE DEMOGRAPHIC CONTEXT OF HFE

The purpose of this analysis was to answer this question: Who is the HFE professional and what does he or she do? The last systematic attempt at such an analysis was the report by Kraft (1970) and that, of course, is now out of date. From time to time, HFES has surveyed its membership about such aspects as salary, but the surveys are not very searching. In that respect, the self-initiated applications to BCPE provide somewhat greater detail.

Results of the Analysis

Most of the sample have doctorates and two thirds have master's degrees; only 5% have no graduate education. Most of these degrees were granted in the 1970s and 1980s, although a few of the older applicants had degrees that go back to the late 1950s. Almost half the sample had degrees in psychology; a quarter, those who were younger, had them in human factors. A further quarter had degrees in something other than behavioral science.

As far as the basic sciences in which applicants received training, there is the usual emphasis on mathematics, physics, and biology. A sign of the changing times is that 35% received at least some computer training and took at least one engineering course. As far as ergonomics-related courses are concerned, one sees the old favorites—experimental design (35%), experimental psychology (50%), general psychology (48%), human factors (55%), and statistics (57%). A few of the courses in human factors at a few schools were quite impressive in terms of their range. However, with that exception, the course listings resemble the sort of psychological training the author had received, which is not a great advance over 1950.

What is surprising is that only 55% of the applicants had any formal training in HFE. Did they get their training, as the early cohort did, on the job? It is possible that if one receives a graduate degree in psychology, there is less likelihood of being trained in HFE than if one receives an advanced degree in industrial engineering or operations research. The importance of statistics to the graduate degree is overwhelming. Statistics and computer training seem to be an integral part of modern HFE, whatever other specific HFE courses are given.

The interesting thing about publications is that 64% of the sample reported at least one refereed journal publication and many others had more than one: 83% reported at least one presented paper, 11% had been involved in writing or editing books, 48% had written at least one government or contract research report, and 19% had at least one chapter publication.

The importance of verbal/written expression to HFE professionals should not be surprising; it is so also for psychologists. However, the data suggest that a great deal of HFE literature is being written by people whom we talk of as practitioners, but who resemble university types in their publication activities. It may be that practitioners, and those who do not publish at all, are a much smaller percentage of all professionals than the author had originally thought. Or the sample, those who wished to be certified, is biased because many of the sample work or worked outside of industry. It is probable that one cannot make a hard and fast distinction between researchers and practitioners—that there is a third category of the researcher/practitioner. Much of the research these people did was related to practice or the data they collected were supposed to be applied in industry. Of course, it is possible that those who did not publish also declined to apply for certification. This would certainly bias the sample, but would not negate the main point—that there is a third group whom one can classify as researcher/practitioners because they do both.

To investigate further, the themes of the journal publications they reported were analyzed. These themes reveal nothing strikingly different from those of journal publications in general. However, the government and contract reports they wrote were much more narrowly focused (e.g., evaluation of a particular system or tactical procedure). Studies like those rarely make it to the general behavioral literature, but there was a small frequency of studies published in scholarly journals. There is a broad range of topics, but little concentration on any one topic.

The effect of report literature stemming from industry-based researchers has not led to any great advances in HFE because it has been content-oriented. Research published from contractor facilities has had a greater impact because much of it has dealt with methodology. For example, task analysis methodology was developed by a member of the American Institute for Research (Miller, 1953). Performance modeling work (e.g., Siegel & Wolf, 1969; Wherry 1976) came out of contractor shops, as did many of the mathematical modeling techniques, such as SAINT (Wortman et al., 1978). A good deal of substantive research directed at specific systems (e.g., sonar, radar, aerospace) has also been performed by specialists working in government laboratories.

The median number of years of work experience was 15, with a range from 4 to 40. Most professionals in the sample worked in government, a government-supported contractor facility, or else aerospace; 23% did part-time teaching or moonlighted as consultants. The range of activities performed by the sample is large. These categories were derived from the applicant's own description; they were not inferred. The most common activities are design, function analysis, design evaluation, management, and test and evaluation. This is understandable because applicants had been told in advance that they would be evaluated on their activities in analysis, design, and evaluation. The median number of activities reported as performed by any single individual was 10, with a range from 4 to 18. This suggests that the HFE professional is somewhat of a *jack of all trades.*

Quite apart from the listing of activities, this author was impressed by the sophistication of the analyses and tests that some of the sample described (examples of these are to be found in the work histories in the final section of this chapter). Of course, they wrote to impress evaluators, but the detail they provided suggested that a sizable proportion of the applicants were very, very good in their jobs.

Finally we come to the product line the HFE professional supported. Military systems are most prominent, followed at some distance by consumer products. When one considers that the preponderance of work by ergonomists is in government and government-supported contract research facilities, together with the importance of aerospace, much of which is funded by the military, it is apparent that government support of HFE has been and still is indispensable to the discipline. Considering the heavy emphasis on computer software training for HFE students in the university, it is a little surprising that only 17% of the sample supported that product line. However, those just getting out of school may be doing more of that sort of thing.

The median number of organizations worked at during the applicant's working life was two, with a range from one to eight. Obviously HFE professionals do not move around frenetically. Of course, the longer one's working life, the more opportunity there is to move from one organization to another. However, the correlation can never be 1 because a number of the applicants had spent 20 to 30 years in the same organization, moving up the ladder as time progressed. When people did move, they tended to stay in the same line of work, although there were some notable exceptions.

HFE EMPLOYMENT

Summary and Introduction

Formal descriptions of HFE activities at work have been provided (Meister, 1987), but they are pale descriptions compared with the actual work histories in this section. HFE work is performed in industry (an engineering environment), government facility (a laboratory or military base), and contractor shop. HFE also takes place in the university, but this topic is not discussed in any detail because it would involve questions of experimental research (already discussed) or university politics, with which the author (not being an academic) is not familiar.

This section is intended to set the stage for the actual work histories of real individuals in the following pages. The conclusions in this summary are based on impressions developed by the author in all three of the work environments over many years, but they are only one specialist's impressions of HFE.[1] As seen in chapter 5, varied impressions of HFE are possible based on individual experiences. Shapiro et al. (1995) provided information about what various job markets require and what professionals may expect in these.

Getting a Job

Before examining the various kinds of HFE work contexts, it is necessary to discuss getting a job, which is the first task the newly graduated HFE professional has. In fact, the effort to secure a position begins long before the diploma is handed over at graduation exercises. Again, this chapter does not discuss the business of landing a university position. That is almost completely a matter of having a network of at least one or two senior professors who will exert themselves in favor of the applicant, activate their network, call other faculty members, write letters, arrange interviews, supply references, and trade on long-standing relationships.

---

[1]The author is indebted to a colleague, Thomas P. Enderwick, for a critical review and comments on the ideas expressed in this section.

For the HFE professional who is not bound for another university (and the number of opportunities in university are quite limited), there are other sources of potential jobs. The network of senior professors and friends can serve just as well for the graduate going into nonuniversity work. In fact, this is the preferred method, assuming that one has knowledgeable friends who can steer one to an appropriate billet. In addition, advertisements from universities, industry, and contractors asking for new personnel can be found in the monthly bulletins of the HFES. Governmental agencies offering new positions are required to advertise these on posters adorning the walls of civil service organizations. These provide information on the type of position, the necessary training, and how to apply.

The HFES offers a job placement service in which potential employers list the jobs they have available; the society also offers a large room at the annual meetings at which applicants can meet potential employers. Advertisements in newspapers are useless; only engineering positions are advertised. If none of the preceding are effective, one can, using the Society's Directory as a guide, write to the organizations listed in its index, asking if any job opportunities exist. This last method is one of desperation. However, if the other methods do not work, it is worth the effort to contact the agency directly. Moroney and Adams (1996) and Moroney, Sottile, and Blinn (1996) provided an analysis of placement opportunities for the novice specialist based on statistics gathered from the HFES Placement Service.

HFE in System Development

There are excellent books about the technical aspects of HFE in system development, some of which the author has written (e.g., Meister, 1987). This section deals with the nontechnical aspects of that work. Additional descriptive material can be found in Gage, Adams, Logan, and Wilson (1996).

The HFE specialist in system development is ordinarily thought of not as a researcher, but as one who works in engineering development. In describing HFE in system development, this is the individual whose activities are described. As was seen in an earlier section, there are researchers in government agencies and contractor facilities who are concerned with problems stemming from system design and operation. These researchers are considered under appropriate sections dealing with government and contractor work.

(Enderwick noted that there are elaborate research facilities—terrain models, flight simulators, etc.—at major aerospace companies like Lockheed-Martin or Boeing, but these and the research performed with them support system development. The reports that are written about this research are usually proprietary and are rarely found in the general HFE literature. A small number of system development personnel do perform contract research on general topics, which may be published in the journals.)

System development involves three major functions: analysis of a design problem; the design itself, which is the solution of the problem; and testing, which is the evaluation of the design to determine if the solution is adequate. System development is a problem-solving situation because the problem is to create a design for something that does not exist at the moment, although the new system may be merely an update of a predecessor system.

The three functions mentioned serve as a conceptual means of organizing the professional's varied work efforts into meaningful units. Another organizing factor is that the HFE specialist's responsibilities involve only the human–machine interface. This interface can be defined broadly or narrowly, but confines those responsibilities to the components and activities involved in the interface.

Another factor that limits the specialist's responsibilities is that he or she is not, in most cases, the primary designer of the system; that is the engineer, although the reader sees in the following work histories that specialists sometimes design in relation to the interfaces that are their responsibility.

In most cases, the specialist is a consultant to the primary designer. Thus, he or she (a) advises on design aspects that affect the operator's capability to operate and/or maintain the equipment, (b) evaluates the engineer's design concepts in terms of how they will influence the operator's performance, (c) translates research conclusions to the designer as these bear on the equipment being designed, and (d) tests the adequacy of the design solution by exposing personnel to the tasks involved in operating and maintaining the system in various phases of its design and completion.

The human–machine interface (henceforth referred to as the *interface*) may be defined in different ways. Generally, from a hardware standpoint, it is the workstation, the cockpit, controls and displays, the working environment (e.g., lighting, noise, temperature), and anything and everything associated with this interface. The computer interface involves the physical terminal as well as a behavioral interface (software), which includes what and how stimuli are presented by the software to the user.

If one were to make a formal list of the common HFE tasks in system development (see also the earlier section of this chapter), it would include: participating in initial (preliminary design) and subsequent planning sessions to develop the design concept; analysis, review, and evaluation of operating and maintenance procedures; recommendations to the designer concerning the design configuration as a whole as well as components (like controls and displays) relative to the interface; training considerations (when no training specialist is available); the development and conduct of component, module, mockup, and other tests where personnel performance is involved; writing of test plans; conduct of the HFE part of the operational system test; providing HFE familiarization training to company personnel; writing reports; making verbal presentations to management and customers; writing the HFE sections of proposals (when so required by the company management); collecting data during test procedures by means of objective measures, observation, and interviews with company personnel; investigating and recommending relative to accident and health-related complaints from company personnel; and so on. The *and so on* includes a great many miscellaneous activities that are too detailed to consider here.

(Enderwick reminds the author that many of these activities are mediated by the computer, design drawings are developed on the computer [computer-assisted design (CAD)], and mockup tests may also be performed on computer as well as in the physical environment. During the time the author worked in system development [roughly 1956 to 1964 and occasionally thereafter], computers were not used in daily engineering work. However, they are much more in evidence now, although there does not appear to be much evidence as to how popular CAD is.)

In the early days, as the author remembers them, a great deal of the work performed by the HFE professional was self-generated because there was no formal HFE program plan generated before a contract was awarded. These days, because of plans established to systematize behavioral efforts in system development, professionals have a more or less specific road map like MANPRINT (Booher, 1990) to guide their activities. This plan mandates certain HFE outputs during system development. The activities needed to produce these outputs require many diverse actions.

The preceding refers to system development performed under a contract given by the government, usually the military. In civilian industry (i.e., when the company for which one works is designing an equipment for general sale to and usage by the general population), it is to be presumed that the rules change somewhat. If a professional is assigned to the development of a new civilian product, management may require the specialist to describe in writing (in detail) what he or she is going to do on the project (a program plan). There are probably certain progress reports that are required, but the mandates come from the engineering management of the company, not the government. (Legal actions against shoddy workmanship also play a role in these mandates.)

The receptivity of engineering to HFE may be more, now that programs like MANPRINT exist, because they serve as a mandate for HFE activities. However, in the early days (see chap. 5), the HFE group (if there was a group) often had to scare up enough work on its own to give the impression of busyness. For military system development, there are one or more (depending on the size of the effort) monitors from the military customer's shop who make periodic visits to the engineering facility to check that required outputs are being produced. The development of formal HFE programs like MANPRINT means that periodic reports must be written to the overseeing agency; this in itself guarantees a certain amount of HFE effort.

As long as the overall project is funded, there is the security of a continuing HFE effort because engineering management is responsive to mandated requirements and the availability of funds. However, if for some reason the funds disappear, so will the HFE effort. As has been pointed out elsewhere, government funding is capricious, particularly in these days of shrinking budgets.

(Enderwick states that funding is not always the problem: "I wrote at least twelve human engineering portions for proposals in the first company I worked for. Some of these proposals resulted in contracts for the company. However, no HFE specialist was ever asked to work on the projects. Junior electronic engineers, with no work assignment, were dubbed 'human engineer specialists' by their supervisors, and given the job.")

There are periods of monetary feast and famine in government-supported work. When the military began to downsize after the disintegration of the Soviet Union, the HFE specialist, like everyone else in system development work, was negatively affected. When a new contract is won by the company, especially if the contract is a large one, HFE is positively affected, but not equally in comparison with engineering because HFE is an auxiliary service in system development and its budget is likely to be a little tight.

The work that HFE performs in system development is, as has been suggested previously, almost entirely applied. There is practically no research performed in system development unless a behavioral problem arises that is so severe that it cannot be solved except by research. Even then it is likely that the work will be contracted out to a research contractor. (One must differentiate HFE research from testing and evaluation of system components. The first is uncommon and the second is an accepted part of system development.)

The inability to perform research means that the HFE specialist is entirely dependent on those who actually do research. Because of this, the specialist working in system development must become a translator of that research into the specifics of the individual design problem. However, comparatively little ongoing behavioral research is translated into design specifics because the relevance of the research to individual design issues is not clear.

What specialists working in system development do that resembles research is testing and evaluation of equipment. This is not research if one defines *research* as being the effort to determine the variables that affect human performance. However, it is measurement; in this respect, the concentration of effort in university on research processes stands the specialist in good stead because much of the research training generalizes to test and evaluation.

To the novice professional, it may appear that the system development process is one of confusion because so many individual projects (or rather parts of projects) are being performed concurrently (this is particularly true if the system being developed is a major one; e.g., a new aircraft, a new warship type). It may appear to the novice that all these projects require his or her assistance. However, unless there is a specific interface relationship in these projects, the specialist can contribute nothing. (Enderwick notes that HFE is also preventive in nature. If it is done, no one will notice. Only if it is not done and things go wrong is the lack of it noticed.)

In addition to the general problem of being a behavioral scientist in a physicalistic environment, HFE in system development faces a number of specific problems:

1. The inadequacy of research-derived design guidelines. It is a common complaint that designers do not make use of HFE data (Meister & Farr, 1967). It is fair to say that HFE specialists make relatively little use of research data, but not for want of trying. This problem has plagued HFE in system development from the beginning of the discipline. The causes are partially within the engineer and specialist, but much more in the inapplicability of the research.
2. Whatever behavioral design guidelines exist in such standards as MIL-STD 1472E (1994) are mostly qualitative and do not describe the increment of improved human performance resulting from use of a guideline.
3. The most common problem in system development is managerial ignorance of HFE (some specialists would say stupidity), which leads to a reluctance to support the HFE effort.
4. The lack of funding has already been mentioned. Time pressures (insufficient time to perform HFE tasks properly) are severe as well, but this affects everyone.
5. There is an inclination to sweep behavioral design deficiencies under the rug because each discrepancy is considered individually, and thus appears to have little importance.
6. The lack of HFE design approval for interface designs has been a continuing problem. This means that, although specialists are allowed to review, analyze, and evaluate such designs, they usually do not have the power to reject a design no matter how deficient.
7. There is also the inability to measure the HFE contribution to design, which makes it easier for management to dismiss the HFE role as unimportant. This inability arises because the final design is a group product. (Enderwick's note: The manager's view—if you are not able to measure it, why fund it?)
8. Sometimes the HFE group is located at so low a level in the engineering organization that its attempts to ensure proper consideration of behavioral factors in design are blunted.

These problems have always existed because the physicalistic orientation of engineering means that HFE, like all minority disciplines, must work harder merely to stay alive. However, there is an increasing awareness of the effects of design on human performance, as represented, for example, by publicity about user-friendly PCs and other consumer equipments. In pointing out these nontechnical problems facing the specialist, the intent is not to suggest that HFE cannot do its job properly; rather the specialist should be aware of what faces him or her. Lest all of the preceding suggest a somewhat gloomy view of behavioral activities in system development, it should be pointed out that for many (most?) system development specialists there is considerable satisfaction in having contributed to the completed physical product.

HFE in Contractor Work

A contractor organization makes its living primarily by performing research under contract to a governmental or civilian agency or company and by occasionally supplying professionals to work on a temporary basis in an engineering facility to support a specific system development.

The major output of the contractor company (occasionally individuals incorporate themselves) is research. Most of that research seeks to answer a specific question asked by the customer. For example, during the Vietnam War, the U.S. military asked for the development of a means of identifying the presence of the enemy by analysis of the odors or waste products left by enemy personnel. Although the research is rarely of an academic nature, much of it is methodological and therefore useful. For example, how long can long-distance truck drivers drive before exhaustion degrades their perceptual and motor capabilities? What graphic characteristics of maps affect map reading performance? What physical components of motion produce sea sickness? How does sea sickness affect display monitoring performance? How could one develop a more effective way to present maintenance instructions in written procedures that will reduce maintenance error? The author once won a contract to write a handbook of human factors for the U.S. Post Office (although why it wanted one was never clearly expressed). In another contract, he evaluated the adequacy of a handbook published by the Air Force.

Obviously the research topics addressed by contractor research vary widely in terms of theme and, although not fundamental in an academic sense (in contrast to questions as general as, how do people learn), are by no means trivial. The questions asked by the (usually) governmental customer are more or less specific, but they are often important questions to the customer (e.g., the effects of fatigue and driving performance are important for the Department of Transportation, desirable map reading characteristics for the U.S. Air Force, sea sickness factors for the U.S. Navy, whether living aboard a cutter instead of onshore is detrimental to work performance for the U.S. Coast Guard). Much excellent work has been and is being performed by contract researchers, some of it requiring data collection in the operational environment (military bases on land, ships at sea, etc.). In a few cases, the military supports the activities of a reasonably large research unit (10–20 people) at one of its bases for a period of years on a continuing contract.

The contractor facility may vary in size from a single individual to organizations of dozens of personnel (e.g., the American Institutes of Research, Dunlap Associates, Anacapa), with branches in several states and even overseas. Outstanding researchers sometimes incorporate themselves (e.g., A.I. Siegel, R. Wherry) and, with one or two assistants, work a limited number and type of projects, in which they specialize (e.g., performance modeling). The larger organizations work on a wide variety of projects—anything that can bring in a *buck*. A recent tendency, as the first generation of specialists retires from industry or government, is for them to set up one-man shops as research contractors.

Research for money is a highly competitive activity. The (usually) governmental customer indicates what it wants researched (in the form of a question or a general topic) in the form of something called the *Request for Proposal* (RFP), which is advertised in a government-published newspaper called the *Commerce Business Daily*. The advertisement is a brief notice (no more than one or two paragraphs) informing the potential contractor of what is wanted and supplying an address from which the contractor can request the RFP (which is often several pages long and occasionally, for a major project, quite lengthy).

Governmental desires for research or other assistance are well publicized because it is a legal mandate. However, the need of a civilian company for contractor assistance may not be known by the contractor until it is approached by the company. If the advertisement sounds inviting (within the presumed expertise of the contractor), the contractor will send for a copy of the RFP. The RFP describes the type of research to be performed, provides specific details, and sets the bounds of the work. After examining the RFP, the contractor decides whether to make a formal proposal.

Much of the work of a contractor is in the writing of proposals to secure new business. This is not the only way to secure new business, but it is a major way. Writing the proposal can be seen as an exercise in problem solving because whether one wins a contract depends, at least in part, on originality in attacking the problem.

Responding to an RFP is the solution of a problem akin to advertising a new commercial product: what unusual approaches to the problem would tempt the proposal reviewer, how much should one promise the reviewer (some exaggeration of proposed outputs is common in all proposal responses, but they must not be obviously a "come-on" or impossible to supply). Who can the contractor promise to perform the proposed research? How can one picture this individual's curriculum vitae in the best light possible?

Other factors involved are the price quoted by the contractor, including overhead (the low bidder usually, but not invariably, wins), the contractor's past experience (track record) in performing similar research projects, the expertise of the people (described by name and resume) who will be assigned to the project if it is won, and its management and resources (the government does not want to wind up with a contractor that, because of its small size, may not have financial resources to complete the job if there is, God forbid, an overrun). Minority and female-owned businesses have some, but not an overwhelming, advantage, which can be canceled out by the other factors mentioned previously.

The competition for contracts creates an atmosphere of uncertainty within the contractor company. As present contracts are completed and word has not yet been received about pending proposals, anxiety begins to suffuse the company, unless the contractor has a backlog of projects sufficient to maintain the company for at least a year.

Because of the uncertainty inherent in contract research, the company cannot rely on responses to RFPs. The company usually sends out members of its senior staff to solicit business from potential funding agencies or least bring the company to the attention of agencies that may eventually issue an RFP. As the nation's capital, Washington is the usual venue of such visitors, but agencies in other parts of the country may often receive visitors.

The contractor professional is often seen visiting one governmental agency or the other (sometimes civilian companies as well), stopping in to request audience with the head of the department, and introducing him or herself (if this is a first visit) to tell the functionary about the capabilities of the contractor organization. It is rare that the visiting salesman pitches a specific item of research unless there is some prior information that the governmental official might be responsive to the contractor idea. The focus of the visit is the contractor's attempt to discover the following:

1. What is the agency interested in in terms of one or more general research areas?
2. What kinds of research is the agency presently supporting?
3. Does it have or does it anticipate having money for research (how much would be nice to discover, but is carefully concealed)?
4. What kind of unsolicited proposal would the agency be responsive to or is the notion of an unsolicited proposal undesirable? (Most agencies dislike such proposals because it means taking time to review and reject them; agencies prefer to fund research *they* think of.)
5. When would it be a good idea to submit something?
6. Is there anything cooking on the front burner that has not been written up as yet in the form of an RFP?

Visits to potential customers are largely detective visits. The governmental official tries to conceal as much as possible, but most feel that they have to be at least polite to the contractor and in the process may reveal some information. If the specialist is a well-known individual and has had a previous relationship (positive, one hopes) with the agency, the detectival process is much easier.

There are agencies that will entertain unsolicited proposals in specific areas of their interest; if they are aware of the proposer's expertise, they will issue a contract for the work. However, in a few instances, an agency will reject an unsolicited proposal and rewrite it in the form of a solicitation, thus depriving the proposer of the benefits of his or her work.

During contract performance, the customer must be satisfied even if the problem to be solved in the research is difficult to attack; the customer may have undue expectations and may in consequence be dissatisfied. Because government contracts are let on the basis of the lowest bid, and are hence very competitive, the contractor in his proposal may promise too much for too little funding (but this must not be obvious; if it is too apparent, the customer may conclude that the contractor does not really understand the problem). The best arrangement is if one is well known to a government agency and the contract is a continuing project, funded over several years, although this is no guarantee of anything.

Enderwick notes: "During the years that I worked for HRB-Singer and Rockwell, once an R&D contract was funded, it remained funded for that year. A year was the nominal length of our contracts. However, the technical preparation of the proposals was extensive and resulted in a formal document. In contrast, my experience with project funding in government service [within the Navy, that is] has been one where funding for a 1 year project can be had with a paragraph or two. However, the funding can be withdrawn by a phone call." "I don't know about the Army or Air Force but the Navy, in addition to block funding, has the 'Navy Industrial Funding' category which recognizes that Naval laboratory researchers do technical marketing much like our counterparts in industry, i.e., Navy research funding is a mixed bag."

The contractor works in tandem with a contract monitor assigned from the funding agency. The monitor is supposed to ensure that the contractor does what it is contracted to do and report progress periodically to the funding agency. At times and over a relatively long period, a close relationship may develop between a contractor and its monitors—to the point that the researcher and monitor collaborate as co-authors of the final research report. In contract research, the monitor is almost always another HFE professional simply because a layman will not understand what the contractor is supposed to do. This is much less likely to be the case in a hardware development project, in which the HFE group is only one of several engineering support efforts.

Depending on the closeness in time of the government's fiscal year completion, when government monies must be expended or returned, the possibility of interesting an agency in a particular line of research increases. Of course, this depends on the general financial environment in which the visit takes place. The first research contract the author won was the result of the fortuitous concatenation of money and interest; on leaving the agency office, he had the promise of a contract for a specified sum. However, this is the exception, not the rule.

The kinds of research the contractor will engage in are: (a) analytical (e.g., development of models, new task analysis methods, how to utilize HFE in the military, new methods of selection or testing of personnel); (b) empirical work, like participating in the testing of a new helicopter system. Some military agencies have field research test facilities at military bases and testing is contracted to civilians; and (c) engineering support—the contractor may be called in to work on a new system development or supply HFE personnel as temporary HFE engineers (the professional equivalent of the office *temp*). The research topic is usually highly circumscribed. The contract researcher does not have a great deal of discretion. That is because the customer cannot monitor the contract research adequately if the problem and its outputs are not specified in some detail.

HFE in the Government R&D Laboratory

HFE activities in a government laboratory, such as the ones in which the author was employed for over 20 years (Navy Electronics Laboratory, Army Research Institute, Naval Personnel R&D Center, Naval Ocean Systems Center), are similar to those in which he engaged when working in a contractor facility (the American Institute for Research):

1. analytical and empirical research (e.g., producing a handbook on human performance related to electronic displays (Meister & Sullivan, 1969) and studies of human factors variables in relation to equipment reliability (Meister, Finley, Thompson, & Hornyak, 1970);
2. monitoring the development of government systems and contractor research; no reports; one does not publish reports on monitoring activities;
3. evaluation of the performance of military systems and personnel (e.g., determination of the performance of Army helicopter pilots when navigating at nap-of-the-earth; Meister, Fineberg, & Farrell, 1976):
4. preparing proposals for new research efforts and justifying expenditure of monies spent in previous HFE efforts (no reports can be published on this).

Government R&D and contractor research have one thing in common: the necessity to *sell* the customer on the particular research to be performed. The R&D specialist is just as dependent as the contract researcher on funding. However, when funds are scarce, the government is capable of creative manipulation of what funding exists, whereas the contractor is much more constrained in selecting funded projects to which unfunded personnel can be assigned.

In the government laboratory, proposals have to be made to funding agencies and there is competition among potential recipients of such funds. Appearances to the contrary, government money is never *flush* for R&D; a Congress that can appropriate millions for dubious hardware projects is much more niggardly when it is a matter of R&D and particularly behavioral R&D, which is essentially *terra incognita* to laymen.

Governmental research is classified in several ways: (a) 6.1 is basic research directed more often at agencies that specialize in advanced projects (e.g., radio-astronomy studies to detect the presence of alien life in space), (b) 6.2 is applied research to solve general problems, (c) 6.3 is research specifically involving applications to system development, and (d) 6.4 is research involved with operational systems. So much money is allocated to 6.1, so much to 6.2, and so on. The research categories (particularly 6.2 and 6.3) are difficult to differentiate. Thus, if 6.2 is not a good category for a certain type of research question, one tries 6.3 and vice versa. In addition, the government laboratory and the departments within it must also justify whatever funding is provided them. Every so often justification is required because of the innate government paranoia for finding waste and abuse.

As the contractor often discovers (and this is true of R&D laboratories as well), government funding is highly erratic, expanding and contracting as higher levels decree. The only difference in this regard between contractors and government workers is that the latter are closer to the sources of perturbation and can more directly lobby them. A primary problem in government funding of behavioral R&D is that it is difficult to maintain the funding agencies' attention for more than a short time (relatively speaking, of course). For example, when the author joined the Army Research Institute in Arlington, Virginia in 1972, one of the significant areas of interest and funding in his department was study of the effects of organizational development and management on soldier performance. By the time the author left in 1975, almost all interest and funding for this work had lapsed—for no reason that was apparent to the researchers. Hence, justification of one's work, requiring some imaginative report writing, is often necessary.

One activity that is peculiar to government R&D is the monitoring of government-supported contract research and system development. As indicated previously, this may lead to a close relationship between the government specialist and contractor professionals.

It is the author's experience (which may be less than wholly representative) that research at a government facility can be a frustrating experience. The requirement that a certain percentage of government research funds (up to 40%) must support contractor research means that often the most interesting research projects are allocated to contractors. It is often necessary to tie government research to a specific system (6.3 research), which means that one is constrained by the nature of the system. Moreover, all research—government and civilian alike—often does not lead to concrete outputs; the effect of the research is unknown.

HFE in a government establishment makes it possible (and often necessary) to become involved in military operations such as field exercises. (This is also true of contractor research, but to a lesser extent.) Involvement with the military usually requires learning the details of operating complex systems. For example, when the author joined the Navy Electronics Laboratory in San Diego in 1951, within 1 week he was enrolled in the Navy's Anti-Submarine Warfare school as a full-time student for 10 weeks, learning the same job and attending the same classes as the novice sonar man (but without having to participate in military formations). At the Army Research Institute, he spent alternate weekends flying the rear seat of a Huey helicopter to supervise a project involving navigation at nap-of-the-earth (heights between 5 and 100 feet above the ground).

The choice of research topic or activity may be no more free in the R&D laboratory than in contractor research. Certain problems are inherent in the nature of a type of system (e.g., detection in sonar and radar) and often this determines the specific research one performs.

HFE specialists are found in a wide variety of government agencies. A glance at the geographical index of the HFES Directory indicates that small numbers of them are in, for example, the Department of Transportation, the General Accounting Office, the Nuclear Regulatory Commission, the Naval Sea Systems Command, and many others, too numerous to list.

HFE in the University

The author has been out of the university for almost 50 years. Except for two evening human engineering seminars he taught 25 years ago, he has no personal experience as a teacher of HFE. For the following discussion, therefore, he is indebted to Dr. Richard C. Pearson, professor in the industrial engineering department of North Carolina State University, which has an outstanding HFE program.

Much of the following is taken from Pearson (1980), which was updated by an unpublished presentation (Pearson, 1994). His investigation of the status of international training programs can be summarized as follows:

1. Most training is given in the United States, Western Europe, and Japan. In 1980, 42% of all programs were in the United States and 40% in Europe, and the situation had not changed materially by 1994.

2. Most programs are located in psychology, industrial engineering, and general engineering departments. As of 1994, worldwide there were 53 programs in psychology, 58 in industrial engineering, and 31 in other engineering areas; 29 programs were explicitly in HFE. Outside the United States, the growth of HFE programs has been largely in engineering rather than psychology. One can speculate on the meaning of this (if any) for further HFE.

3. Titles of the programs differ. Those in the United States, Canada, and Australia most frequently carry the title of *human factors* or *engineering psychology*. European programs are called *ergonomics, occupational psychology*, and *work physiology*. American programs emphasize system design and applied research. European ones emphasize worker concerns, selection, safety, and satisfaction.

Pearson summarized his 1994 study by reporting that "a course in "Man–Machine Systems Design" is clearly the most popular subject taught. This is the rather traditional course, which covers systems development, displays, controls, and panel layout. Growing in popularity in the Americas and Asia is a course in "Workplace Ergonomics," which focuses on the industrial and/or office workstation, seating, VDTs, and perhaps the cumulative trauma disorders. Popular courses not identified 15 years ago include cognition, human–computer interaction, product design, sociotechnical systems, and human reliability. On a relative basis, courses in work physiology, work design, hygiene, and health appear to be on the decline. Course popularity in human information processing, environmental stress, and human performance appear to have leveled off. Those gaining in popularity include biomechanics, safety, and research methods. (See also Williges, 1992.)

WHAT THE HFE SPECIALIST SHOULD LEARN AND KNOW

The following description is biased because the author received his training in a psychology department and he professes little knowledge of what is taught in operations research or other engineering departments. A number of papers about HFE training presented at annual meetings of the HFES describe efforts to provide hands-on training for graduate students. These center around the solution of a design problem and follow the general process of system development, starting with analysis of an RFP, leading to the development of a proposed design solution, and so on. Students work either individually or in small teams. The interested reader may find additional detail in Moroney and Cameron (1995); Moroney, Green, and Konz (1996); Sind-Prunier (1996); and Sojourner, Olson, and Serfoss (1995).

The following description assumes that the reader will imagine he or she is or once was a graduate student in a university department that awards an advanced degree in HFE or some discipline related to it. The HFES distributes a monograph (Wilson, 1997) that lists universities that provide such degree programs.

If one is going to be a researcher, then probably all the psychology courses one ordinarily takes in a psychology graduate department (e.g., experimental psychology and experimental design, statistics, principles of learning, sensation and perception) are necessary if only as background material. In addition, no graduate student today can escape a hefty dose of computers because everything—regardless of whether it is warranted—is being computerized. The graduate student will at the very least use a computer and one or more software programs to analyze his or her thesis or dissertation data. The author's generation used a Friden electronic calculator to analyze data because the computer did not as yet exist (in university departments). With this exception, the courses cited earlier are much the same as those to which he and his student colleagues were exposed.

If there is such a course as basic engineering taught in university, the behaviorally oriented student would do well to take it for familiarization with the other major discipline involved with HFE. For someone who expects to work in system development, the more engineering courses the better (within reason, of course).

Whether one anticipates research or practice as one's final job destination, every HFE student must have at least one course that includes system development processes, including its stages and what is done at each stage, the design problems encountered, and so on (see Czaja, 1997). He or she should also be exposed to the history of HFE and its more mundane problems (e.g., job getting). This course should also describe the specialties into which HFE fractionates because, whatever the core HFE curriculum presents, the student will eventually be faced with specialties. These should include industrial ergonomics and macroergonomics as described in chapter 7.

Most important is HFE methodology (e.g., task analysis and its variations, graphic analytic methods like operational sequence diagrams, common instrumentation for measuring environmental factors, the use of simulators). Undoubtedly, the department will provide one or more HFE laboratory courses in which students will be presented with realistic design problems and asked to solve these on the basis of what they have learned in more theoretical coursework. This is a practicum, in effect. The laboratory should emphasize, as does HFE in system development, analysis, design solutions and their evaluation, and mockup, system, and usability testing. The laboratory is only a simulation of reality. Hence, if it is possible to arrange, the advanced student should pursue at least a 6-month internship in an actual system development facility. An internship in a research facility not linked to system development problems will probably add nothing to the research, which the university requires of the student for graduation. The emphasis in the training recommended by the author is familiarization with real-world HFE activities and problems.

As to what HFE specialists should know, it is necessary to introduce the subject within the context of their activities. The author believes that everything the specialist does, whether it is research or applied work, presents a problem that must be solved. What specialists should know is how to solve these problems. Therefore, they should be able to:

1. Analyze the problem in terms of elements and the factors that are involved in the problem. The specialist should be able to determine which of the problem elements will impact operators and their performance, and to what extent. Manifestly, the specialist will only be interested in serious problems. For example, for the operator required to make sound detections using very weak (near threshold) signals, it will be necessary to boost these. Then the question becomes, to what strength? If it is not possible to boost signal strength, perhaps one can transform the auditory signals into visual ones that would be more perceptible. This is the kind of analysis that is required in the initial analytic stage of system development. The same analysis is also required of the researcher, although if the research is self-initiated the specialist has more freedom to specify the parameters of the research.

2. Specialists should know where to go to secure the behavioral information they need to help solve the problem. There are many potential sources: books, published papers, governmental standards, and company test data. Specialists must know to which of these resources they should address themselves. If the problem is one involving the development of a new system, the HFE professional must determine where relevant technological information exists. Unless one has total recall, it will be always necessary to access informational sources.

Any thought that the specialist must memorize any extensive set of facts (e.g., minimum resolution required for TV viewing) should be dropped for two reasons: Most specialists are unlikely to be able to do so and it is unnecessary if they know where to access the information. For this reason, graduate training in HFE should emphasize basic concepts, methodology, and the way to attack problems.

3. One of the most important skills the specialist must have is knowledge of the methods and instrumentation needed and available; and how to set up, run the instruments or use the methods, collect the data, and analyze it. Hands-on, performance-based knowledge is much more useful than acquaintance through books alone.

4. Specialists should know (or be in a position to find) what the most recent HFE literature has to say that bears on their problem. In particular,

they should know or learn what functional capabilities (perceptual, cognitive, motoric) the human has to have to do the job, and whether these capabilities will be stressed by routine operations. (If the system is involved in an emergency, obviously the operator will be stressed, regardless of anything the designer can do to help.) Knowledge of human capabilities to perform a particular job is important, but unfortunately is often lacking. Ideally the specialist would have a catalogue of operational tasks with corresponding error probabilities and other human response characteristics, but such a catalogue is not available, although there is a little error data in the form of quantitative probability statistics in a few databases like Swain and Guttmann (1983).

In a system development project, almost all the information the specialist will have at the start of the project is what the customer requires the system to do. If the project is one on which the company has been working, there may be test data, analytic reports, and so on, but do not bet on this. From this point on, the specialist must be able to determine which physical factors in systems operation will have an effect on the operator.

Problem analysis is common to both research and application. When we come to the design phase of system development, specialists should know which aspects of the human–machine interface should be considered in analyzing and evaluating design outputs like drawings of cockpit displays and what the preferred design configuration is (i.e., behavioral design guidelines). In the test and evaluation stage of system development, the specialist should know what kind of tests are required (mockup, system, usability) and how to conduct these to secure the desired data.

Hence, the emphasis is on how to analyze a problem, determine its elements, anticipate the effects of physical and operational factors on human performance, anticipate what the operator will have to do, and what level of performance (in terms of error, time, or other indexes) one can anticipate in operator activity. In HFE training, a great deal of attention must be given to methodology because if professionals do not know what techniques they must or can use they will be inept. The total amount of material to be remembered need not be excessive, but practical knowledge (where to get information, how to use techniques) is important. Unless one is a researcher, fundamental laws of perception and learning need not be remembered in detail, but it is necessary to know basic paradigms like S–O–R for use in analyzing problem situations.

WORK HISTORIES

A few words are required to put the following work histories into perspective. These descriptions are derived from applications of individuals to the Board of Certification in Professional Ergonomics. These are the same individuals whose demographic data were described in a preceding section. The intent in providing these descriptions is to present a more illustrative picture of HFE applications than can be found in formal texts like the author's (Meister & Rabideau, 1965).

This is not as useful as actually observing HFE specialists in their routine activities because the authors of the following work statements are looking back in time and formalizing their statements, which leads to a certain artificiality. It would also be useful to expose a sample of highly skilled specialists to a set of standard design problems, for example, and have them verbalize their responses and reasons for these. To the extent that there are subjective elements in what all specialists do, studies of their performance under standard conditions could lead to a better understanding of HFE. Many years ago, the author and a colleague (Meister & Farr, 1966) performed a study of how HFE and industrial design specialists analyzed and evaluated control panel layouts. The results were illuminating but somewhat distressing in terms of the great response variability found.

The individuals who wrote their work histories were asked to categorize them into three types of activity: analysis of a design-related problem, design as a solution to the problem, and testing of the solution to ensure its adequacy. The need to formalize the episodes reported may make the activities involved appear somewhat static, whereas actual system development activities are usually quite dynamic.

In the course of their activities, the work histories show that HFE professionals perform a wide variety of activities: They go out to sea or to factories, observe operations and interview personnel, study manuals, work both with engineers and independently, and do other things that the work histories describe much better than any author can. In reproducing these work experiences, all identifications of the individuals involved and their organizations have been either eliminated or disguised. These histories were selected at random from the 101 available to the author. Because of space limitations, only 4 of the 101 resumes have been included; to have included them all would have required another book. Repetitive material was eliminated to shorten the descriptions. Other than editorial changes, the following words are those of the professionals.

Work History 1

System Involved: Supermarket Check Stands

I have been involved in several projects aimed at improving the occupational health and safety of cashier operators in supermarkets. This began with my thesis in university identifying the health problems of cashiers, and later involved the design of checkstands with adjustments, modular equipment components, and sit/stand options.

To identify these health problems a sequential task analysis was carried out. With each task element there was a physical assessment of the cashier's posture, particularly body positions (head, shoulder, elbow, wrist, and back) deviating from the neutral position. Cashiers were surveyed through questionnaires and a postural fatigue scale. Changes in hand strength and leg edema measurements were used as indicators of physical fatigue. Anthropometric norms were compared with dimensions of checkstands presently available.

Negative features of checkstand design documented included protruding objects, excessive register keyboard heights, and low bag-wells, resulting in poor working postures. Postural fatigue was linked to lower back, neck, left shoulder, and arm discomfort. There were significant changes in calf size linked to continuous standing.

A second supermarket study compared different cashiering methods, in relation to health complaints and work performance. A checkstand design with two different work methods, touch-checking versus electronic scanning, and two different bagging methods (scan & bag versus scanning followed by bagging) were evaluated.

Different problems were identified with the various equipment components and work methods. Using physiological assessment tools, postural analyses and questionnaire forms, [these] identified electronic scanning and scanning followed by bagging methods as the most favorable.

I have also had the opportunity to design and evaluate several checkstand designs. A systems approach was taken, which required the development of ergonomic criteria for work procedures, equipment, work station, software, training, and so on. A number of interviews with end-users provided the material for these criteria. In design emphasis was placed on designing as many different work methods as possible into a single checkstand, the philosophy being that no one work posture should be maintained for any great length of time. This meant designing a checkstand in which one was able to perform the task in a number of different ways, including standing, sit-standing, with or without a bagger. When the prototype checkstand was evaluated against existing checkstand designs, the prototype was always an improvement.

System Involved: Meat Processing Industry

The meat processing industry ranks fourth largest of all industries and employs approximately 50,000 employees. The industry has experienced a wide range of technological changes, from sausage making machines to assembly-line cutting and packaging of meat. However, manual operations are still prevalent. One tool which has not undergone much change, yet is frequently in use, is the meat cutter's knife. Technological development in the meat cutting sector has increased task specialization and increased productivity. At the same time, a growing number of cumulative trauma type injuries related to the back and upper limbs have been reported.

Although few studies have specifically examined the meat processing industry, a number of workplace factors have been identified in the literature as contributing to cumulative trauma disorders. These include repetitive hand/arm movements and forces, awkward workstation layout, cold temperatures, and the use of gloves.

To obtain an overall understanding of the problems in the industry a questionnaire survey of 500 retail and wholesale workers was carried out. To determine whether cold environmental temperatures were a contributing factor in the incidence of upper limb disorders the hand surface temperatures of 20 wholesale meat cutters were monitored throughout a work day. To evaluate different knife designs, forearm muscle forces were measured using electromyographic recording techniques (EMG), including an analysis of the different hand motions cutters utilize to do their job. To evaluate different workstation lay-outs reach dimensions and heights were measured.

The industry is divided into wholesale (beef or pork) and retail operations. The work environments, tasks, and work procedures differ significantly between wholesale plants and retail supermarket stores where both beef and pork are prepared for the customer. The study indicated that different workplace factors were responsible for the occupational problems in these different areas.

In retail stores excessive work surface dimensions (size of display cases), cold drafts, and work procedures were the primary factors associated with back and upper limb disorders in deli workers, meat cutters, fish workers, and meat wrappers.

In the beef industry, hand/wrist disorders were most prevalent amongst beef boners and cutters. Workplace factors which were associated with these disorders included cold hands, wearing a mix of cotton and rubber gloves on the knife holding hand, and the interaction between the knife and the cut which frequently placed the hand in a deviated position.

The pork industry reported the highest prevalence of hand/wrist disorders. A pork plant differs from a beef plant in that the meat is greasier; hence, making the entire workplace greasier, and also that the environmental temperatures are considerably warmer. The majority of workers do not wear gloves on their knife holding hand. The major workplace factors which had to be addressed are knife handles which reduce slippage and hence, gripped forces, possibly introducing the use of a glove for the knife holding hand, and examining the effects of hand tool vibration among knife users.

Both beef and pork workers frequently assumed static work postures for prolonged periods of time, repeatedly stressing the same joints; mini-stretch/massage pauses and job rotation were therefore recommended. During the project, I had the opportunity to evaluate and design a meat cutters knife that was more suited to the various meat cutting tasks than existing knives. Studies were conducted to examine the use of existing knives in the industry. One knife study included developing and testing a prototype knife design at +/–30° angles from the center of the handle. The combination of handle design and a positive knife angle reduced muscular effort and increased productivity. The second knife study examined the vibration levels of knives which are used to trim fat from pork cuts. Three different anti-vibration devices were developed and evaluated. An anti-vibration sleeve cover slipped over the knife was the most effective device.

Work History 2

System Involved: Coefficient of Friction (COF) Tester

As a forensic human factors consultant, each of the hundreds of cases I have been retained on requires that I analyze the systems involved in the incident. The majority of cases have also required many forms of testing and evaluation including illumination testing, coefficient of friction testing, perception/lines of sight evaluation, reaction times, various force measurements and slope measurements. At times, bio-mechanical testing has also been required. My principal contribution in the area of ergonomics design has been in the development of an automatic coefficient of friction tester. I have also redesigned offices to maximize comfort and efficiency of the employees. While at ——— Aircraft Company, I performed a needs analysis and developed spec. sheets to automate the monthly reporting system.

As a Safety and Human Factor Associate, I was asked to contribute to the design of a coefficient of friction (COF) tester which is currently going through the patenting process. I was responsible for providing human factors input regarding the capabilities of the instrument, as well as the type of information included on the control/display panel.

I have laid out the panel so that the LED displays are on the upper portion with the control buttons below the corresponding display. The four basic control buttons on the bottom of the panels are also placed in sequence of use. The "zero" button has been moved up near the test number so an operator does not accidently press it until the end of the run when s/he wants to zero out the test number.

System Involved: Forensic Investigation

Several cases illustrate different areas of Human Factors analysis and testing that I have performed.

Case One. The case involved a nurse rupturing a patient's ear drum and shattering the ossicular canal bones when applying too much force during an ear lavage (ear wash). The defense contended that the injury was a result of the patient jerking his head. I was retained by the plaintiff's attorney to evaluate the reasonableness of conduct of both the plaintiff and the nurse, the selection of the instrument used, the adequacy of training, and the credibility of the testimony.

This case required both a bio-mechanical analysis, testing of the ear lavage instrument, analysis of design problems with the instrument, and evaluating whether there was a more ergonomically designed instrument readily available to the defendants. As part of the testing and evaluation for this case, I interviewed two doctors (retained to testify as medical experts) regarding the instruments they typically use, the proper procedure for the lavage and any criticisms they had of the instrument used in this case. I also interviewed two nurses in an allergist's office and did some bio-mechanical testing with them in their lab on a sample instrument.

Because the defense's contention was that the patient jerked his head prior to the rupture of his ear drum, I wanted to test whether it was possible for the incident to have occurred as the nurse testified. Therefore, during my testing, I asked one nurse in the lab to hold the instrument as though she was going to perform the lavage on the other nurse using her nondominant hand (as was done in the actual incident). I asked the other nurse to jerk her head as the defendant demonstrated in her video deposition. As expected, the nurse holding the instrument moved her hand when the other nurse jerked her head. I then repeated this procedure using several people as "patients" and "nurses" in order to have a significant sampling. While I found a small learning effect within subjects, no "nurse" was able to keep her hand perfectly still. I measured a total of 15 subjects, with 3 samples from each. The range of responses was between 3/4" from the opening of the ear, to 3-1/2." Therefore, no one actually got the tip back aligned with the meatus (opening of the ear). I confirmed that it was unlikely that the nurse (who in the actual incident would not have been trying to keep her hand level as were the people in the study) would have had the instrument in position to reinter the ear after the patient jerked.

Secondly, I wanted to demonstrate that even if the nurse somehow managed to align the tip with the opening, the anatomy of the ear would block entrance of the tip without doing significant damage to the ear. For this, I referred to my interviews with the 2 doctors as well as *Gray's Anatomy*. The tragus blocks the entrance to the ear; the ear canal itself is not straight. Therefore, there would have been evidence of damage (scratches/cuts) from the metal tip of the instrument if the instrument jerked into the ear (there was no such evidence).

My opinions in this case were: 1) that the plaintiff did not move his head *prior* to the rupture of his ear drum, and 2) that the incident was caused by a combination of human error and the selected instrument. The basis for my first opinion was my testing of the instrument, that the plaintiff knew not to move, he was comfortable with the procedure and had it done numerous times; that people do not move into pointed objects (it is not consistent with a reasonable/reflexive response), that adults in general do not move during this procedure and that the plaintiff jerked away from the syringe.

My second opinion is based on the fact that the defendant did not follow standard procedures. S/he used her nondominant hand, she put the tip of the instrument into the ear canal (it is supposed to remain outside), and the instrument itself was heavy, awkward and difficult to use (based on my experimentation and interviews). There was also a more ergonomically designed instrument already available in the defendant's office (both the disposable syringe and the children's tip which was too short to reach the ear drum but functional for the procedure). Both of these instruments eliminate the chance for operator error.

Case Two. In this case, the plaintiff fell while walking through a yard to reach her unit in a duplex. I was retained by the plaintiff's attorney to evaluate the condition of the ground in the yard to determine whether it constituted an unsafe condition. I was also requested to analyze whether there was adequate lighting at the accident site at the time the plaintiff fell.

On visual inspection, it was clear that the ground was uneven dirt with clips of grass, weeds, rocks, and fruit from a nearby tree. Furthermore, there was no permanent pathway from the gate of the yard to the doorway of the unit.

To evaluate the level of lighting, I performed illumination testing. The only notable source of lighting came from a porch light that was partially obstructed by plants. Testing was conducted in both the "with light" and "without light" conditions. I also insured that the test was performed with the same moon phase as was present on the night of the incident.

The results of both conditions yielded extremely low levels of illumination, .00 foot candles (ft-C) for "without light" and an average of .03 ft-C for "with light" at ground level. What's more, the .03 ft-C is a conservative figure. Because the back unit was inaccessible, lighting from the front unit, which was closer to the ground and blocked by the plants, was used for the testing.

I was also requested to evaluate the reasonableness of conduct of the plaintiff. Based on my inspection, testing, and a review of the discovery materials, I found that it was both foreseeable and reasonable to use the path from the gate to the door of the back unit of the duplex. This path should either have been paved, or at a minimum, had stepping stones. It was also my opinion that there should have been at least one foot candle of light directly and uniformly lighting this pathway.

Work History 3

System Involved: Control Room Design

This three-year Detailed Control Room Design Review (DCRDR) project was a human factors review of a large control room at a nuclear power generating plant. I served as the Principal Investigator and Program Manager throughout the project, including its successful completion, defense, and acceptance by the Nuclear Regulatory Commission (NRC). It resulted in a refitting of many areas in the control room, including the central work space, panel components, panel layouts, and the modification of procedures used in operations.

As mandated by the NRC, this DCRDR covered a broad range of ergonomic design and system issues using multiple investigation methods. The issues included control panel components and panel layout; panel configuration; work space furniture, CRT monitors, keyboards, lighting, sound, communications, environment; Safety Parameter Display System for the computerized presentation of critical parameters; alarm systems; and Emergency Operating Procedures.

The methods used in the control room work space review included interviews with operating personnel, research of the operating experience event records, a detailed evaluation of all control room components against the NUREG-0700 human factors design guidelines and plant-specific design conventions, and a survey of the control room work space.

Our review also included a system review and task analysis (SRTA) of the plant's emergency operating procedures. This included a systematic analysis of the procedures, tracing incipient conditions through the detailed operator tasks and the availability and adequacy of the control room displays and controls. This analysis was followed by a complete simulator walk-through and a series of simulated off-normal events at the utility's simulator facility using control room personnel. In these simulator tests, we attempted to confirm our review and task analysis as well as observe new real-time problems and crew interactions. The SRTA was a thorough effort that lasted over six months.

After the work space, component, and operating procedures evaluation, we were engaged in a year-long redesign effort that focused on the repositioning, relabelling, or replacement of control panel components. A new system of control room conventions were developed and documented for system layout, component selection, and labeling. Working with the utility personnel, we completed a detailed control room redesign and a full-scale color mockup of the control room panels as part of this effort.

The mockup was used to facilitate review by operations personnel. Using a walk-through of standard operating procedures, we identified where the modified components and layout effectively solved the original human factors problems and did not introduce new design deficiencies, and where another design was required. The redesign effort included the physical work space where all the furniture, computers, and work surfaces were completely refitted with custom units, following our new designs and configurations. Two large panel retrofit projects were completed shortly after the completion of the DCRDR.

System Involved: Thermostat

This extensive test of a candidate home control thermostat was completed at the request of a product design team. My role as the human factors specialist was to develop an evaluation that would resolve issues of control labeling, display design, programming, physical operation, and product aesthetics.

Using a computer simulation of the product, I developed a usability test that addressed the engineers' issues. The usability test included a pretest qualification, pretest interview, simulator familiarity session, scenario-based tests of all thermostat functions with reallocations, post-test questionnaires and interviews, and a session with physical thermostat models to test operation and consumer preference. For each task, we analyzed time-on-task and the details of a computer-generated file of user actions. We were interested in both overall performance and understanding users' mental models of the product and its features.

The usability testing was separated into several phases with approximately eight participants per phase, with periodic usability test summary reports, design team review, and redesign in an attempt to improve the design through iteration. Although some of these redesigns were not completely successful, the process was very successful in identifying solutions to consumer problems with the product. We human factors specialists provided most of the redesign recommendations that were implemented on the prototype.

The end result was a thermostat design that was modified in many respects, including control labeling, display layout and character size, operating instructions, and programming logic. We did not escape the need for the operating manual, but we minimized its importance, confirmed the success of the design, and outlined the human factors design challenges for subsequent thermostat product development efforts.

System Involved: Home Security User Interface Panel

A large interdivisional team was selected to design and build a new home automation system. This fast-track development program had a human factors position that included working closely with the design team to define the system functions and user interface. I was selected to serve that position.

After completing a function analysis, the design team developed a basic design configuration including a wall-mounted control panel, auxiliary components for temperature sensing and control, and appliance and lighting controls. As the human factors specialist on the team, I promoted the concept of "house modes" that would allow the homeowner to configure the house with a minimum number of keystrokes. This was based on an analysis completed during previous internal research indicating that patterns of climate, lighting, and security settings could be developed that would simplify the operation of equipment in a home, though these patterns would differ between households. We developed the product concept around these "modes" and prepared a scheme for custom installation to support this design requirement.

I also developed requirements and candidate designs for the keyboard, display hardware, display font, and a menu-based interface. This design was first documented on paper and subsequently implemented on a fully functional screen-based prototype. The prototype was used to make preliminary design decisions within the design team. Because of the fast development time, I was unable to test the system with typical homeowners, but the heuristic tests completed in-house proved to be very successful at identifying potential interface problems. One of my most important roles as a human interface designer was enforcing a consistent approach to the product's user interface design. This was a continuing problem as the project drew the attention of various marketing, management, and engineering staff.

A significant feature of the product was telephone access to the home automation system from both inside and outside the home. This feature necessitated a separate human factors design task, coordinated with the first, to develop an interface using a telephone keypad for system input and a digitized voice module for output to the user. Although a simulator was not available for prototyping this interface, we did successfully implement the interface using a "Wizard of Oz" technique and identified several potential voice feedback issues that were remedied early in the design process.

Work History 4

System Involved: All Weather Landing Monitor

This project was to determine what type of information pilots really needed in the presentation of display symbology in order to appraise approach and landing progress and make consistently correct decisions to land or go around during low-to-zero visibility automatic landings.

From the time of the first ILS landing, it had been accepted that visual acquisition was necessary for the decision to land. No one knew what the real requirements were, that is, the cues that the pilot actually used. The operational solution has always been to get the pilot down to some reasonable level for visual acquisition, then he would use direct visual cues to land. To extend from this point, we knew we would have to create a different approach from anything that preceded this effort. We chose to work as though the problem had never been examined.

Human Factors efforts were started by (1) completing a detailed system analysis of the landing operation, then (2) creating an experimental design that would permit cost effective testing without compromising the principles of statistical design of experiments, and finally (3) performing a series of simulated automatic landings for a number of incrementally sophisticated display symbology concepts.

Display simulations started with a sketchy outline of the runway similar to what would result if a runway were outlined by 8 equally spaced runway lights (4 each side), a configuration that could be provided by small radio frequency transmitters or radar corner reflectors. Display information on relative aircraft position and projected status was systematically added one element at a time and a series of simulated automatic approaches was completed for each configuration. Pilot subjects appraised present and projected status based on the information display, and accuracy of display use was measured by judgments of approach conditions and estimated 'aim point' from decision height. Accordingly, display contribution to pilot judgments of approach progress and aim point at decision height would demonstrate display adequacy for each of the various configurations.

The test procedure was to complete a series of 'simulated automatic landing' approaches for each display configuration, using predetermined mixes of glide path, cross track/crab and aim point conditions. Test conditions ranged through a wide selection of approach paths, including high through low glide paths, alternative cross track conditions (that could relate to winds or approach irregularities), and a selection of longitudinal and lateral aim points—from short of runway threshold to well past the desired touchdown zone, and from near to far sides of the runway. The path selected was changed in a random order from one approach to another. Reasonable levels of turbulence and cross wind conditions were included to provide realism in the dynamics of the display adequacy for each of the various configurations.

The statistical matrix for presentation order was a balanced Latin Square design with counterbalanced, nested variables. The matrix of the design was created for the purpose of providing a wide mixture of approach conditions, and minimizing number of conditions from the thousands that would be needed for a traditional factorial design covering the range of desired conditions. With it, we could appraise main effects and first order interactions, even though accepting confounding of higher order interactions—any implications of subtle effects would be reason for more specific explorations. This alternative was acceptable in that a vast amount of information could be collected within time and budget constraints.

Pilots participating in the project were training pilots. Experimental engineering pilots served as project advisors.

Changes in display features showed progressive improvements in judgments that were clearly related to selected changes in display symbology formats, most notably by moving runway "lights" to designate a touchdown zone, and adding symbology for (a) projected flight path, and (b) alignment cues for glide slope and lateral position.

Significant improvements occurred with the addition of meaningful runway reference points (aim point, touchdown zone, and centerline extension), relative aircraft alignment status and projected flight path. Interestingly, "errors" in judgment tended in the "desired" direction, that is, path alignment judged reasonable and aimpoint safely on the runway, even though the approach and touch down conditions might be short, long, cross track or to the side of the runway.

The tests confirmed features then under consideration for an Electronic Attitude Director Indicator, and added some—most specifically to enhance appraisal of alignment for glide slope and lateral position. While some were of the opinion that lateral information was less needed, interactions effects existed. Later tests applied these principles while exploring refueling director lights; they showed an improvement in overall performance when lateral alignment information was given to the pilots. Lateral cues reduced workload, demonstrated by less control activity, through less need to dedicate attention to sorting out cross track conditions, i.e., track, cross wind drift and crab.

Conclusions

Work Histories 1 and 2 illustrate what industrial ergonomists, whose work seems tied to production processes, do. Histories 3 and 4 illustrate HFE associated with system development. The distinction is not firm, however; occasionally both sets of activities are to be found in the same history.

In all cases, what initiates the HFE activity is a problem. For industrial ergonomists, the problems are usually worker complaints of muscular trauma or excessive effort requirements. In system development, the problem is to design a new or updated system or review the already existent design of an operational system. One can schematize the process as: problem → analysis → design → test → evaluation → problem solution (or at least a partial solution).

In many, if not all, cases, the nature of the problem must be explored. This is often done by interviewing and observing system experts—those who have worked extensively with the same or similar system. There is a great deal of documentary review and analysis. Design is iterative; preliminary sketches are drawn up, submitted to criticism by others, and then revised until a consensus of approval is reached. The final stage in the process is when the design is tested, either by elimination of worker complaints as in industrial ergonomics or, more positively, by exposure of the design to usage in a test.

A range of methods is available to the HFE specialist; prominent is the walkthrough, which represents a cheap and effective partial simulation of the operation of the system under examination. Standards are available: MIL-STD 1472E (1994), those of NIOSH, and, in the case of industrial ergonomics, even physical performance models. There is little or no mention of models in system development HFE, despite the large number of models that have been developed over the years. Are these models inadequate? Too expensive to use? Too little known to HFE system development specialists? It would be interesting to explore these questions.

HFE, whether in industrial ergonomics or system development, is not a rapid process, usually requiring months, if not years, to complete the system. HFE is essentially a team-oriented process because the basic design is usually performed by engineers and because of the multiple aspects of most HFE projects, which demand additional personnel. The need to document HFE efforts, to write required reports, requires at least minimal communication capabilities. University training, which demands much writing skill, would automatically eliminate those who are feeble in communication, although in the early days the author once had a chap working for him who was unintelligible in writing, although understandable in conversation.

One of the interesting features of the industrial ergonomics activity is that it indicates that it is possible to perform good research, as in the Work History 1 checkstand study, that is purely applied and quite useful. One could sneer at such research as dealing with trivialities, but even if one accepts this point of view (which the author does not), this research is at least useful, which is more than can be said of much other HFE research dealing with *fundamental* issues. The ideal would be fundamental HFE research with much greater applicability than that research has displayed (see chap. 6 and Singleton, 1994).

The characteristics of the more applied research are that (a) a much more concrete problem initiated the research than is usually the case with fundamental research; (b) it was concerned with producing an improved output rather than with expanding the knowledge base; and (c) the utility of the research could be determined almost immediately, whereas any utility of the more fundamental HFE research can be determined only after years and perhaps never.

One of the characteristics of the applied research described in these work histories is that it was research, not experimentation. Although some HFE specialists may confuse the two, implying that only experimentation qualifies as research, they are distinct phenomena. The more applied research may involve accessing the knowledge structure of operators, their attitudes, and feelings, rather than performing an experiment, in which the subject's attitudes and opinions are ignored or rigidly suppressed.

The author does not suggest that fundamental HFE research be eliminated. He merely wishes to draw the reader's attention to something that should be obvious—that there is more than one type of research (experimentation) directed at more than one type of product (knowledge). At the same time, one must ask why these applied studies cannot supply knowledge-type information about human performance if those who perform these studies were to look at what they are doing for more than the immediate goals of the research. Admittedly this sounds almost impossibly vague, but an example might help. If one is doing system development and the goal is to develop an improved trainer, perhaps the specialist should ask this question: What do the characteristics of the improved trainer tell us about how people in general, or the trainees in particular, respond to the trainer features that were improved?

Another characteristic of the HFE activity described in these work histories is that HFE specialists do look at the literature to gain helpful hints, but often the literature does not have much to say. Perhaps someone should examine this phenomenon by finding out why the literature did not prove more useful. This could be done by asking those who consulted the literature on a particular project how well they were satisfied with the aid the literature provided and why. The obvious hypothesis that is used as an excuse is that the specificity of the question the specialist asks does not permit the literature (which is much more general) to be as relevant as it should be, but there may be other factors that could make HFE literature more useful.

By raising these points, the author is implicitly asking whether the ways in which the HFE specialist does what he or she does is not a legitimate subject for research on its own. This part of the oversight function, which the author has suggested in the initial chapters, should be one of the things that the discipline should routinely do.

9

A Commentary on the Big Issues

The reader should think of this chapter as a series of essays expanding or commenting on the themes discussed previously. These essays are based on the notion that the study of HFE as a discipline in toto, rather than of its molecular specialty aspects, can be as intellectually stimulating and satisfying as, for example, the study of theology. In both, *big* questions are addressed (in theology, what is God?; in HFE, what are the nature and parameters of the human–technology relationship?). The really big questions in HFE are listed in Table 9.1. The list may not be exhaustive and others may ask different questions. The significant thing about these questions is that, to a great extent, they are not experimental in nature, although experiments will help answer them.

This biggest question of all is Number 7: What are the parameters of the human–technology relationship? This is arguably the one question susceptible to empirical test. However, answers to the other questions must be found before Question 7 can be fully answered. The answers to these questions depend primarily on analysis, which may pose a serious problem for HFE professionals: It is easy to perform an experiment, but much harder to think about a problem. That is why the HFE literature has comparatively little to say about them.

Because these questions are overarching and abstract, they cannot be directly attacked. Each question requires the performance of a number of what may be called *implementing* or lower level studies. For example, to develop a quantitative prediction methodology, one requires first a quantitative database. However, such a database requires a preceding analysis of available HFE data, which in turn requires a review of the HFE research literature as a whole.

TABLE 9.1
The Big HFE Questions

(1) What is the nature, scope, and parameters of HFE?
(2) What are the purposes and goals of HFE in general? Of HFE research in particular?
(3) What research topics should HFE study?
(4) What should the relationship between HFE research and HFE application be? What should one do with these research results? How can research results be translated into guidelines for the design of technological entities?
(5) How do system characteristics affect human and system performance individually? How are human and system performance related to each other?
(6) How can quantitative prediction of human performance be performed? How can it be associated with various characteristics of technology? What should the HFE database consist of? How can one develop such a database?
(7) What are the parameters of the human–technology relationship? How can behavioral principles be translated into hardware/software equivalents and vice versa?
(8) Does the history of HFE show progress toward accomplishment of special goals? What criteria can be applied to measure that progress?
(9) How does technology affect human attitudes and responses? How can the human impress human characteristics on technology?
(10) What are the characteristics of the operational environment? How can these characteristics be incorporated into experimental research? How does one validate conclusions gathered by research in nonoperational environments?
(11) Can a theory/model of human–technology relationships be developed? What would its outlines be?
(12) How do the concept structures of the HFE professional and discipline develop? What is their effect on each other and on the discipline as a whole?

The starting point for study of these questions is the analysis and discussion of them to agree that they are important for HFE and to develop in outline form the necessary implementing studies for each question.

Of course, there are *little* or lesser questions, but these cannot be listed because of their number. All the reader has to do is thumb through any volume of *Human Factors* and he or she will see examples of little questions. To call them *little* is not to denigrate them, but to recognize that these questions are not as comprehensive as the larger ones. A lesser question is simply less complex and involves fewer variables than a bigger one.

Solutions to lesser questions should logically produce answers to bigger questions, providing one can analyze those lesser solutions properly. The process is not a simple linear one, however. One must think about those more abstract questions at the same time one is attempting to answer more concrete ones. It is also necessary to ask what the relation is of the more concrete to the more abstract. We are dealing here with two distinct intellectual domains, which, like the physical and behavioral domains, are not easily crossed. In fact, it is possible that one could spend a lifetime studying visual perception or workload and find no answers to the bigger questions. Indeed, it is likely that performing research on the lesser questions will not automatically supply answers to the greater ones. For example, the kinds of experimental research as found in *Human Factors* may answer Question 7 in Table 9.1 about the parameters of the human–technology relationship, but do nothing to answer the other questions in that table. However, Questions 1 and 2 in Table 9.1 may significantly affect lower level experimental research by indicating what the topics or themes of that research should be. Therefore, there is a relationship between more molecular research and the bigger questions of Table 9.1: The latter may determine the former, but the former may not inform the latter.

Of course, one cannot perform experiments at random and have them be meaningful in terms of the big questions. Rather, one must start with the big questions and develop a series of studies that are related by their purpose to these questions. As long as HFE research is performed without some sort of master plan derived from thinking about the big questions, the research results will not be very meaningful in terms of the big questions—meaningful perhaps in terms of answering the immediate research question, but not in leading to big question answers. This is the situation in which we find ourselves at the present time because HFE research is essentially anarchic (in the sense that everyone is doing his or her own thing) and hence, to a certain extent, disorganized.

The scientific culture (Zeitgeit) and HFE supply a certain direction to its research, but only a gross one. Therefore, it is necessary to keep the big questions constantly in mind. Perhaps before any researcher begins an experiment, he or she should stop to review them to see how the experiment fits into them. When the experiment is over, perhaps the questions can be reviewed again to see how the research results illuminate these questions. The big questions are points of reference for the researcher. Without them, research to answer lower level questions is somewhat uncoordinated. However, it must be admitted that few professionals are interested in the big questions; most say, we have a hard enough time trying to get answers to lower level specialty questions. They are right, of course, but this condemns them (and the discipline) always to be at a lower level.

CONCLUDING COMMENTS

One can ask whether any of the activities described in this book accomplish the HFE goals we started with in chapter 1. In system development, if the specialist prevents the designer from making egregious errors, it could be maintained that a primary HFE function was accomplished. Of course, this is a far cry from what specialists would like to be able to do in system development, but HFE is only one of the influences (and not the most important) on that development. Moreover, the state of the HFE art is weak.

It may be easier to achieve HFE goals in research because these goals are so much more tenuous. If the goal is to provide knowledge about the human–technology relationship, then almost anything one learns from a study presumably satisfies that goal. What the research adds to the store of knowledge one can never know unless one defines the knowledge contribution in operational terms, and no one can or would be willing to do that. Publication may be considered a concrete proof, but at least some of the HFE research that is published is probably irrelevant to HFE.

If the goal of HFE is a strong discipline (criteria for which are certain capabilities, such as quantitative prediction and design guidelines), we are not doing so well. The saving grace for HFE is that hardly anyone applies criteria of competence to the discipline, although one encounters among colleagues a certain feeling of dissatisfaction.

If one can suppose that one criterion of HFE success is popularity or at least acceptance, then we fail that standard also. It is a paradox, at which the author wonders, that there is opposition to HFE. No one opposes physics, biology, or chemistry as sciences and in their practical manifestations. It may be that there is something about the human–technology relationship that makes HFE appear in an equivocal light. What that element is is unclear, but the observation seems to be on target. At this writing, manufacturers and businessmen in California are vigorously fighting a new state ergonomics regulation involving repetitive trauma disorders. Nor is HFE popular with engineering designers, many of whom resent HFE as a constraint on their creative freedom. (Of course, all we have is anecdotal data for this conclusion.) The term *ergonomics* is becoming more familiar every day, but as used by advertising companies to tout their products, it means nothing to the layman.

In the meantime, technology rolls on, irreversibly, carrying with it in its train certain unpleasant side effects. Many times that technology is much more complex than it need be; consequently, people cannot use it efficiently. Often it is not effective because it is not adapted to its human partners. Technology is tied to the profit motive; consequently, it has little appreciation of the human.

It is possible to view HFE in several ways, all of them important. As science, it is needed to understand the ramifications of the human–technology relationship. Note that the word *human* precedes *technology*; that is because technology should be the servant, not the master, although in too many instances these roles are reversed. In the area of design and application, HFE does two things: It helps make equipment and systems more effective and productive, and it helps to adjust technology to respond to human limitations and capabilities.

As propagandists for the human in relation to technology, HFE professionals may be performing a job of work as or even more important than the preceding functions. It is possible to view technology as a ravaging beast that needs to be tamed. To say this is not to be a modern Luddite, but merely to recognize that certain effects of technology are undesirable. For example, we would not have aircraft accidents if there were no aircraft. To say this is not to suggest the elimination of aircraft technology, but the necessity for controlling it. That is why, for example, we design more effective displays so that air traffic controllers can do a better job.

It is necessary to recognize that the human–technology relationship has both positive and negative elements. Technology as a means of accomplishing many things otherwise impossible to humans is a boon. Technology, especially the most recent technology, which is becoming increasingly complex, has costs associated with it. These are responsible for the tension inherent in the relationship. Many people are uncomfortable with that technology because they feel they are incapable of mastering it; they feel inept and at a disadvantage with it. Newspaper stories threaten those without computer literacy with becoming second class citizens in a technological civilization. Computerization, as it impinges directly on the information processing of the individual, as one sees it on the Internet, for example, threatens a degraded, mindless culture.

In this connection, HFE can be conceptualized as having a mediating role—as attempting to mitigate the worse excesses of unbridled technology. Most HFE professionals might reject this because it smacks of cultural activism and elitism. They would probably prefer an HFE as an engineering specialty whose purpose is limited to working around human limitations in system development or as a purely academic research activity. These goals are fine as far as they go, but fall far short of what HFE could conceivably become. (If HFE is to fulfill a higher vocation, it must examine itself more fully.)

One would not wish, even if one could, to reverse technology. However, one might seek to ameliorate its effects. For that reason, it appears to the author that in the future, whether it is HFE or some other avatar, there will always be a science directed at the human–technology relationship (see also Meister, 1992). HFE may change, improve, or even retrogress, but the need for it will continue.

This has been as much a voyage of exploration for the author as he hopes it has been for the reader. At its culmination, he realizes that one concept animates the entire structure of HFE: *transformation*—in one direction, the transformation of physical elements (the equipment) into human (behavioral) performance; in the other direction, the translation of behavioral principles derived from human performance into physical structures (the design of equipments and systems).

As in the Renaissance, when the mind–body problem was discovered, so in the 20th century the human–technology problem has been discovered. The investigation of the human–technology relationship will continue in the new century simply because we now live in a world not only of physical forces, but also of technological entities. The world of technology is now so overwhelming that the human must find means of controlling it. One way of controlling technology is to imbue it with human qualities by translating behavioral processes into physical ones. This is the quintessential problem for HFE—it is what makes HFE unique among the sciences.

It is also necessary to keep in mind that HFE is a cultural and historical phenomenon as well as a technological and conceptual one. HFE represents, among other things, a recognition that maximization of technology as represented by Taylorism demands as a compensatory weighting force the need to humanize that technology, lest we become its intellectual and conceptual slaves. Those military and civilian managers who introduced psychologists in America and Britain into World War II research did not know the forces they were unleashing, nor did the psychologists themselves. However, their work had certain unintended consequences (which is always the case in history). In some ways, World War II was a watershed for technological America and the West in the 20th century, just as the American Civil War was a watershed for political America in the 19th century.

Events become redefined in terms of new concepts, as logical (or unintended, if one prefers) consequences of those events. That is what we have at the end of 50 years of HFE development. HFE can no longer be considered a purely technological practice or psychological concept; it is a cultural concept that must be recognized as going well beyond its origins. HFE must be considered as a reaction to the dominance of technology—as a means of trying to control that dominance. It is unlikely that such a view will find complete acceptance within (or without) the discipline. Those who consider themselves primarily as scientists will see this viewpoint as profoundly nonscientific; those who are primarily technologists will see it as antitechnological.

However, every technological phenomenon has cultural and historical consequences. What would we be today if the atom bomb and the computer revolution had not occurred? One can ignore cultural and historical consequences if one wishes; the consequences will still occur. Is it better to ignore these or examine them for what they have to tell us about our discipline?

References

Unless otherwise indicated, the term *Proceedings* refers to reports of papers presented at the annual meetings of the Human Factors Society.

Adams, J. A. (1989). *Human factors engineering.* New York: Macmillan.

Anokhin, P. K. (1935). *The problem of center and periphery in the physiology of higher nervous activity.* Gorky: Gorky Publishers.

Anokhin, P. K. (1955). Features of the afferent apparatus of the conditioned reflex and their importance in psychology. *Problems of Psychology, 6,* 16–38.

Anonymous (1967). *Proceedings, Symposium on Human Performance Quantification in System Effectiveness.* Washington, DC: Naval Material Command and National Academy of Engineering.

Apostolakis, G. E., Mancini, G., van Otterloo, R. W., & Farmer, F. R. (Eds.). (1990). Special issue on human reliability analysis. *Reliability Engineering & System Safety, 29(3).*

Appleby, J., Hunt, L., & Jacob M. (1994). *Telling the truth about history.* New York: Norton.

Askren, W. B., & Regulinski, T. L. (1971). *Quantifying human performance reliability. Report AFHRL-TR-71-22.* Brooks AFB, TX: Air Force Human Resources Laboratory.

Bailyn, B. (1994). *On the teaching and writing of history.* Hanover, NH: Dartmouth College.

Baker, K., Olson, J., & Morrisseau, D. (1994). Work practices, fatigue, and nuclear power plant safety performance. *Human Factors, 36,* 244–257.

Barnes, R. M. (1980). *Motion and time study, design and measurement of work* (7th ed.). New York: Wiley.

Bedny, G. Z., & Meister, D. (1997). *The Russian theory of activity: Current applications to design and learning.* Mahwah, NJ: Lawrence Erlbaum Associates.

Beringer, D. B. (1984). Calculator watches as data entry devices: The fundamental things apply as time goes by. *Proceedings,* 253–257.

Berliner, D. C., Angell, D., & Shearer, J. W. (1964). Behaviors, measures and instruments for performing evaluation in simulated environments. *Proceedings,* Symposium on Quantification of Human Performance, Albuquerque, NM.

Bernshtein, N. A. (1935). The problem of the relationship between coordination and localization. *Archives of Biological Science, 38(1),* 1–34.

Bittner, A. C., & Champney, P. C. (Eds.). (1995). *Advances in industrial ergonomics and safety-VII*. London: Taylor & Francis.

Boehm-Davis, D. (1994). Human-computer interface techniques. In J. Weimer (Ed.), *Research techniques in human engineering* (pp. 268–301). Englewood Cliffs, NJ: Prentice-Hall.

Boff, K. R., & Lincoln, J. E. (1988). *Engineering data compendium: Human perception and performance*. Dayton, OH: Wright-Patterson AFB.

Booher, H. R. (Ed.). (1990). *MANPRINT, An approach to systems integration*. New York: Van Nostrand Reinhold.

Brady, J. S., & Daily, A. (1961). *Evaluation of personnel performance in complex systems. Report GM 6300.5 - 1431*. Los Angeles, CA: Space Technology Laboratories.

Broen, N. L., & Chiang, D. P. (1996). Braking response times for 100 drivers in the avoidance of an unexpected obstacle as measured in a driving simulator. *Proceedings*, 900–904.

Caldwell, L. S. (1970). Decrement and recovery with repetitive maximal muscular exertion. *Human Factors, 12*, 547–552.

Campbell, J. L., Carney, C., & Kantowitz, B. H. (1997). Design guidelines for advanced traveler information systems (ATIS): the user requirements analysis. *Proceedings*, 954–958.

Casey, S. M. (1997). The business of human factors. *HFES Bulletin, 40*(4), 1–2, 6.

Chapanis, A. (1996). *Human factors in systems engineering*. New York: Wiley.

Chapanis, A., Garner, W. R., & Morgan, C. T. (1949). *Applied experimental psychology*. New York: Wiley.

Chapanis, A., Garner, W. R., Morgan, C. T., & Sanford, F. H. (1947). *Lectures on men and machines: An introduction to human engineering*. Baltimore, MD: Systems Research Laboratory.

Chapanis, A., & Lindenbaum, L. E. (1959). A reaction time study of four control-display linkages. *Human Factors, 1*, 1–7.

Checkland, P. B. (1981a). Rethinking a systems approach. *Journal of Applied Systems Analysis, 8*, 3–14.

Checkland, P. B. (1981b). *Systems thinking, systems practice*. New York: Wiley.

Christensen, J. M. (1962). The evolution of the systems approach in human factors. *Human Factors, 5*, 7–16.

Christensen, J. M., Topmiller, D. A., & Gill, R. T. (1988). Human factors definitions revised. *HFES Bulletin, 31*, 7–8.

Cole, M., & Maltzman, I. (1969). *A handbook of contemporary Soviet psychology*. New York: Basic Books.

Comer, M. K., Kozinsky, E. J., Eckel, J. A., & Miller, D. P. (1983). *Human reliability databank for nuclear power plant operation: Vol. 2. A databank concept and system. Report NUREG/CR-2744/2 of 2, RX, AN*. Albuquerque, NM: Sandia Laboratory.

Comer, M. K., Seaver, D. A., Stillwell, W. G., & Gaddy, C. D. (1984a). Generating human reliability estimates using expert judgment (Vol. 1, Main Report. Report NUREG/CR-3688/1 of 2, SAND84-7511, X, RX). Columbia, MD: General Physics Corporation.

Comer, M. K., Seaver, D. A., Stillwell, W. G., & Gaddy, C. D. (1984b). Generating human reliability estimates using expert judgment (Vol. 2, Appendexes. Report NUREG/CR-3688/2 of 2, SAND84-7115, X, RX). Columbia, MD: General Physics Corporation.

Conant, R. C. (1996). An information-theoretic method for revealing system structure. *Proceedings*, 169–172.

Craik, K. W. (1947). Theory of the human operator in control systems: Vol. 1. The operator as an engineering system. *British Journal of Psychology, 38*, 56.

Craik, K. W. (1948). Theory of the human operator in control systems: Vol. 2. Man as an element in a control system. *British Journal of Psychology, 38*, 142.

Cumming, G., & Corkindale, K. (1969). Human factors in the United Kingdom. *Human Factors, 11*, 75–79.

Czaja, S. J. (1997). System design and evaluation. In G. Savendy (Ed.), *Handbook of human factors and ergonomics* (pp. 17–40). New York: Wiley.

Dandavate, V., Sanders, B. V., & Stuart, S. (1996). Emotions matter: User empathy in the product development process. *Proceedings,* 415–418.

Davies, D. R., Taylor, A., & Dorn, L. (1992). Aging and human performance. In A. P. Smith & D. M. Jones (Eds.), *Handbook of human performance: Vol. 3. State and trait* (pp. 25–61). San Diego, CA: Academic Press.

Dempsey, C. A. (1985). *Fifty years research on man in flight.* Dayton, OH: Wright-Patterson AFB.

Dhillon, B. S. (1988). *Human reliability with human factors.* New York: Pergamon.

Dockeray, F. C., & Isaacs, S. (1921). Psychological research in aviation in Italy, France and England. *Journal of Comparative Psychology, 1,* 115–148.

Dougherty, E. M., & Fragola, J. M. (1988). *Human reliability analysis.* New York: Wiley.

Driskell, J. E., & Olmstead, S. (1989). Psychology and the military. *American Psychologist, 44*(1), 43–54.

Edholm, O. G., & Murrell, K. F. H. (1973). *The ergonomics research society: A history, 1949–1970.* Winchester, Hampshire, England: Warren & Son.

Edwards, W. (1987). Decision making. In G. Salvendy (Ed.), *Handbook of human factors* (pp. 1061–1104). New York: Wiley.

Embrey, D. E. (1983). *The use of performance shaping factors and quantified expert judgment in the evaluation of human reliability, an initial appraisal.* Report NUREG/CR-2986, BNL-NUREG-51591. Upton, NY: Brookhaven National Laboratory.

Embrey, D. E., Humphreys, P., Rosa, E. A., Kirwan, B., & Rea, K. (1984). *SLIM-MAUD: An approach to assessing human error probabilities using structured expert judgment: Vol. 1. Overview of SLIM-MAUD.* Report NUREG/CR-3518, BNL-NUREG-5716. Upton, NY: Brookhaven National Laboratory.

Emery, F. E., & Trist, E. L. (1960). Sociotechnical systems. In C. W. Churchman & M. Verhulst (Eds.), *Management science: Models and techniques II.* Oxford, England: Pergamon.

Endsley, M. R. (1995). Toward a theory of situation awareness. *Human Factors, 37,* 32–64.

Ericsson, K. A., & Simon, H.A. (1980). Verbal reports as data. *Psychological Review, 81,* 215–251.

Fischetti, N. A., & Truxal, C. (1985). Simulating "the right stuff." *IEEE Spectrum,* 38–47.

Fitts, P. M. (Ed.). (1947). Psychological research on equipment design. (Research Report 19). Washington, DC: U.S. Army Air Force Aviation Psychology Program.

Fitts, P. M., & Jones, R. E. (1947). Psychological aspects of instrument display. Analysis of 270 "pilot error" experiences in reading and interpreting aircraft instruments. (Report TSEAA-694-1-12A). Dayton, OH: Aeromedical Laboratory, Air Material Command.

Flanagan, J. C. (1954). The critical incident technique. *Psychological Bulletin, 51,* 327–358.

Fleishman, E. A., & Quaintance, M. K. (1984). *Taxonomies of human performance: The description of human tasks.* Orlando, FL: Academic Press.

Forbes, T. W. (1939). The normal automobile driver as a traffic problem. *Journal of General Psychology, 20,* 471–474.

Forbes, T. W. (Ed.). (1981). *Human factors in highway traffic safety research.* Malabar, FL: Robert E. Krieger.

Fowler, B. (1994). P300 as a measure of workload during a simulated aircraft task. *Human Factors, 36,* 670–683.

Fu, K.-S. (1987). Artificial intelligence. In G. Salvendy (Ed.), *Handbook of human factors* (pp. 1106–1129). New York: Wiley.

Gage, M., Adams, E., Logan, R., & Wilson, J. (1996). Human factors in the product development process. *Proceedings,* 398–400.

Galaktionov, A. I. (1978). *The fundamentals of engineering in the psychological design of ASGTP.* Moscow: Energy Publishers.

General Accounting Office. (1981). *Effectiveness of U.S. forces can be increased through improved weapon system design* (Report PSAD-81-17). Washington, DC: Author.

Gennaidy, A. M., & Asfour, S. S. (1987). Review and evaluation of physiological cost prediction models for manual materials handling. *Human Factors, 29*, 465–476.

Goodstein, L. P., Andersen, H. B., & Olsen, S. E. (Eds.). (1988). *Tasks, errors, and mental models.* London: Taylor & Francis.

Green, D. M., & Swets, J. A. (1966). *Signal detection theory and psychophysics.* New York: Wiley.

Hahn, H. A. (1996). Using electronic dialogue to augment traditional classroom instruction. *Proceedings,* 454–458.

Hall, R. H., Haas, J. E., & Johnson, N. J. (1967). Organizational size, complexity, and formalization. *Administrative Science Quarterly, 12*, 72–91.

Hancock, P. A. (1997). *Essays on the future of human-machine systems.* Minneapolis: Author.

Hannanman, G. W., Spurgin, A. J., & Lukic, Y. D. (1984). *Human cognitive reliability model for PRA analysis* (Draft report NUS-4531). San Diego, CA: NUS Corporation.

Hanson, B. L. (1983). A brief history of applied behavioral science at Bell Laboratories. *Bell Systems Technical Journal, 62*, 1571–1590.

Harper, W. H., Knapik, J. J., & de Pontbriand, R. (1997). Equipment compatibility and performance of men and women during heavyload carriage. *Proceedings,* 604–608.

Hart, S. G., Hauser, J. R., & Lester, P. T. (1984). Inflight evaluation of four measures of pilot workload. *Proceedings*, 945–949.

Heisenberg, W. (1971). *Physics and beyond: Encounters and conversations.* New York: Harper & Row.

Hendrick, H. W. (1984). Cognitive complexity, conceptual systems, and organizational design and management: Review and ergonomic implications. In H.W. Hendrick & O. Brown (Eds.), *Human factors in organizational design and management* (pp. 467–478). Amsterdam: North Holland.

Hendrick, H. W. (1991). Ergonomics in organizational design and management. *Ergonomics, 34,* 743–756.

Hendrick, H. W. (1995). Future directions in macroergonomics. *Ergonomics, 38*, 1617–1624.

Hendrick, H. W. (1996a). All member survey: Preliminary results. *HFES Bulletin, 36*(6), 1, 4–6.

Hendrick, H. W. (1996b). Human factors in ODAM: An historical perspective. In O. Brown, Jr., & H. W. Hendrick (Eds.), *Human factors in organization design and management* (Vol. 5, pp. 429–434). Amsterdam: North Holland.

Hendrick, H. W. (1997a). Organizational design and macroergonomics. In G. Salvendy (Ed.), *Handbook of human factors and egonomics* (pp. 594–636). New York: Wiley.

Hendrick, H. W. (1997b). Good ergonomics is good economics. *Ergonomics in Design, 5*, 1–14.

Henneman, R. L., & Rouse, W. B. (1983). Human performance in monitoring and controlling hierarchical large scale systems. *Proceedings,* 685–689.

Henneman, R. L., & Rouse, W. B. (1987). Human problem solving in dynamic environments: Understanding and supporting operators in large scale, complex systems. (ARI Research Note 87-51.) Atlanta, GA: Georgia Institute of Technology.

James, P., & Thorpe, N. (1994). *Ancient inventions.* New York: Ballentine.

Jamieson, G. A. (1996). Using the Conant method to discover and model human-machine structures. *Proceedings,* 173–177.

Kalawsky, R. S. (1993). *The science of virtual reality and virtual environments.* Wokingham, England: Addison-Wesley.

Kanigel, R. (1997). *The one best way; Fredrick Winslow Taylor and the enigma of efficiency.* New York: Viking.

Kanis, H. (1994). On validation. *Proceedings,* 515–519.

Kantowitz, B. H. (1989). The role of human information processing models in system development. *Proceedings,* 1059–1063.

Kao, H. S. R. (1973). Human factors in penpoint design. *Proceedings*, 503–504.

Karwowski, W., & Mital, A. (1984). Validation of a fuzzy model for the assesment of human operator responses to manual lifting tasks. *Proceedings,* 408–412.

Kelly, C. R. (1969). What is adaptive training? *Human Factors, 11,* 547–556.

Kelso, B. J. (1965). Legibility study of selected scale characteristics for moving tape instruments. *Human Factors, 7*, 545–554.

Kelvin, A. E. (1973). Human factors in rehabilitation and health care-transportation for the handicapped. *Proceedings,* 337–344.

Kemeny, J. R. (1979). *Report of the President's Commission on Three Mile Island.* Washington, DC: Government Printing Office.

Kim, H.-K. (1990). *Development of a model for combined ergonomics approaches in manual materials handling tasks.* Unpublished doctoral dissertation, Texas Tech University, Lubbock.

Kirwan, B. (1994). *A guide to practical human reliability assessment.* London: Taylor & Francis.

Kirwan, B. (in press). Trends in human reliability analysis, Section 4.2.15. In W. Karwowski (Ed.), *Handbook of occupational ergonomics.* Boca Raton, FL: CRC Press.

Kirwan, B., Kennedy, R., & Taylor-Adams, S. (1996). The validation of three human reliability quantification techniques—THERP, HEART, and JHEDI: Part II. Results of validation exercise. *Applied Ergonomics.*

Klein, G. A. (1993). A recognition-primed decision (RPD) model of rapid decision making. In G. A. Klein, J. Orasanu, R. Calderwood, & C. E. Zambo (Eds.), *Decision making in action: Models and methods* (pp. 138–147). Norwood, NJ: Ablex.

Kline, T. J. B., & Beitel, G. A. (1994). Assessment of push/pull door signs: A laboratory and field study. *Human Factors, 36*, 684–699.

Koffka, K. (1935). *Principles of Gestalt psychology.* New York: Harcourt & Brace.

Konz, S., & Streets, B. (1984). Bent hammer handles: Performance and preference. *Proceedings,* 438–440.

Koonce, J. M. (1984). A brief history of aviation psychology. *Human Factors, 26*, 499–508.

Kraft, J. A. (1970). Status of human factors and biotechnology in 1968–1969. *Human Factors, 12*, 113–151.

Kuhn, T. S. (1962). *The structure of scientific revolutions.* Chicago: University of Chicago Press.

Kumashiro, M., Hasegawa, T., Mikami, K., & Saito, K. (1984). Stress and aging effects on female workers. *Proceedings,* 751–755.

Laughery, K. R., Jr., & Corker, K. (1997). Computer modeling and simulation. In G. Salvendy (Ed.), *Handbook of human factors and ergonomics* (pp. 1375–1408). New York: Wiley.

Laughery, K. R., Sr., & Schmidt, J. K. (1984). Scenario analysis of back injuries in industrial accidents. *Proceedings*, 471–475.

Lehto, M. R. (1997). Decision making. In G. Salvendy (Ed.), *Handbook of human factors and ergonomics* (pp. 1201–1248). New York: Wiley.

Leuba, H. R. (1967). Information transmission in operator reports of equipment malfunctions (Report AMRL-TR-67-44). Wright-Patterson AFB, OH: Aerospace Medical Research Laboratories.

Lewin, K. (1936). *Principals of topological psychology.* New York: McGraw-Hill.

Lindsey, P. H., & Norman, D. A. (1992). *Human information processes* (2nd ed.). San Diego, CA: Harcourt-Brace Jovanovich.

Lipshitz, R., & Strauss, O. (1996). How decision makers cope with uncertainty. *Proceedings,* 189–193.

Lomov, B. F., & Bertone, C. M. (1969). The Soviet view of engineering psychology. *Human Factors, 11*, 69–74.

Lukas, W. J., Jr., Lettieri, V., & Hall, R. E. (1982). *Initial quantification of human error associated with specific instrumentation and control system components in licensed nuclear power plants* (Report NUREG/CR-2146). Washington, DC: Nuclear Regulatory Commission.

Lund, A. M. (1997). Expert ratings of usability maxims. *Ergonomics in Design, 5*(3), 15–20.

Mackworth, N. H. (1950). *Researchers on the measurement of human performance.* (Special Report Series, No. 268.) London: Medical Research Council.

Mane, A., & Donchin, E. (1989). The Space Fortress game. *Acta Psychologia, 71,* 17–22.

McCalpin, J. P. (1974). *Development of a methodology for obtaining human performance reliability data for a machine gun system. Technical Memorandum 6-74.* Aberdeen Proving Ground, MD: Human Engineering Laboratory.

McCloskey, M. J. (1996). An analysis of uncertainty in the Marine Corps. *Proceedings,* 194–198.

Medvedev, G. (1991). *The truth about Chernobyl.* New York: Basic Books.

Meister, D. (1971). *Comparative analysis of human reliability models.* Report L0074-1U7, Bunker Ramo Corporation, Westlake Village, CA.

Meister, D. (1984). Alternative approaches to human reliability analysis. In R. A. Waller & V. T. Covello (Eds.), *Low probability/high consequence risk analysis* (pp. 319–333). New York: Plenum.

Meister, D. (1985). The two worlds of human factors. In R. E. Eberts & C. G. Eberts (Eds.), *Trends in ergonomics/human factors—II* (pp. 3–11). Amsterdam: Elsevier.

Meister, D. (1987). System design, development and testing. In G. Salvendy (Ed.), *Handbook of human factors* (pp. 17–42). New York: Wiley.

Meister, D. (1989). *Conceptual aspects of human factors.* Baltimore, MD: Johns Hopkins University Press.

Meister, D. (1991). *The psychology of system design.* Amsterdam: Elsevier.

Meister, D. (1992). Some comments on the future of ergonomics. *International Journal of Industrial Ergonomics, 10,* 257–260.

Meister, D. (1993a). Human reliability database and future systems. *Proceedings,* Annual Reliability and Maintability Syposium, 276–280.

Meister, D. (1993b). A critical review of human performance reliablility predictive models. *IEEE Transactions in Reliability, R22*(3), 116–123.

Meister, D. (1993c). Non-technical influences on human factors. *CSERIAC* (Crew Systems Ergonomics Information Analysis Center) *Gateway, IV*(1), 1–3.

Meister, D. (1995a). Cognitive behavior of nuclear reactor operators. *International Journal of Industrial Ergonomics, 16,* 109–122.

Meister, D. (1995b). Human factors—the early years. *Proceedings,* 478–480.

Meister, D. (1996a). The effects of automation on ergonomics R&D. *Proceedings,* 47th International Industrial Engineering Conference and Exposition, 315–322.

Meister, D. (1996b). A new theoretical structure for developmental ergonomics. *Proceedings,* 4th Pan Pacific Conference on Occupational Ergonomics, Taiwan, 688–690.

Meister, D. (1996c). History and characteristics of human factors research. *Proceedings,* 541–545.

Meister, D. (1997). The value of research topics. In S. Casey (Ed.), *The practice of ergonomics, reflections on a profession* (pp. 95–100). Bellingham, WA: Board of Certification in Professional Ergonomics.

Meister, D., & Farr, D. E. (1966). *The methodology of control panel design* (Report AMRL-TR-66-28). Wright-Patterson AFB, OH: Aerospace Medical Research Laboratories.

Meister, D., & Farr, D. E. (1967). The utilization of human factors information by designers. *Human Factors, 9,* 71–87.

Meister, D., Fineberg, M. L., & Farrell, J. F. (1976). *NOE navigation: An overview of ARI experiments.* Arlington, VA: Army Research Institute. (AD- AO76 824/2)

Meister, D., Fineberg, M. L., & Farrell, J. P. (1976). *An assessment of the navigation performance of Army aviators under Nap of the Earth conditions* (Research Report 1195). Arlington, VA: Army Research Institute.

Meister, D., Finley, D. L., Thompson, D. E., & Hornyak, S. J. (1970). *The effect of operator performance variables on airborne electronic equipment reliability* (Report RADC-TR-70-140). Rome, NY: Rome Air Development Center.

Meister, D., & O'Brien, T. G. (1996). The history of human factors testing and evlauation. In S. Charlton & T. G. O'Brien (Eds.), *Handbook of human factors testing and evaluation* (pp. 3–11). Mahwah, NJ: Lawrence Erlbaum Associates.

Meister, D., & Rabideau, G. F. (1965). *Human factors evaluation in system development.* New York: Wiley.

Meister, D., & Sullivan, D. J. (1969). *Guide to human engineering design for visual displays* (Final report, Contract N00014-68-C-0278). Washington, DC: Office of Naval Research.

Miller, J. G. (1965). Living systems: Basic concepts. *Behavioral Science,* 10, 193–237, 380–411.

Miller, R. B. (1953). *A method for man-machine task analysis* (Report 53-137). Wright-Patterson AFB, OH: Wright Air Research and Development Command.

Mital, A. (1995). Industrial ergonomics. In J. Weimer (Ed.), *Research techniques in human engineering* (pp. 302–331). Englewood Cliffs, NJ: Prentice-Hall.

Moray, N. (1993, October). Technosophy and humane factors: A personal view. *Ergonomics in Design*, 33–37, 39.

Moray, N. (1994). De maximis non curat lex [How context reduces science to art in the practice of human factors]. *Proceedings,* 526–530.

Moray, N. (1997). Human factors in process control. In G. Salvendy (Ed.), *Handbook of human factors and ergonomics* (pp. 1944–1971). New York: Wiley.

Moroney, W. F. (1995). The evolution of human engineering: A selected review. In J. Weimer (Ed.), *Research techniques in human engineering* (pp. 1–19). Englewood Cliffs, NJ: Prentice-Hall.

Moroney, W. F., & Adams, C. M. (1996). Placement opportunities for human factors engineering and ergonomics professionals: Part I. Industry, government/military and consulting positions. *Proceedings*, 436–440.

Moroney, W. F., & Cameron, J. A. (1995). Using simulations in teaching human factors: Bridging the gap between the academic and the world of work. *Proceedings*, 399–403.

Moroney, W. F., Green, P. A., & Konz, S. (1996). Providing the team experience to human factors and ergonomics students. *Proceedings*, 445–449.

Moroney, W. F., Sottile, A., & Blinn, B. (1996). Placement opportunities for human factors engineering and ergonomics professionals: Part II. Academic and internship positions. *Proceedings*, 441–444.

Mosier, K. L., Skitka, L. J., & Heers, S. T. (1996). Automation, bias accountability, and verification behaviors. *Proceedings*, 204–208.

Munger, S., Smith, R. W., & Payne, D. (1962). An index of electronic equipment operability: Data Store (Report AIR-C43-1/62-RP[1]). Pittsburgh, PA: American Institute for Research.

National Research Council. (1949). *A survey report on human factors in undersea warfare.* Washington, DC: Author.

Neibel, B. W. (1972). *Motion and time study.* Homewood, IL: Irwin.

Nickerson, R. S. (1992). *Looking ahead: Human factors challenges in a changing world.* Hillsdale, NJ: Lawrence Erlbaum Associates.

Nickerson, R. S. (Ed.). (1995). *Emerging needs and opportunities for human factors research.* Washington, DC: National Academy Press.

Norman, D. A. (1981). Categorization of action slips. *Psychological Review, 88,* 1–15.

Norman, D. A. (1995, April). On differences between research and practice. *Ergonomics in Design,* pp. 35–36.

Okawaga, K. (1993). The role of design guidelines in assisting the interface design task. *Proceedings,* 272–276.

Park, K. S. (1997). Human error. In G. Salvendy (Ed.), *Handbook of human factors and ergonomics* (pp. 150–173). New York: Wiley.

Parsons, H. M. (1972). *Man-machine system experiments.* Baltimore, MD: Johns Hopkins University Press.

Parsons, H. M (1974). What happened at Hawthorne? *Science, 183,* 922–933.

Payne, D., & Altman, J. W. (1962). *An index of electronic equipment operability, Report of development* (Report AIR-C-43-1/62-FR). Pittsburgh, PA: American Institute for Research.

Pearson, R. G. (1980). Educational programmes in ergonomics: A world-wide profile. *Ergonomics, 23,* 797–808.

Pearson, R. G. (1994, August). *Educational programmes in ergonomics: A world-wide profile*—revised. Paper presented at the International Ergonomics Association meeting in Toronto, Canada.

Perrow, C. (1967). A framework for the comparative analysis of organizations. *American Sociological Review, 32,* 194–208.

Petho, F. C. (1993). A history of Naval aviation psychology during World War Two. *Proceedings,* Military Testing Association, 1–6.

Proctor, R. W., & Van Zandt, T. (1994). *Human factors in simple and complex systems.* Boston: Allyn & Bacon.

Rasmussen, J. (1983). Skills, rules, knowledge; signals, signs, and symbols; and other distinctions in human performance models. *IEEE Transactions in Systems, Man, and Cybernetics,* SMC-13, 257–266.

Rasmussen, J. (1985). Trends in human reliability analysis. *Ergnomics, 28,* 1185–1195.

Rasmussen, J., Pettersen, A. M., & Goodstein, L. P. (1994). *Cognitive systems engineering.* New York: Wiley.

Reason, K. J. (1991). *Human error.* Cambridge, England: Cambridge University Press.

Regulinski, T. L. (Ed.). (1973). Special issue on human performance reliability. *IEEE Transactions in Reliability, R22*(3).

Regulinski, T. L., & Askren, W. B. (1969). Mathematical modeling of human performance reliability. *Proceedings,* Annual Symposium on Reliability, 5–11.

Robbins, S. R. (1983). *Organization theory: The structure and design of organizations.* Englewood Cliffs, NJ: Prentice-Hall.

Roscoe, S. N. (1968). Airborne displays for flight and navigation. *Human Factors, 10,* 321–332.

Rouse, W. B. (1990). Designing for human error: Concepts for error tolerant systems. In H. R. Booher (Ed.), *MANPRINT, An approach to system integration* (pp. 237–255). New York: Van Nostrand Reinhold.

Rouse, W. B., & Rouse, S. H. (1983). Analysis and classification of human error. *IEEE Transactions on Systems, Man, and Cybernetics, SMC-13, 4,* 539–549.

Rouse, W. B., & Boff, K. R. (1987). *System design: Behavioral perspectives on designers, tools, and organizations.* New York: Elsevier.

Sanderson, P. M., Reising, D. V. C., & Augustiniak, M. J. (1995). Diagnosis of multiple faults in systems: Effect of fault difficulty, expectancy, and prior exposure. *Proceedings,* 459–463.

Sartor, N. B., & Woods, D. D. (1995). From tool to agent: The evaluation of (cockpit) automation and its impact on human-machine coordinates. *Proceedings,* 79–83.

Sartor, N. B., Woods, D. D., & Billings, C. E. (1997). Automation surprises. In G. Salvendy (Ed.), *Handbook of human factors and ergonomics* (pp. 1926–1943). New York: Wiley.

Schroder, H. M., Driver, M. J., & Streufert, S. (1967). *Human information processing.* New York: Holt, Rinehart & Winston.

Seaver, D. A., & Stillwell, W. G. (1983). Procedures for using expert judgement to estimate human error probabilities in nuclear power plant operations (Report NUREG/CR-2743, SAND82-7054, AN, RX). Falls Church, VA: Decision Science Consortium.

Seminara, J. L. (1975). Human factors in Roumania. *Human Factors, 17,* 477–487.

Seminara, J. L. (1976). Human factors in Bulgaria. *Human Factors, 18,* 33–44.

Shapiro, R. G., Brown, M. L., Fogelman, M., Goldberg, J. H., Granda, R. E., Hale, J. P., & Sanders E. B.-N. (1995). Preparing for the human factors/ergonomics job market. *Proceedings,* 379–383.

Sheridan, T. B. (1985). Forty-five years of man-machine systems: History and trends. *Proceedings,* 2nd IFAC Conference, 5–13.

Siegel, A. I., & Wolf, J. (1969). *Man-machine simulaton models.* New York: Wiley.

Siegel, A. I., Wolf, J., & Lautman, M. R. (1974). *A model for predicting integrated man-machine system reliability; model logic and description.* Wayne, PA: Applied Psychological Services.

Sind-Prunier, P. (1996). Bridging the research/practice gap; human factors practitioners' opportunity for input to define research for the rest of the decade. *Proceedings,* 865–867.

Singleton, W. T. (1994, July). From research to practice. *Ergonomics in Design,* pp. 30–34.

Smith, J. H. & Smith, M. J. (1994). Community ergonomics: an emerging theory and engineering practice. *Proceedings,* 729–733.

Smith, P. C., & Kendall, L. M. (1963). Retranslation of expectations: An approach to the construction of unambiguous anchors for rating scales. *Journal of Applied Psychology, 47,* 149–155.

Smolensky, M. W. (1995). Should human factors psychology and industrial/organizational psychology be re-integrated for graduate training? *Proceedings,* 775–778.

Sojourner, R. K., Olson, W. A., & Serfoss, G. L. (1995). Performing the system design process: An intelligent way to learn. *Proceedings,* 394–398.

Stearns, P. N. (1993). *The Industrial Revolution in world history.* Boulder, CO: Westview.

Stein, J. (Ed.). (1980). *The Random House College Dictionary.* New York: Random House.

Stubler, W. F., & O'Hara, J. M. (1996). Human factors challenges for advanced process control. *Proceedings,* 992–996.

Swain, A. D. (1967). *Field calibrated simulation. Proceedings,* Symposium on Human Performance and Quantification in Systems Effectiveness. Washington, DC: Naval Material Command and the National Academy of Engineering, pp. IV-A-1–IV-A-21.

Swain, A. D. (1968). *Some limitations in using the multiplicative model in behavior quantification* (Report SC-R-68-1697). Albuquerque, NM: Sandia Laboratories.

Swain, A. D. (1987). *Accident sequence evaluation program, human reliability analysis procedure* (Report NUREG/CR-4772). Washington, DC: Nuclear Regulatory Commission.

Swain, A. D. (1989). *Comparative evaluation of methods for human reliability analysis* (Report GRS-71). Germany: Gesellschaft for Reaktor-Sicherheit (GRS) mbh.

Swain, A. D., & Guttmann, H. E. (1983). Handbook of human reliability analysis with emphasis on nuclear power plant applications. Sandia National Laboratories, NUREG/CR-1278, Washington, DC: Nuclear Regulatory Commission.

Taylor, F. W. (1919). *Principles of scientific management.* New York: Harper.

Taylor, H. L. (Ed.). (1994). *Division 21 members who made distinguished contributions to engineering psychology.* Washington, DC: American Psychological Association.

Taylor, H. L., & Alluisi, E. A. (1993). Military psychology. In V. S. Ramachandran (Ed.), *Encyclopedia of human behavior.* San Diego, CA: Academic Press.

Tolman, E. C. (1932). *Purposive behavior in animals and men.* New York: Century.

Topmiller, D. A., Eckel, J. S., & Kozinsky, E. J. (1982). *Human reliability databank for nuclear power plant operations: Vol. 1. A review of existing human reliability databanks* (Report NUREG/CR-2744/1 of 2, SAND82-7057/1 of 2). Columbia, MD: General Physics Corporation.

Van Cott, H. P. (1984). From control systems to knowledge systems. *Human Factors, 26,* 115–123.

Van Gigch, J. F. (1974). *Applied general system theory.* New York: Harper & Row.

Vicente, K. J. (1997). Heeding the legacy of Meister, Brunswik, and Gibson: Toward a broader view of human factors research. *Human Factors, 39,* 323–328.

Vigotsky, L. S. (1962). *Thought and language.* Cambridge: MIT Press.

Viteles, M. (1945a). *The aircraft pilot: Five years of research.* Washington, DC: Civil Aeronautics Administration.

Viteles, M. (1945b). The aircraft pilot: Five years of research—A summary of outcomes. *Psychological Bulletin, 42.*

War Department. (1941). Notes on psychology and personality studies in aviation medicine (Report TM 8-320). Washington, DC: Author.

Wickens, C. D. (1992). *Engineering psychology and human performance.* New York: HarperCollins.

Wickens, C. D., & Goettle, B. (1984). Multiple resources and display formation: The implications of task integration. *Proceedings,* 722–726.

Wickens, D. O., Sandry, D. L., & Vidulich, M. (1983). Compatibility and resource competition between modalities of input, central processing, and output. *Human Factors, 25,* 227–248.

Williams, H. L. (1958, July 4). Assigning a value to human reliability. *Mechanical Design*, pp. 102–110.

Williges, B. H., & Williges, R. C. (1984). Dialogue design considerations for interactive computer systems. In F. A. Muckler (Ed.), *Human factors review—1984* (pp. 167–208). Santa Monica, CA: The Human Factors Society.

Williges, R. C. (1984). The tide of computer technology. *Human Factors, 26,* 109–114.

Williges, R. C. (1992). Human factors specialists' education and utilization. In H. P. Van Cott & B. M. Huey (Eds.), *Human factors specialists' education and utilization* (pp. 36–43). Washington, DC: Committee on Human Factors of the National Research Council.

Williges, R. C., Williges, B. H., & Elkerton, J. (1987). Software interface design. In G. Salvendy (Ed.), *Handbook of human factors* (pp. 1416–1449). New York: Wiley.

Wilson, D. L. (Ed.). (1997). *1997 Directory of human factors/ergonomics graduate programs in the United States and Canada.* Santa Monica, CA: Human Factors and Ergonomics Society.

Wilson, J. R. (1996). Ergonomics in the year 2000 and beyond. *Proceedings,* 4th Pan Pacific Conference on Occupational Ergonomics, Taiwan, 5–8.

Wise, J. A., Guide, P. C., Abbott, D. W., & Ryan, L. (1993). Automated corporate cockpits: Some observations. *Proceedings,* 6–10.

Woods, D. D., Roth, E. M., & Hanes, L. F. (1986a). *Models of cognitive behavior in nuclear power plant personnel. Report NUREG/CR-4532: Vol. 1. A feasibility study, summary of results.* Pittsburgh, PA: Westinghouse R&D Center.

Woods, D. D., Roth, E. M., & Hanes, L. F. (1986b). *Models of cognitive behavior in nuclear power plant personnel: Vol. 2. A feasibility study, main report* (Report NUREG/CR-4532). Pittsburgh, PA: Westinghouse R&D Center.

Woodson, W. R. (1954). *Human engineering guide for equipment designers.* Berkeley, CA: University of California Press.

Wortman, D. B., Duket, S. D., Seifert, D. J., Hann, R. I., & Chubb, A. P. (1978). *Simulation using SAINT: A user-oriented intruction manual* (Report AMRL-TR-77-61). Wright-Patterson AFB, OH: Aerospace Medical Research Laboratory.

Zarakovsky, G. M. (1966). *Psychophysiological analysis of work activity: Logical probability approach.* Moscow: Science.

Author Index

T

U

V

W

Y

Z

Subject Index

D

E

F

G

H

I

J

K

L

M

N

Printed in the United States
by Baker & Taylor Publisher Services

Printed in the United States
by Baker & Taylor Publisher Services